Assisted Reproduc[tion in] Wild Mammals of South America

I0034125

Editor

Alexandre Rodrigues Silva
Federal University of Semiarid Region–UFERSA
Mossoró, Brazil

CRC Press
Taylor & Francis Group
Boca Raton London New York

CRC Press is an imprint of the
Taylor & Francis Group, an **informa** business

A SCIENCE PUBLISHERS BOOK

Cover credit: A Jaguar, photo by Luciana Chiyo, DVM (E-mail: luchiyo@uol.com.br)

First edition published 2024
by CRC Press
2385 NW Executive Center Drive, Suite 320, Boca Raton FL 33431

and by CRC Press
4 Park Square, Milton Park, Abingdon, Oxon, OX14 4RN

Library of Congress Cataloging-in-Publication Data (applied for)

ISBN: 978-1-032-13961-6 (hbk)
ISBN: 978-1-032-13963-0 (pbk)
ISBN: 978-1-003-23169-1 (ebk)

DOI: 10.1201/9781003231691

Typeset in Times New Roman
by Shubham Creation

Foreword

Biodiversity in South America is under unprecedented pressure due to human activities. Like in the rest of the World, the number of wild animal and plant species has been decreasing rapidly. More than ever, saving species in the 21st century can be summarized by: understanding and sustaining a biodiverse and changing planet.

To preserve animal biodiversity, it is crucial to maintain genetic diversity and high numbers of individuals in each population (in the wild as well as in breeding centers and zoos). Besides the high-priority efforts in the protection of natural habitats, knowing the biology of species is fundamental to ensure their survival. Among the scientific disciplines involved in animal conservation, reproductive biology is key. Although the reproduction of <5% of mammalian species has been thoroughly characterized, we already know that there is a rich diversity in terms of anatomy and physiology. Importantly, understanding basic traits in reproduction has enabled the development of assisted reproductive technologies (semen collection, ovarian activity stimulation, artificial insemination, etc.). These techniques can potentially multiply the number of individuals at a much faster rate than natural mating and enhance the genetic management of the populations. One of the goals is to build sustainable populations in breeding centers that can then be reintroduced in their natural habitat.

Even if some are more common than others (like the rodents), mammalian species presented in the present book have a high conservation value as they belong to different hotspots of biodiversity in South America. The book fills a knowledge gap by focusing on the assisted reproduction of understudied and underreported species while reminding the reader about their reproductive physiology. It perfectly illustrates the rich diversity in reproductive mechanisms and the need to use species-specific approaches to assist their reproduction. Specifically, the chapter on Cervidae shows that the development of assisted reproduction in South American species has benefited from studies in semi-domesticated deer species in other parts of the world. However, advanced techniques that are currently used highlight the need for species-specific adaptations. Interestingly, the chapter on

Tayassuidae demonstrates that assisted reproduction is not like in domestic pigs. The chapters on Canidae and Procyonidae contain a lot of information about the huge diversity in reproductive biology and the complexity to develop assisted reproduction in those carnivores. The chapters on large Felidae (jaguar and puma) report impressive advances in assisted reproduction, which still remain behind the achievements in small Felidae (in the next chapter). Regarding primates, the chapters on Aotidae, Atelidae, Callitrichidae, Saimiri, and Sapajus prove that human assisted reproduction has inspired the approaches and methods in wild primates (including advanced techniques like *in vitro* culture of follicles). The impressive progresses made in the recent years justify separate chapters for each primate family. Even if laboratory rodents have been extensively used as models to study and develop assisted reproduction, chapters on the wild counterparts Like Agoutis, Capybaras, Cavies, Pacas, and Vizcachas illustrate the difficulties to make substantial progress in those wild species. Lastly, fascinating chapters on Xenarthra (Anteaters, Sloths, and Armadillos) shed new light on those least studied species in terms of assisted reproduction. Overall, Chapters 19 and 20 report more information in male than female assisted reproduction. They also emphasize the need to freeze and store biomaterials (especially gametes and somatic cells) in combination with the development of assisted reproduction.

When I started my career as a research veterinarian in French Guyana in the 1990s, I had the opportunity to work on some of the species mentioned above. Several years (or decades!) later, it is a real pleasure to write the preface of a great book sharing so many exciting advances that we could not even imagine when I began my studies. The Editor of the book, Dr. Alexandre Rodrigues Silva, is one of the most prominent scientists in wildlife reproduction in South America. He has been highly productive for many years by training students, organizing workshops, conducting cutting-edge research, and extensively publishing his work. He has gathered an excellent group of contributors who also are actively involved in wildlife research. This book is a good example of their profound dedication to animal conservation. It will be highly valuable to the next generation of conservationists who will learn that reproductive science is as important as the protection of natural habitats.

The book also opens many doors and shows the next directions in the discipline. In addition to the optimization of the current techniques, more collaborative and multi-disciplinary research as well as more resources are needed to integrate the assisted reproduction (and the associated biobanking) into the population management and future reintroductions. Unfortunately, this has not been achieved for a vast majority of species yet. However, looking at all the progress made in the past years, we can only be optimistic about the years to come, with more tools and more options to save species and preserve our planet.

Pierre Comizzoli, DVM, PhD
Senior Program Officer for Science
Smithsonian Institution

Preface

South America has one of the greatest riches in the world in terms of biodiversity, including hosting the Amazon biome, in which the largest tropical forest in the world is located. Despite all this richness, recent reports have raised alarm about different threats to this biodiversity. Thus, society, and the scientific community in particular, have been extremely concerned in order to develop strategies to reduce this loss of biodiversity that will certainly affect the life of the entire humankind.

Currently, we have been spectators of the scientific advances achieved in terms of mastering assisted reproduction techniques (ARTs) applied to farm animals, to provide quality protein to feed the human society. However, these ARTs can also be applied to the propagation and conservation of wild mammals, to reduce the effects of this imminent biological loss. In fact, in many countries in Europe, Asia, Oceania and even North America, the adoption of wildlife conservation programs using ARTs is already a reality. In South America, however, efforts to develop or even apply ARTs for the conservation of its biodiversity are still in a nascent stage.

We, for example, have been coordinating with the team at the Laboratory of Animal Germplasm Conservation at the Federal University of Semiarid Region (LCGA, UFERSA), located into the Caatinga biome, in Northeastern Brazil, since 2006. Day by day, we have challenged ourselves to learn more and more about reproductive physiology of hitherto unknown wild species. With this knowledge in hand, we have tried to establish efficient protocols for the conservation of South American species through different ARTs, as gamete and gonadal tissue conservation, pharmacological control of the reproductive cycles, and artificial insemination.

In this sense, this book embodies the idea of dreaming that we can indeed walk, albeit in small steps, towards a future in which wildlife can be respected and preserved. For it to take shape, we brought together a team of great specialists from different regions of Brazil, and from other countries, such as Argentina, Peru, the United States, and Spain, who have been carrying out research aimed at the conservation of wild species using ARTs. Thus, this book compiles recent and advanced information that shows the scientific progress achieved by these research groups in favor of South American biodiversity, thus serving as an important reference resource for professionals and students interested in wildlife conservation.

For a better organization, the book is divided into five parts that correspond to different orders: Arctiodactyla, Carnivora, Primata Rodentia, and Xenarthra, which precisely correspond to the groups of species where the greatest number of research has advanced in this field. Unfortunately, this also gives us an idea of so many other groups of unexplored South American species that we still need to know and help to preserve. We are just beginning our journey.

Good reading everyone!

Alexandre Rodrigues Silva
Editor

Contents

Part V: XENARTHRA
(*Part Editor*: Alexandre Rodrigues Silva)

Chapter 1

Wildlife in South America: The Use of Assisted Reproduction Techniques as Conservation Tools

Alexandre Rodrigues Silva[1]* and Alexsandra Fernandes Pereira[2]

[1]Laboratory for the Conservation of Animal Germplasm,
Federal Rural University of Semi-Arid (UFERSA),
Av. Francisco Mota, 572, Costa e Silva, Mossoró, RN, Brazil, 59625900.
Email: alexrs@ufersa.edu.br

[2]Laboratory of Animal Biotechnology,
Federal Rural University of Semi-Arid (UFERSA),
Av. Francisco Mota, 572, Costa e Silva, Mossoró, RN, Brazil, 59625900.
Email: alexsandra.pereira@ufersa.edu.br

South America, the fourth largest continent on the globe, represents 13% of the emerged lands, being inhabited by 6.5% of the world's population (IUCN 2019). The vast territory that comprises this continent belongs to 12 countries, which have a significant diversity characterized by their geography, biodiversity, economic, cultural, and social aspects. In substantial numbers and information, South America has five (Brazil, Colombia, Peru, Venezuela, and Ecuador) of the 17 megadiverse countries on the planet, in addition to the Amazon as the largest forest ecosystem in the world, still presenting a quarter of the net water resources (approximately 26%), the most biodiverse savanna (the Brazilian Cerrado) and a rich variety of ecosystems (IUCN 2019).

This significant diversity is also observed in the large number of its biomes, which comprise tropical, alpine, desert, savannah, grassland, chaparral, freshwater,

*Corresponding author: alexrs@ufersa.edu.br

marine and deciduous forest environments. These richly diverse biomes also result in a very diverse fauna (Silva et al. 2016), distributed in important ecosystems that have extraordinarily high levels of endemism and species diversity (IUCN 2019). South America, because it has the Amazon, is the region on the planet that comprises the largest number of known species, with Brazil being the most important country in terms of biodiversity on this continent (Feldmann 2021). It is estimated that 99% of the species that once existed are extinct, and information about them can only be obtained through fossil records. The current biodiversity of mammals in the New World tropics consists of 12 orders, among which the following can be highlighted: Artiodactyla, Carnivora, Primates, Rodentia, and Xenarthra (MacFadden 2006).

Environmental indicators related to global warming, deforestation, pollution, and species extinction have also been in demand in South America. According to data from the IUCN Annual Report, deforestation, fragmentation of ecosystems, advance of the agricultural frontier, irresponsible mining, population growth, climate circulation, desertification processes, water pollution represent a historical constant on the continent without many signs of improvement (IUCN 2019). Therefore, seeking to ensure a sustainable future associated with human progress and the preservation of nature, conservation strategies aimed at protecting wild fauna are also a concern of many research centers and conservation institutions.

The maintenance of animal biodiversity occurs mainly through the balance in the interactions of individuals, which can interfere in the vegetation through their nutrition. Wild animals have high scientific and ecological relevance and the importance of preserving the planet's biodiversity has been proposed. In this sense, different conservation strategies have been indicated for the maintenance of the global ecosystem (Souza et al. 2012).

Interestingly, an efficient conservation strategy would imply the preservation of an entire habitat and all the species in it, called *in situ* conservation. Representatives of South American countries have been working on proposals for the conservation of the continent's biodiversity. In recent data, South America had 4,868 protected areas through national protection systems, resulting in 24.4% of its territory, as well as 23.4% of its marine areas under some protection category (IUCN 2019). In addition, *ex situ* conservation strategies have been suggested, aiming to restore the genetic resources of the animals of interest through the formation of Biological Resource Banks or BRBs (Holt et al. 1996, Holt and Comizzoli 2021) and use in Assisted Reproduction Techniques or ARTs (Hildebrandt et al. 2021).

These ARTs include artificial insemination, multiple ovulations, and embryo transfer and/or sexing, *in vitro* and *in vivo* embryo production, and somatic cell nuclear transfer cloning. Some actions can complement these ARTs, such as monitoring hormonal changes, estrous cycle variations, understanding physiological principles, measuring gonads, micromanipulation of gametes and embryos, cryopreservation of tissues, cells, DNA, and formation of BRBs. All these tools are examples of non-invasive techniques that complement ARTs and understanding these techniques will directly reflect on the success of conservation (Andrabi and Mawxell 2007).

Among the ARTs mentioned, those currently most studied with the aim of conserving native species are artificial insemination, embryo transfer combined with *in vitro* production of embryos by *in vitro* fertilization, micromanipulation of embryos, cryopreservation, and cloning. The application of artificial insemination for the conservation of wild animals stands out for being an optimized method and used in several species, with the obtaining of born offspring (Swanson and Brown 2004, Comizzoli 2020). However, for the success of this technique, it is important to have well-defined estrous cycle monitoring protocols.

The *in vitro* production of embryos by *in vitro* fertilization is a technique that requires a controlled environment for the co-incubation of spermatozoa and oocytes, also requiring the steps of collection, selection, oocyte maturation, and sperm capacitation. After fertilization, *in vitro* development/culture occurs until the formation of blastocysts. Finally, these produced embryos favor the development of two forms of interest for the conservation and perpetuation of the species: the transfer of embryos to synchronized recipients or even the cryopreservation of these blastocysts until their transfer or developmental competence studies (Andrabi and Mawxell 2007).

Cryopreservation consists of one of the techniques for conservation of genetic material that can vary from its method to amounts and types of cryoprotectants, in addition to temperature variations and equilibrium conditions. In general, the type of material to be cryopreserved will promote protocol modifications and other existing variations (Comizzoli and Holt 2014). Cryopreservation can be classified into slow freezing, rapid freezing, and vitrification. All these techniques are essentially differentiated by temperature variations and amount of cryoprotectants used.

The reintroduction of genes into populations of animals of high genetic value with a view to conserving biodiversity is one of the goals that strengthen research and the creation of BRBs capable of storing gametes, embryos, and somatic cells. In general, some of the difficulties encountered when it is desired to cryopreserve genetic material from reproductive cells is the existence of a series of factors that limit obtaining the material, or even its use, such as the age of the donor or recipient animals, the age, and the phase of the estrous cycle of females, among others. Regarding somatic cells, however, these limitations become practically nonexistent, and genetic material can be obtained from any animal regardless of age or reproductive stage (Praxedes et al. 2018).

This book proposes to present, by species group, the main advances achieved in the development of ARTs applied to wild mammals in South America. It brings advances in terms of reproductive monitoring and application of estrus control methods, either for germplasm multiplication or for suppression of reproductive activity, when necessary. In addition, the book presents the main tools developed for the formation of biobanks of South American species, including protocols for the collection and cryopreservation of gametes, gonads, and somatic recourses. Finally, it presents the first attempts at *in vitro* production of embryos and cloning.

We try to contemplate the largest possible number of groups of species, in particular, those in which ARTs are more developed, even if only experimentally at the moment. We therefore hope that this book will serve as an incentive for professionals, researchers and students in the area to dedicate themselves to knowing the reproductive physiology of more and more species in South America, and with that they can improve strategies for their conservation. We even hope that we can envision the ARTs not only in the context of scientific experimentation, but mainly with their practical application in efficient programs for the conservation of wild species not only from South American ecosystems, but from all over the globe.

REFERENCES

Andrabi, S.M.H. and W.M.C. Maxwel. 2007. A review on reproductive biotechnologies for conservation of endangered mammalian species. Anim. Reprod. Sci. 99: 223–243.

Comizzoli, P. 2020. Birth of a giant panda cub after artificial insemination with frozen-thawed semen: a powerful reminder about the key role of biopreservation and biobanking for wildlife conservation. Biopreserv. Biobank. 18: 349–350.

Comizzoli, P. and W.V. Holt. 2014. Recent advances and prospects in germplasm preservation of rare and endangered species. Adv. Exp. Med. Biol. 753: 331–356.

Feldmann, P.R. 2021. África e América do Sul: O futuro passa pela biodiversidade. Estud. Av. 35: 111–123.

Hildebrandt, T.B., R. Hermes, F. Goeritz, R. Appeltant, S. Colleoni, B. Mori et al. 2021. The ART of bringing extinction to a freeze e History and future of species conservation, exemplified by rhinos. Theriogenology 169: 76–88.

Holt, W.V., P.M. Bennett and V. Volobouev. 1996. Genetic resource banks in wildlife conservation. J. Zool. 238: 531–544.

Holt, W.V. and P. Comizzoli. 2021. Genome resource banking for wildlife conservation: promises and caveats. CryoLetters 42: 309–320.

IUCN – International Union for the Nature Conservation. 2019. South America Regional Office Annual Report 2018, Quito, Equador.

Macfadden, B.J. 2006. Extinct mammalian biodiversity of the ancient New World tropics. Trends. Ecol. Evol. 21: 157–165.

Praxedes, E.A., A.A. Borges, M.V.O. Santos and A.F. Pereira. 2018. Use of somatic cell banks in the conservation of wild felids. Zoo. Biol. 37: 258–263.

Silva, A.R., A.F. Pereira, G.L. Lima, G.C. Peixoto and A.L.P. Souza. 2016. Assisted reproductive techniques on south american wild mammals. pp. 39–66. *In*: Rita Payan Carreira (ed.). Insights from Animal Reproduction. IntechOpen Limited.

Souza, J.M.G., R.I.T.P. Batista, L.M. Melo and V.J.F. Freitas. 2012. Reproductive biotechnologies applied to the conservation of endangered ruminant—Past, Present and Future. Rev. Port. Ciênc. Vet. 110: 31–38.

Swanson, W.F. and J.L. Brown. 2004. International training programs in reproductive sciences for conservation of Latin American felids. Anim. Reprod. Sci. 82: 21–34.

Part I
Artiodactyla

Part Editor: *Alexandre Rodrigues Silva*

Laboratory on Animal Germplasm Conservation
Federal University of Semiarid Region - UFERSA
Mossoró, Brazil
alexrs@ufersa.edu.br

Chapter 2

The South American Deer

Luciana Diniz Rola[1], David Javier Galindo[2],
Vicente José de Figueirêdo Freitas[3]
and José Mauricio Barbanti Duarte[4]*

[1]Reproduction of Organisms Laboratory, Federal University of Paraíba (UFPB),
Rodovia 12, PB-079, Areia - PB, 58397-000.
Email: luciana.diniz@academico.ufpb.br

[2]Laboratory of Animal Reproduction, Faculty of Veterinary Medicine,
National University of San Marcos (UNMSM),
Av. Circunvalación, 28, San Borja, Lima, Peru, 15021.
Email: dgalindoh@unmsm.edu.pe

[3]Laboratory of Physiology and Control of Reproduction,
Faculty of Veterinary, State University of Ceará (UECE),
Av. Dr Silas Munguba, 1700, Fortaleza-CE, Brazil, 60714-903.
Email: vicente.freitas@uece.br

[4]Deer Research and Conservation Centre
(NUPECCE – Núcleo de Pesquisa e Conservação de Cervídeos),
Faculty of Agricultural and Veterinary Sciences (FCAV),
State University of São Paulo (UNESP),
Department of Animal Science, Jaboticabal, São Paulo, Brazil, 14884-900.
Email: mauricio.barbanti@unesp.br

INTRODUCTION

South America contains several biodiversity hotspots and cover an impressive range of biomes, being considered one of the richest biogeographic regions in the world for deer diversity (Myers et al. 2000). Among the 15 species of South American deer currently described, two categories are included: species of small and spike-like antlers, weighing less than 30 kg, adapted to forests and closed vegetation (genera *Mazama, Subulo, Passalites* and *Pudu* – Figure 2.1);

*Corresponding author: mauricio.barbanti@unesp.br

and species with branched antlers, weighing more than 30 kg, inhabit open-field environments (genera *Odocoileus, Hippocamelus, Ozotoceros* and *Blastocerus*– Figure 2.1) (Merino and Rossi 2010, Gutierrez et al. 2017, Peres et al. 2021, Bernegossi et al. 2022).

Figure 2.1 Photos of some of the South American species identified by their scientific name and popular name: **(A)** *Blastocerus dichotomus* (Marsh deer) **(B)** *Mazama americana* (Red brocket deer) **(C)** *Mazama jucunda* (Small red brocket deer) **(D)** *Mazama nana* (Dwarf brocket deer) **(E)** *Mazama rufa* (Southern red brocket deer) **(F)** *Odocoileus virginianus* (White-tailed deer) **(G)** *Ozotoceros bezoarticus* (Pampas deer) **(H)** *Passalites nemorivagus* (Amazonian brown brocket deer) **(I)** *Subulo gouazoubira* (Brown brocket deer).

Many South American deer species have shown a declining trend, such as *Blastocerus dichotomus, Hippocamelus antisensis, Hippocamelus bisulcus, Mazama nana, Mazama jucunda, Mazama rufina, Mazama chunyi, Ozotoceros bezoarticus, Pudu mephistophiles,* and *Pudu puda* (Duarte and Gonzalez 2010). The reductions are estimated to range from 40 to 90% (Weber and Gonzalez 2003), which represents a conservation status considerably worse when compared to other mammals in the world. Globally in mammals, there is a 25% rate of threatened species and 15% are data deficient (Schipper et al. 2008). Data on the South American deer species shows that 5 species are Vulnerable (33%), 1 Endangered (7%), 2 Near Threatened (13%), 2 Data Deficient (13%), 1 not evaluated (*M. rufa*), and only are 3 Least Concern (IUCN 2022).

Wide deficiencies are found about information that refers to the evolution and taxonomy of the Cervidae family (Cifuentes-Rincón 2020, Gonzalez and Duarte 2020, Peres et al. 2021, Sandoval et al. 2022, Bernegossi et al. 2022). South American deer show great karyotypic variability (2n = 32–70), but especially in the genus *Mazama* (2n = 32–54), which has shown chromosomal differences but no clear morphological distinction (Duarte et al. 2008). Considering this information, it is fundamental for the taxonomy of deer that genetic studies are used to unravel the complex of cryptic species that exist (Gonzalez and Duarte 2020, Peres et al. 2021). One of the species considered as a superspecies or a system of cryptic species is *M. americana*. *Mazama americana stricto sensu* was described by Cifuentes-Rincón (2020) and revealed the big challenge to describe the other species inside the *M. americana sensu lato*; *Mazama jucunda*, *M. temama* and *M. rufa*, originally considered as junior synonymous of *M. americana*, were validated by integrative studies (Duarte and Jorge 2003, Peres et al. 2021, Mantellatto et al. 2022, Sandoval et al. 2022).

Some species of *M. americana* complex are very similar morphologically, but distinct cytogenetically. The chromosomal differences impose an efficient post zygotic reproductive barrier between these cytotypes from two chromosome pairs of difference, resulting in infertile or subfertile progenies (Cursino et al. 2014, Salviano et al. 2017, Galindo et al. 2021). Because of this, it is extremely important to evaluate all animals cytogenetically before they are part of artificial reproduction technologies (ART) programs. Thus, only animals with similar karyotype should be crossed.

In this context, in well-defined species, ART can be used to avoid genetic diversity losses in small populations, enabling the equal contribution of all individuals to the next generations (Duarte 2005, Jabbour et al. 2009). The collection and cryopreservation of gametes and embryos can considerably reduce the need to capture free living individuals or to transport captive animals between institutions (Wildt and Wemmer 1999). It is indicated that the sample collection is carried out not only in threatened species but also for those that are not yet at risk of extinction, in an attempt to retain the genetic diversity that is generally lost in rare species, after bottleneck and genetic drift (Duarte and Garcia 1995).

REPRODUCTIVE BIOLOGY

Most South American deer species have shown weaker seasonality or are completely aseasonal (Lincoln 1985), being accepted that reproduction of these species is more influenced by local climatic factors like annual rainfall patterns activity (Mishra and Wemmer 1987, Bubenik et al. 1991, Monfort et al. 1993) or food supply (Bubenik et al. 1991, Monfort et al. 1993) than by photoperiod. These climatic factors can result in some synchrony between reproduction of individuals (peak fawning), but conceptions and fawning can occur at any time of the year (Putman 1988).

Exceptions where the photoperiod seems to have some influence on South American deer reproduction can be found, as is the case of the Southern pudu (*Pudu puda*). They are animals whose reproduction is partially controlled by the photoperiod, but also respond to other environmental signals, such as variation of food supply. Also, they present postpartum estrus, which reflects a greater flexibility of the species to the photoperiod than we observe in other deer from temperate regions (Blanvillain et al. 1997). Studies have shown that, like the Northern pudu (*Pudu mephistophiles* — Wemmer 1998, Bubenik 2006), the Southern Pudu appears to have two reproductive periods per year or to reproduce for a longer period than previously documented (Vidal et al. 2012). In addition to the two previously mentioned species, the only South American deer females that seem to be seasonally polyestric are the North Andean deer (*Hippocamelus antisensis*) and Huemul (*Hippocamelus bisulcus*) (Merkt 1987, Blanvillain et al. 1997, Wemmer 1998).

Since the majority of the South American deer can reproduce at any time of the year, females of the genera *Blastocerus, Ozotoceros, Subulo, Passalites* and *Mazama*, are considered as continuous polyestric (Pereira et al. 2010). The occurrence of estrous cycles throughout the year has already been confirmed for brown brocket deer females (*Subulo gouazoubira*) kept in captivity (Santos et al. 2001). Also, captive males of the same species were monitored throughout the year, demonstrated aseasonality, with no morphological (testis size and antler cycle), endocrine (testosterone and cortisol levels), or seminal (volume, concentration, motility, and sperm morphology) variations correlated with environmental variations (temperature, photoperiod, and rainfall) (Barrozo et al. 2001).

SEMEN COLLECTION AND CRYOPRESERVATION

Male gametes are easier to obtain and cheaper to store than female gametes or embryos, but the knowledge to carry out their collection and freezing is still limited or non-existent for most wild species (Fickel et al. 2007). Moreover, sperm characteristics among related species have shown considerable differences in the viability, function, and success of fertilization (Leibo and Songsasen 2002). Regarding deer species, semen collection is obtained by three main methods: electroejaculation (EE), artificial vagina (AV), and post-mortem epididymal recovery (Asher et al. 2000). Seminal characteristics from South American deer species and different semen collection protocols are described in Table 2.1.

The most used method to obtain ejaculate from deer, whether in captive or free-living animals, is the EE. However, since chemical containment is necessary, successive collections with an interval of less than one week between procedures are not recommended (Duarte and Garcia 1997). Anatomical species-specific characteristics may influence the choice of the electrodes used during EE and the protocols may also require adaptations based on individual responses to electrical stimuli (Castelo and Silva 2014). Some studies performing EE in South American Deer are based on transrectal electrical stimuli ranging between 250 mA to 750 mA, administered in a three-second stimulation cycle followed by a three-second rest for a total of 10 stimuli. After one to two-minute resting period, the stimulation cycle is repeated attempting at least three stimulation cycles per session for successful semen collection (Duarte and Garcia 1995). The EE has been reported as an invasive method that affects several aspects of animal welfare, even under the effect of general anesthesia (Damián and Ungerfeld 2011, Fumagalli et al. 2012).

Table 2.1 Seminal characteristics and cryoprotectants used in the cryopreservation of semen from South American deer

Species	Collection	Volume (mL)	Concentration (sptz/mL)	Cryopreservation/ cryoprotectant	Motility Pre-freezing (%)	Motility Post-thawing (%)	References
Blastocerus dichotomus	EE	0.22 – 1.20	2×10^9	–	–	–	Duarte and Garcia 1995
	EE	0.37 ± 0.12	$2.68 \pm 0.81 \times 10^9$	Yes			Favoretto et al. 2012
				Tris-yolk (10%)	70	16.33	
				Tes-Tris-Yolk (20%) [a]	65	5.44	
				Tes-Tris-Yolk (20%)-Equex	70	24.66	
	EE	0.39 ± 0.14	$2.44 \pm 0.99 \times 10^9$	–	69.60 ± 8.92	–	Rola et al. 2013
Mazama americana	VA	0.14 ± 0.03	$4.32 \pm 1.17 \times 10^9$	Yes			Alvarez et al. 2020
				Tris-citric acid-egg yolk (20%) + Glycerol	66.30 ± 6.61	55.31 ± 7.39	
				Tris-citric acid-egg yolk (20%) + Ethylene glycol	66.30 ± 7.77	48.13 ± 2.39	
				Tris-citric acid-egg yolk (20%) + Dimethylformamide	66.90 ± 6.25	55.94 ± 2.77	
Mazama bororo jucunda	EE	0.06 – 0.40	$1.28 - 5.63 \times 10^9$	–	–	–	Unpublished data[b]
	EE	0.15 – 0.40	2×10^9		–	–	Duarte and Garcia 1995
Mazama nana				Yes			Abreu 2006
				Tris-citric acid-egg yolk (10%)		33.20	
				Tris-citric acid-egg yolk (10%) + Vitamin E		38.70	
	EE	0.02 – 0.25	$1.54 \pm 0.35 \times 10^9$	–	70.00 ± 8.16	–	Abreu et al. 2009

(Contd.)

Table 2.1 Seminal characteristics and cryoprotectants used in the cryopreservation of semen from South American deer.

Species	Collection	Volume (mL)	Concentration (sptz/mL)	Cryopreservation/ cryoprotectant	Motility Pre-freezing (%)	Motility Post-thawing (%)	References
	EE	0.42 ± 0.19	0.14 ± 0.11 × 10^9	—	—	—	Cursino and Duarte 2016
Passalites nemorivagus	EE	0.14	2.27 × 10^9	Yes Tris-citric acid-egg yolk (10%)	83	20	Orjuela et al. 2013
	EE	1.22 ± 0.13	1.59 ± 0.38 × 10^9	Yes Egg yolk and milk	83	36	Stewart et al. 2016
Odocoileus virginianus	EE	0.10 – 0.4	0.67 – 2.60 × 10^9	Biladyl-glicerol (4%)[c]	82	25	Duarte and Garcia 1995
				Biladyl-glicerol (6%)[c]	80	25	
				Biladyl-glicerol (8%)[c]	80	30	
				Triladyl[c]	81	40	
				Ovine Red[c]	80	40	
				Andromed[c]	80	40	
				Andromed[d]	61	50	
				Ovine Red[d]	72	50	
				Optixcell[d]	63	60	
	EE	0.41 ± 0.05	0.79 ± 0.10 × 10^9	—	—	—	Beracochea et al. 2014
Ozotoceros bezoarticus	EE			Yes Andromed	75	35	Remedi 2014
				Triladyl	75	50	
	VA	0.05 – 0.25	4.80 × 10^9	—	—	—	Soto, unpublished data

Species	Collection	Volume (mL)	Concentration (sptz/mL)	Cryopreservation/ cryoprotectant	Motility Pre-freezing (%)	Motility Post-thawing (%)	References
	EE	0.08 ± 0.03	0.41 × 10^9	Yes Tris SB milk	82.8 ± 2.8 80.6 ± 3.3	53.4 ± 2.6 49.4 ± 3.7	Muñoz-Toledo 2008
Pudu puda	EE	0.10 – 0.40	0.46 × 10^9	–	–	–	M. Fabry, unpublished data
	EE	0.05 – 1.50	0.17 – 0.22 × 10^9	–	–	–	Barrozo et al. 2001
	EE	0.46 ± 0.24	0.76 ± 0.38 × 10^9	–	–	–	Cursino and Duarte 2016
Subulo gouazoubira	post-mortem epididymal recovery	3.00^e	1.62 × 10^9	Yes/Botu-bov®	80%	50%	Assumpção and Santos 2017

[a] with modifications – without Equex STM paste, Nova Chemical SalesInc., Scituate, Massachusetts, USA.

[b] Data from the Deer Research and Conservation Center (NUPECCE).

[c] Experiment 1. Biladyl: egg yolk-based extender; Triladyl: egg yolk-based extender; Ovine Red: egg yolk-based extender; Andromed: soybean-based extender.

[d] Experiment 2. Andromed: soybean-based extender; Ovine Red: egg yolk-based extender; Optixcell: liposome-based extender.

[e] Volume of recovery medium used for sperm collection.

The use of AV offers the collection of an ejaculate with similar quality of the coitus, reducing the animal's stress (Duarte and Garcia 1989), although a previous training of the animals is required to minimize excitement and distress during handling (Spindler and Wildt 2010). Whenever possible, females in estrus can be used as a live mannequin, performing the lateral deviation of the penis towards AV. The success rates of collection by AV vary in the literature from 75 to 100% in *S. gouazoubira* (Rola et al. 2012) and from 50 to 70% in *M. americana* (Rola et al. 2012, Alvarez et al. 2020). When no females are available, it is also possible to use imprinted animals for seminal collections by AV, where the males perform the mounts directly on the operator's knee. This type of collection has already been described in *O. bezoarticus* (Soto et al. 1995), in *S. gouazoubira* (Giraldi and Duarte 2004) and *M. americana* (Alvarez et al. 2020).

Post-mortem sperm recovery acts as a last resort to obtain genetic material from animals that died unexpectedly or were euthanized for medical reasons, and are collected from vas deferens and/or the tail of the epididymis. However, one of the main points for the success of this technique is to perform seminal collection as soon as possible, and it is recommended that it occur before 24 h after death (Strand et al. 2016). A report with *S. gouazoubira* showed the viability of post-mortem epididymal recovery around 5 h after death, presenting motility of 80% and sperm vigor with score 3 (Assumpção and Santos 2017) (Table 2.1).

A unique characteristic observed in *Passalites nemorivagus* semen is the reddish coloration when collected by EE (Peroni et al. 2012) or pinkish when collected by VA (Rola, unpublished data). However, no blood cells or hemoglobin were found in the seminal plasma samples. Moreover, the presence of clusters of a red pigment was observed, with different sizes and formats that were not bound to the spermatozoa (Peroni et al. 2012). Therefore, it is not yet known what role this element plays in the fertility of this species.

ARTIFICIAL INSEMINATION (AI)

Among the assisted reproduction technologies, AI is considered one of the most promising for the conservation of wild species. This technique can be performed by monitoring the female's natural cycle of estrus, by synchronizing her estrus, or even with a fixed time where it is not necessary to follow the receptive behavior of females (see in Table 2.2). It could also be performed by several techniques, including intravaginal, intracervical, and intrauterine deposition of semen. However, the application of each one of them to different species of deer will depend on their reproductive anatomy. If passage through the cervix is not possible and semen deposition were performed vaginally, a greater number of sperm will be required for pregnancy to be achieved. The use of frozen semen with vaginal deposition was tested for *S. gouazoubira*, but it did not result in pregnancy (Duarte and Garcia 1995). In *O. virginianus*, pregnancy rates ranging between 50 and 100% have been reported after intravaginal/intracervical insemination with frozen semen (Jacobson et al. 1989, Magyar et al. 1989). Vaginal deposition of fresh semen was also used to obtain an offspring of *O. bezoarticus*

Table 2.2 Artificial insemination protocols and techniques used in different species of South American deer

Species	Protocol	Time of AI (h)*	AI technique	Semen	% Pregnancy (pregnancy/AI)	References
Odocoileus virginianus	Natural estrus	10-12	TCAI	Thawed	73 (33/45)	Jacobson et al. 1989
	Natural estrus	0, 12 or 18	TCAI	Thawed	75	Magyar et al. 1989
	Natural estrus	6, 24 or 30	TCAI	Thawed	60	Gentry et al. 2012
	CIDR® 14 days + eCG	60	TCAI	Thawed	59 (21/34)	
	CIDR® 14 days	60	TCAI	Thawed	43 (17/40)	
	CIDR® 14 days + eCG	60	TCAI	Fresh	40 (8/20)	Mellado et al. 2013
	CIDR® 14 days + eCG	60	LAI	Fresh	45 (113/251)	
Mazama americana	CIDR® 10 days + cloprostenol sodium (day 10) + eCG (day 10) + LH (together with IA)	80	LAI	Fresh	100 (1/1)	Zanetti et al. 2009
Passalites nemorivagus	Natural estrus	8h	TCAI	Refrigerated	100 (1/1)	Oliveira et al. 2016
Mazama jucunda	CIDR® 8 days + cloprostenol sodium (day 8) + LH (together with IA)	–	LAI	Thawed	100 (1/1)	Rola and Duarte, unpublished data
Ozotocerus bezoarticus	Natural estrus	–	Vaginal	Fresh	100 (1/1)	Rola and Duarte, unpublished data
Subulus gouazoubira	MPA for 14 days	–	Vaginal	Thawed	0 (0/1)	Duarte and Garcia 1995
	CIDR® 8 days + cloprostenol sodium (day 8) + LH (together with IA)	64 – 112h	LAI	Thawed	50 (4/8)	Peroni 2013

*time between the end of protocol and AI procedure or after the end of natural estrum, MPA = medroxyprogesterone acetate; CIDR® = "Controlled Internal Drug Release"; eCG = equine chorionic gonadotropin; LH = luteinizing hormone; TCAI = Transcervical intrauterine artificial insemination; LAI = Laparoscopic intrauterine artificial insemination

species, since the male exhibited imprinting behavior with humans and had no sexual interest in the female (Rola and Duarte, unpublished data).

An important point to achieve cervical transposition is the use of adequate instruments. In the case of the species *P. nemorivagus,* three attempts to perform AI were carried out, but cervical traction was not possible due to the inadequacy of the instruments used, the semen being deposited vaginally. In this attempt, no pregnancy was obtained. After developing specific material to attend the anatomy of this species, cervical traction was possible, and refrigerated semen was deposited inside the uterus, resulting in pregnancy (Oliveira et al. 2016).

Besides technique of intrauterine AI with cervical transposition being relatively simple, low cost, and minimally invasive, applying this technique can be difficult for many South American species. This difficulty is due to its anatomy which includes a small cervical opening, the length of the vagina and the large number of cervical rings (Duarte and Garcia 1995). When cervical transposition of the cervix is not an option, intrauterine deposition can be performed surgically, laparoscopy being the most reliable technique for AI (Jabbour et al. 1997, Duarte 2005). This has been the technique of choice for most non-domestic species, including small deer (Leibo and Songsasen 2002), as it allows placement of relatively small amounts of semen into the uterine lumen and use of frozen semen effectively (Asher et al. 2000). However, the technique has some limitations such as the need for animal anesthesia, technical training and equipment costs (Willard et al. 1998).

Multiple Ovulation and Embryo Transfer

Multiple ovulation, known as superovulation (SOV), is one of the factors that will determine the success of embryo collection. A great variation in the response to hormonal protocols for SOV was observed in deer species, poor responses in wild ungulates being associated with stress, which can be induced by intense manipulation during the procedures (Comizzoli et al. 2000). The treatments proposed for SOV in deer involve the use of progesterone devices associated with a single injection of equine chorionic gonadotropin (eCG) and/or multiple injections of follicle-stimulating hormone (FSH) that are administered a few days before or at the moment of progesterone withdrawal (Jabbour et al. 2009) as shown in Table 2.3. Despite obtaining better SOV responses in treatments where higher doses of eCG are used, some authors report that it can cause a premature regression of the CL, indicating possible poor follicle quality produced (Zanetti et al. 2014).

For embryo collection, it is necessary to carry out a uterine flushing using a catheter, which can be inserted via transcervical or by laparoscopy/laparotomy techniques. Although the collection by transcervical method has already been successful in *O. virginianus* (Magyar et al. 1988), this technique is unsuitable for most South American deer species due to difficulties in transposing the catheter through the cervix. Among the techniques, laparoscopy is less invasive and presents lower chances of adhesions in the reproductive tract, which could

Table 2.3 Superovulation protocols used in different species of South American deer

Species	Superovulation protocols	Ovulation (CL)	Total stimulus	Fertilization method	Collection and/ or CL counting method	Collection rate (%)	References
Subulo gouazoubira	MPA sponge for 13 or 14 days + 700 iu eCG (day 13 or 14) + hCG day 16, 17 or in the moment of the AI	–	–	Natural copulation	Laparotomy	–	Duarte and Garcia 1995
	oCIDR for 8 days, 0.5 mg of EB (D0), 600 iu of eCG (D4) and 250 µg of cloprostenol sodium (D8)	3.40 ± 0.68	4.80 ± 1.02	Natural copulation	Videolaparoscopy	–	Zanetti and Duarte 2012
	oCIDR for 8 days, 0.5 mg of EB (D0), 300 iu of eCG (D4) and 250 µg of cloprostenol sodium (D0)	1.40 ± 0.24	1.80 ± 0.37	Natural copulation	Videolaparoscopy	–	
	oCIDR for 8 days, 0.5 mg of EB (D0), 250 mg of FSH dissolved in polyvinylpyrrolidone (D4) and 250 µg of cloprostenol sodium (D8)	0.80 ± 0.49	1.40 ± 0.60	Natural copulation	Videolaparoscopy	–	
	oCIDR for 8 days, 0.5 mg of EB (D0), 700 iu of eCG (D4) and 250 µg of cloprostenol sodium (D8)	6.0 ± 1.7	13.2 ± 2.6	Natural copulation	Videolaparoscopy	–	Zanetti et al. 2014
	oCIDR for 7.5 days, 0.5 mg of EB (D0), 130 mg FSH (D3.5/D7 – each 12h) and 250 µg of cloprostenol sodium (D7.5)	2.0 ± 0.3	10.4 ± 1.1	Natural copulation	Videolaparoscopy	–	
Odocoileus virginianus	MPA sponge for 10 or 14 days; or cloprostenol sodium (14 d post estrus) + 1000 iu eCG	3.1	–	Natural copulation	Laparotomy	68	Waldham et al. 1989

(Contd.)

Table 2.3 Superovulation protocols used in different species of South American deer (*Contd.*)

Species	Superovulation protocols	Ovulation (CL)	Total stimulus	Fertilization method	Collection and/ or CL counting method	Collection rate (%)	References
	Single injection of cloprostenol sodium (25 mg) given 11 days post estrus	Three animals were used but just one was analyzed for the number of CLs that was 3 structures	–	Artificial insemination	Transcervical	One blocked, one poor quality, two of good quality	Magyar et al. 1988
Blastocerus dichotomus	0.03 mg progesterone + 2 mg EB (D0); 1500 UI FSH (D4); 0.53 mg cloprostenol sodium (D7)	0 (4 animals)	0.25	Natural copulation	Laparotomy	–	Duarte, unpublished data
	oCIDR for 8 days + 1 mg EB (D0); 1200 UI FSH (D4); 0.53 mg cloprostenol sodium (D8)	0.2 (5 animals with just 1 animal with 2 CLs)	0.4	Natural copulation	Laparotomy	50 (1 degenerated)	
	oCIDR for 8 days + 1mg EB (D0); 1200 UI FSH (D3); 0.53 mg cloprostenol sodium (D7)	0.66 (3 animals were 2 presenting 1 CL each)	5.33	Natural copulation	Laparotomy	50 (1 blocked)	
	CIDR for 7 days + 0.5 mg EB + 0.11 mg gonadorelin tetrahydrate diacetate (D0); 800 UI eCG (D5); 0.53 mg cloprostenol sodium (D7); 0.11 mg gonadorelin tetrahydrate diacetate (after estrum was detected)	2 (1 animal)	2	Natural copulation	Laparotomy	–	Galindo et al. 2015

Species	Superovulation protocols	Ovulation (CL)	Total stimulus	Fertilization method	Collection and/ or CL counting method	Collection rate (%)	References
	CIDR for 8 days + 0.5 mg EB + 0.11 mg gonadorelin tetrahydrate diacetate (D0); 1200 UI eCG (D5); 0.53 mg cloprostenol sodium (D7); 0.11 mg gonadorelin tetrahydrate diacetate (after estrum was detected)	1 (1 animal)	9	Natural copulation	Laparotomy	–	Galindo et al. 2015
	CIDR for 8 days + 0.5 mg EB + 0.11 mg gonadorelin tetrahydrate diacetate (D0); 1200 UI eCG (D5); 0.53 mg cloprostenol sodium (D7); 2.5 mg LH (between 12–16 hours after estrum was detected)	8 (3 animals)	10	Natural copulation	Laparotomy	37.5 (2 blocked, 4 degenerate, 2 viable, 1 regular)	Galindo et al. 2015 Duarte, unpublished data

CL = corpus luteum; MPA = medroxyprogesterone acetate; eCG = equine chorionic gonadotropin; CIDR® = "Controlled Internal Drug Release"; EB = estradiol benzoate; FSH = follicle stimulating hormone; LH = luteinizing hormone.

lead to reduced fertility and embryo collection. Surgical procedures are applied seven/eight days after insemination/copulation, showing success in embryo collection (Jabbour et al. 2009). Using laparotomy, a greater collection efficiency in *S. gouazoubira* species was obtained, resulting in up to eight ova and embryos (Duarte and Garcia 1995). For the surgical procedure in *B. dichotomus*, laparotomy was used to flush the uterus and tubas (Galindo et al. 2015), and the two viable embryos obtained were vitrified and stored in liquid nitrogen. Subsequently, these embryos were rewarmed and transferred into a free-living female belonging to a population that had been reintroduced, and from which it was known to need genetic reinforcement (Duarte, unpublished data).

In vitro Embryo Production

Among the available ARTs, *in vitro* embryo production (IVEP) is considered the most efficient for the propagation of small populations (Comizzoli et al. 2000). The IVEP technology has advantages over AI and *in vivo* embryo production, such as total flexibility in pairing between males and females, the possibility of obtaining a greater number of embryos than *in vivo* collection per donor and allows the cryopreservation of oocytes and embryos.

Despite all the advantages, this technique is considered expensive and includes the following steps: oocyte recovery, *in vitro* maturation (IVM), *in vitro* fertilization (IVF), and *in vitro* development of embryos (IVD). The low success rates of IVEP are associated with failure in some of these steps, such as the ability to obtain good quality oocytes, the lack of knowledge of the optimal conditions for the maturation of these gametes and embryos, and uncertainties about their subsequent transfer (Reviewed by Rola et al. 2021a).

For *S. gouazoubira*, the only neotropical species where the oocyte recovery was performed using laparoscopic *ovum pick up* (LOPU), an oocyte recovery rate of 52.84% was obtained, which resulted in the collection of an average of 7.15 ± 3.72 oocytes/female. The recovered oocytes were submitted to IVM for 24 h and stained with Hoechst 33342 to evaluate the nuclear status; 64.5% reached metaphase II and 16.1% were spontaneously activated (Rola et al. 2021b). As can be seen from the results obtained so far, much remains to be done in order to improve the effectiveness of IVEP in neotropical deer.

THE POTENTIAL OF SOMATIC CELL SOURCES

Somatic cells are biological materials considered very promising, representing today a major cryopreserved source for South-American deer species. The high representativeness of this samples in relation to the maintenance of gametes and embryos is due to the greater ease in obtaining and cryopreservation of tissues and cells (Rola et al. 2021a).

Somatic Cell Nuclear Transfer (SCNT)

The ability of cytoplasmic machinery within the oocyte to reprogram differentiated somatic cell nuclei has provided potential strategies to generate embryos and living offspring, making this technique extremely interesting for endangered wild species. Regarding the difficulty in obtaining oocytes to use as cytoplasts, interspecific cloning (iSCNT) is an alternative (Lanza et al. 2000). The generation of animals through iSCNT also presents several obstacles such as compatibility of genomic/mitochondrial DNA, embryonic genomic activation of the donor nucleus (karyoplast) by the recipient oocyte (cytoplast), and the availability of receptors for the reconstructed embryos. To increase the chances of compatibility between karyoplast and cytoplast, the use of oocytes of phylogenetically similar domestic species is recommended (Lagutina et al. 2013).

Among the initial steps for the development of cloning is the evaluation of the cell line to be used as a karyoplast. In studies using *S. gouazoubira* fibroblasts, it was observed that as more passages in culture were performed, the greater the number of fibroblasts with fragmented DNA. Passages four and seven demonstrated good viability after frozen/warmed; however, low quality of 10th passage was reported, demonstrating that its use was not suitable for iSCNT (Magalhães et al. 2017).

Among neotropical deer species, iSCNT has already been reported in *S. gouazoubira* and *P. puda*. For *P. puda*, iSCNT was performed using fibroblasts taken from the ear as karyoplast and bovine oocytes as cytoplasts. It was possible to obtain embryos up to the blastocyst stage with normal morphology and karyotype. Embryos reconstructed from 4th passage fibroblasts produced only structures up to the 8–16 cells (9.5%). Using 6th passage fibroblasts, better results were obtained, with 53% reaching the 8–16 cell phase and where two embryos (4%) reached the blastocyst stage (Venegas et al. 2006).

For *S. gouazoubira*, goat (Magalhães et al. 2020) and cattle (Melo et al. 2022) cytoplasts were used. The reconstructed deer/goat embryos were able to develop until the morula stage (12.5%) whereas, for the reconstructed deer/cattle embryos, 5.9% of them reached the blastocyst stage (Figure 2.2). To assess the influence of heteroplasmy on these embryos, the mitochondrial activity was measured through the levels of expression of some of its genes (*ATP6*, *COX3*, and *ND5*). These studies indicate that both goat and cattle oocytes can reprogram the somatic cells of *S. gouazoubira* demonstrating the potential of iSCNT in the preservation and dissemination of genetic diversity from small and fragmented populations of rare South America deer species.

USE OF STEM CELLS

Several potential strategies for the use of stem cells have been developed recently, which offer the opportunity to produce individuals from the reprogramming of terminally differentiated cells in pluripotent lines (iPSC) or the transdifferentiation of adult stem cells into gametes. In *B. dichotomus*, three adult stem cell lines (antler, fat, and skin) were cultured and characterized and showed to be promising sources to assist in the construction of germplasm banks due to their plasticity, high rate of proliferation, as well as ease of obtaining. In all cell types, the multipotency of cells can be proven by the ability to differentiate into adipocytes, osteocytes, and chondrocytes. They also have rapid cell doubling time (25.96 h for antler, 32.24 h for fat, and 33.32 h for skin) and have expressed pluripotency markers such as *OCT4, SOX2, NANOG, REX1*, and *LIN28* (Rola et al. 2020a).

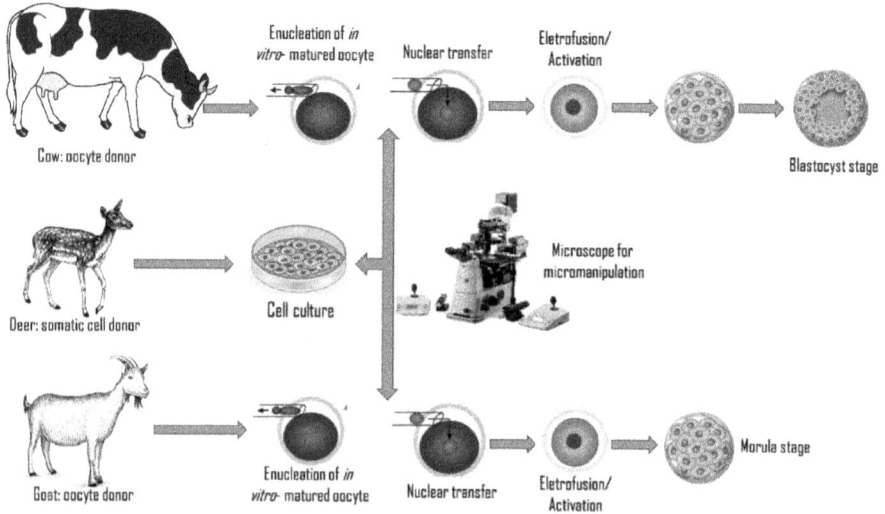

Figure 2.2 The different steps for cloning of *Mazama gouazoubira* deer using interspecific somatic cell nuclear transfer technique. The use of bovine oocytes resulted in better embryonic development (up to the blastocyst), while goat oocytes only reached the morula stage.

The iPSC reprogramming efficiency varies according to the cell types used in a similar way to what occurs in the SCNT, being lower in terminally differentiated cells and faster and more efficient in less differentiated or undifferentiated cells (Liebau et al. 2013). Multipotent stem cells from antler, fat, and skin of *B. dichotomus* were tested to be reprogrammed into iPSC. Nonetheless, resistance to input the reprogramming transcription factors into the interior of the cells was observed, and also the presence of colonies with partial reprogramming that did not develop (Rola et al. 2021c). Despite that, a preliminary study showed that when these same multipotent lineages were directly subjected to transdifferentiation into PGCs, fat and skin responded well to the protocol, and showed morphological changes, expression of surface proteins, and transcripts consistent with that of PGCs (Rola et al. 2020b).

Other interesting sources to be maintained in the germplasm banks are the gonad tissues and the spermatogonial stem cells (SSCs). A study with *Odocoileus virginianus* obtained post-mortem testis tissue collected from a sexually immature male, and fragments of the tissue were grafted under the back skin of immunodeficient recipient mice. This fragment resulted in successful testicular maturation and development of spermatogenesis, where 12 months post-grafting fully formed sperm was observed (Abbasi and Honaramooz 2012).

Final Considerations

Assisted Reproductive Technology may help to preserve the genetic diversity in captive and free ranging populations. These techniques are useful to permit gene flow between institutions from *ex situ* programs, and also between *ex situ* and *in situ* populations. Techniques like embryo transfer and artificial insemination would be used for this purpose. It is an urgent goal to implant banks of South American deer cells (skin, fibroblasts, semen, embryos) to preserve the allelic pool for future use if necessary. The advance of cloning and reprograming technology will permit to use the somatic cell bank to produce animals and help the sustainability of the endangered species.

REFERENCES

Abbasi, S. and A. Honaramooz. 2012. Feasibility of salvaging genetic potential of post-mortem fawns: production of sperm in testis tissue xenografts from immature donor white-tailed deer (*Odocoileus virginianus*) in recipient mice. Anim. Reprod. Sci. 135(1–4): 47–52.

Abreu, C.O. 2006. Efeito da vitamina E na criopreservação de sêmen de veado-bororó-do-sul (*Mazama nana*). Ph.D. Thesis, Universidade Estadual Paulista, Jaboticabal, São Paulo, Brazil.

Abreu, C.O., A.C. Martinez, W. Moraes, J.C. Juvenal and N. Moreira. 2009. Características reprodutivas de veado-bororó-do-sul ou veado-mão-curta (*Mazama nana*). Pesq. Vet. Bras. 29(12): 993–998.

Alvarez, M.C.L., L.D. Rola and J.M.B. Duarte. 2020. Comparison between Three Cryoprotectants in the Freezing of *Mazama americana* Semen Collected by Artificial Vagina. Biopreserv. Biobank. 18(5): 351–357.

Asher, G.W., D.K. Berg and G. Evens. 2000. Storage of semen and artificial insemination in deer. Anim. Reprod. Sci. 62(1–3): 195–211.

Assumpção, T.I. and A.L.Q. Santos. 2017. Cryopreservation of genetic material collected post-mortem from male gray brocket deer *Mazama gouazoubira* Fischer, 1814. Int. J. Curr. Sci. Tec. 5: 510–512.

Barrozo, L.A., G.H. Toniolo, J.M.B. Duarte, M.P. Pinho and J.A. Oliveira. 2001. Padrão anual de variação da testosterona sérica, volume testicular e aspectos seminais de veados-catigueiros (*Mazama gouazoubira*) em cativeiro. Rev. Bras. Reprod. Anim. 25(2): 210–211.

Beracochea, F., J. Gil, A. Sestelo, J.J. Garde, J. Santiago-Moreno, F. Fumagalli, et al. 2014. Sperm characterization and identification of sperm sub-populations in ejaculates from pampas deer (*Ozotoceros bezoarticus*). Anim. Reprod. Sci. 149(3–4): 224–230.

Bernegossi, A.M., C.H.S. Borges, E.D.P. Sandoval, J.L. Cartes, H. Cernohorska, S. Kubickova, et al. 2022. Resurrection of the genus *Subulo* Smith, 1827 for the gray brocket deer, with designation of a neotype. J. Mammal. gyac068.

Blanvillain, C., J.L. Berthier, M.C. Bomsel-Demontoy, A.J. Semperé, G. Olbricht and F. Schwarzenberg. 1997. Analysis of reproductive data and measurement of fecal progesterone metabolites to monitor the ovarian function in the pudu, *Pudu puda* (Artiodactyla, Cervidae). Mammalia 61(4): 589–602.

Bubenik, G.A., R.D. Brown and D. Schams. 1991. Antler cycle and endocrine parameters in male axis deer (*Axis axis*): seasonal levels of LH, FSH, testosterone and prolactin and results of GnRH and ACTH challenges tests. Comp. Biochem. Physiol. 99A: 645–650.

Bubenik, G.A. 2006. Seasonal versus non-seasonal reproduction in deer: From the arctic to the tropics. Advances in Deer Biology, 16.

Castelo, T.S. and A.R. Silva. 2014. Eletroejaculação em mamíferos silvestres: principais fatores que afetam sua eficiência. Rev. Bras. Reprod. Anim. 38: 208–213. http://www.cbra.org.br/pages/publicacoes/rbra/v38n4/pag208-213%20(RB529).pdf

Cifuentes-Rincón, A. 2020. Caracterização morfológica, citogenética e molecular de *Mazama americana* (Artiodactyla: Cervidae) a partir de um topótipo atual. Dissertação (Mestrado em Genética e Melhoramento Animal), Universidade Estadual Paulista Júlio de Mesquita Filho, Jaboticabal, Brasil.

Comizzoli, P., P. Mermillod and R. Mauget. 2000. Reproductive biotechnologies for endangered mammalian species. Reprod. Nutr. Dev. 40: 493–504.

Cursino, M.S., M.B. Salviano, V.V. Abril, E.S. Zanetti and J.M.B. Duarte. 2014. The role of chromosome variation in the speciation of the red brocket deer complex: the study of reproductive isolation in females. BMC Evol. Biol. 14: 40.

Cursino, M.S. and J.M.B. Duarte. 2016. Using sperm morphometry and multivariate analysis to differentiate species of gray *Mazama*. R. Soc. Open Sci. 3(11): 160–345.

Damián, J.P. and R. Ungerfeld. 2011. The stress response of frequently electroejaculated rams to electroejaculation: hormonal, physio-logical, biochemical, haematological and behavioural parameters. Reprod. Domest. Anim. 46(4): 646–650.

Duarte, J.M.B. and J.M. Garcia. 1989. Colheita e criopreservação do sêmen de veado-catingueiro (*Mazama gouazoubira*). Ciênc. Vet. 3: 8–9.

Duarte, J.M.B. and J.M. Garcia. 1995. Reprodução assistida em Cervidae brasileiros. Rev. Bras. Reprod. Anim. 19(1–2): 111–121.

Duarte, J.M.B. and J.M. Garcia. 1997. Tecnologia da reprodução para propagação e conservação de espécies ameaçadas de extinção. pp. 228–238. *In*: J.M.B. Duarte (ed.). Biologia e Conservação de Cervídeos Sul-americanos: *Blastoceros*, *Ozotoceros* e *Mazama*. Fundação de Estudos e Pesquisas em Agronomia, Medicina Veterinária e Zootecnia, Jaboticabal, São Paulo, Brasil.

Duarte, J.M.B. and W. Jorge. 2003. Morphologic and cytogenetic description of the small red brocket (*Mazama bororo* Duarte, 1996) in Brazil. Mammalia 67(3): 403–410.

Duarte, J.M.B. 2005. Coleta, conservação e multiplicação de recursos genéticos em animais silvestres: o exemplo dos cervídeos. Agrociencia 9: 541–544.

Duarte, J.M.B., S. Gonzalez and J.E. Maldonado. 2008. The surprising evolutionary history of South American deer. Mol. Phylog. Evol. 49: 17–22.

Duarte, J.M.B. and S. Gonzalez. 2010. Neotropical Cervidology, Biology and Medicine of Latin American Deer. Funep/IUCN, Jaboticabal/Gland, p. 394.

Favoretto, S.M., E.S. Zanetti and J.M.B. Duarte. 2012. Cryopreservation of red brocket deer semen (*Mazama americana*): comparison between three extenders. J. Zoo. Wildl. Med. 43(4): 820–827.

Fickel, J., A. Wagener and A. Ludwig. 2007. Semen cryopreservation and the conservation of endangered species. Eur. J. Wildl. Res. 53: 81–89.

Fumagalli, F., M. Villagrán, J.P. Damián and R. Ungerfeld. 2012. Physiological and biochemical parameters in response to electroejac-ulation in adult and yearling anesthetized pampas deer (*Ozotoceros bezoarticus*) males. Reprod. Domest. Anim. 47(2): 308–312.

Galindo, D.J., J.M. Garcia, M.E.F. Oliveira and J.M.B. Duarte. 2015. Superovulation with equine chorionic gonadotropin in Marsh deer (*Blastocerus dichotomus*). Spermova 5: 163–167.

Galindo, D.J., G.S. Martins, M. Vozdova, H. Cernohorska, S. Kubickova, A.M. Bernegossi, et al. 2021. Chromosomal polymorphism and speciation: The case of the genus *Mazama* (Cetartiodactyla; Cervidae). Genes 12(2): 165.

Gentry, G.T., J. Lambe, W. Forbes, B. Olcott, D. Sanders, K. Bondioli, et al. 2012. The effect of equine chorionic gonadotropin (eCG) on pregnancy rates of white-tailed deer following fixed-timed artificial insemination. Theriogenology 77: 1894–1899.

Giraldi, T. and J.M.B. Duarte. 2004. Implantação e Avaliação de Métodos não Invasivos para Colheita de Sêmen de Veado-Catingueiro (*Mazama gouazoubira*). Proc. Congr ABRAVAS, Jaboticabal, São Paulo, Brazil 8: 83.

Gonzalez, S. and J.M.B. Duarte. 2020. Speciation, evolutionary history and conservation trends of Neotropical Deer. Mastozool. Neotrop. 27: 37–47.

Gutiérrez, E.E., K.M. Helgen, M.M. McDonough, F. Bauer, M.T.R. Hawkins, L.A. Escobedo-Morales, et al. 2017. A gene-tree test of the traditional taxonomy of American deer: the importance of voucher specimens, geographic data, and dense sampling. ZooKeys 697: 87–131.

The IUCN red list of threatened species 2022. Available online: https://www.iucnredlist.org/ (accessed on 29 november 2022).

Jabbour, H.N., V. Hayssen and M.W. Bruford. 1997. Conservation of deer: contributions from molecular biology, evolutionary ecology, and reproductive physiology. J. Zool. 243(3): 461–484.

Jacobson, H.A., H.J. Bearden and D.B. Whitehouse. 1989. Artificial insemination trials with white-tailed deer. J. Wildl. Manage. 53: 224–227.

Lagutina, I., H. Fulka, G. Lazzari and C. Galli. 2013. Interspecies somatic cell nuclear transfer: advancements and problems. Cell. Reprogram. 15: 374–384.

Lanza, R.P., J.B. Cibelli, F. Diaz, C.T. Moraes, P.W. Farin, C.E. Farin, et al. 2000. Cloning of an endangered species (*Bos gaurus*) using interspecies nuclear transfer. Cloning 2: 79–90.

Leibo, S.P. and N. Songsasen. 2002. Cryopreservation of gametes and embryos of non-domestic species. Theriogenology 57(1): 303–326.

Liebau, S., P.U. Mahaddalkar, H.A. Kestler, A. Illing, T. Seufferlein and A. Kleger. 2013. A hierarchy in reprogramming capacity in different tissue microenvironments: what we know and what we need to know. Stem Cells Dev. 22: 695–706.

Lincoln, G.A. 1985. Seasonal breeding in deer. In the Biology of Deer Production Eds PF Fennessy and KR Drew. Bulletin No. 22, Royal Society of New Zealand, Wellington 165–179.

Magalhães, L.C., J.V. Cortez, M.H. Bhat, A.C.N. Sampaio, J.L. Freitas, J.M.B. Duarte, et al. 2020. *In vitro* development and mitochondrial gene expression in brown brocket deer (*Mazama gouazoubira*) embryos obtained by inter-specific somatic cell nuclear transfer. Cell. Reprogram. 22: 208–216.

Magalhães, L.C., M.H. Bhat, J.L. Freitas, L.M. Melo, D.I.A. Teixeira, L.C. Pinto, et al. 2017. The effects of cryopreservation on different passages of fibroblast cell culture in brown brocket deer (*Mazama gouazoubira*). Biopreserv. Biobank. 15: 463–468.

Magyar, S.J., C. Hodges, S.W.J. Seager and D.C. Kraemer. 1988. Successful nonsurgical embryo collection with surgical transfer in captive white-tailed deer. Theorigenology 29: 273.

Magyar, S.J., T. Biediger, C. Hodges, D.C. Kraemer and S.W.J. Seager. 1989. A method of artificial insemination in captive white-tailed deer (*Odocoileus virginianus*). Theriogenology 31(5): 1075–1080.

Mantellatto, A.M.B., S. González and J.M.B. Duarte. 2022. Cytochrome b sequence of the *Mazama americana jucunda* Thomas, 1913 holotype reveals *Mazama bororo* Duarte, 1996 as its junior synonym. Genet. Mol. Biol. 45(1): e20210093.

Mellado, M., C.G. Orta, E.A. Lozano, J.E. Garcia, F.G. Veliz and A. Santiago. 2013. Factors affecting reproductive performance of white-tailed deer subjected to fixed-time artificial insemination or natural mating. Reprod. Fertil. Develop. 25: 581–586.

Melo, L.M., S.B. Silva, L.C. Magalhães, J.V. Cortez, S. Kumar, J.M.B. Duarte, et al. 2022. The use of somatic cell nuclear transfer to obtain interspecific cloned embryos from brown brocket deer karyoplast and bovine cytoplast: Embryo development and nuclear gene expression. Theriogenology Wild 1: 100001.

Merino, M.L. and R.V. Rossi. 2010. Origin, Systematics, and Morphological Radiation. pp. 2–11. *In*: J.M.B. Duarte and S. Gonzalez (eds). Neotropical Cervidology, Biology and Medicine of Latin American Deer. Funep/IUCN, Jaboticabal/Gland.

Merkt, J.R. 1987. Reproductive seasonality and grouping patterns of the North Andean deer or taruca (*Hippocamelus antisensis*) in Southern Peru. pp. 388–401. *In*: C. Wemmer (ed.). Biology and Management of the Cervidae. Smithsonian Inst. Press, Washington D.C., USA.

Mishra, H.J. and C. Wemmer. 1987. The comparative breeding ecology of four cervids in Royal Chitwan National Park. p. 577. *In*: C. Wemmer (ed.). Biology and management of the cervidae. Washington D.C.: Smithsonian Inst. Press.

Monfort, S.L., J.L. Brown, M. Bush, T.C. Wood, C. Wemmer, A. Vargas, et al. 1993. Circannual inter-relationships among reproductive hormones, gross morphometry, behaviour, ejaculate characteristics and testicular histology in Eld's deer stags (*Cervus eldi thamin*). J. Reprod. Fertil. 98: 471–480.

Myers, N., R.A. Mittermeier, C.G. Mittermeier, G.A.B. Fonseca and J. Kent. 2000. Biodiversity hotspots for conservation priorities. Nature 403: 853–858. http://doi. org/10.1038/35002501.

Muñoz-Toledo, S.B. 2008. Congelación de semen de Pudú (Pudu pudu): Efecto de los diluyentes Tris y leche SB sobre semen diluido. Bachelor Thesis, Universidad Austral de Chile, Valdivia, Chile.

Oliveira, M.E.F., E.S. Zanetti, M.S. Cursino, E.D.F.C. Peroni, L.D. Rola, M.A.R. Feliciano, et al. 2016. First live offspring of Amazonian brown brocket deer (*Mazama nemorivaga*) born by artificial insemination. Eur. J. Wildl. Res. 62(6): 767–770.

Orjuela, L.L.C., M.D. Christofoletti, M.S. Cursino and J.M.B. Duarte. 2013. Comparação de dois diluentes para congelação de sêmen de *Mazama nemorivaga*. Proc. Congr ABRAVAS, Salvador, Bahía, Brazil 16: 194–195.

Pereira, R.J.G., E.S. Zanetti and B.F. Polegato. 2010. Female reproduction. pp. 51–64. *In*: J.M.B. Duarte and S. Gonzales (eds). Neotropical cervidology: Biology and Medicine of Latin American Deer. FUNEP: Jaboticabal, Brasil.

Peres, P.H.F., D.J. Luduvério, A.M. Bernegossi, D.J. Galindo, G.B. Nascimento, M.L. Oliveira, et al. 2021. Revalidation of *Mazama rufa* (Illiger 1815) (Artiodactyla: Cervidae) as a distinct species out of the complex *Mazama americana* (Erxleben 1777). Front. Genet. 12: 1–18.

Peroni, E.F.C. 2013. Inseminação artificial intrauterina por videolaparoscopia em veado catingueiro (*Mazama gouazoubira*) com sêmen congelado. Dissertação (mestrado)– Universidade Estadual Paulista, Faculdade de Ciências Agrárias e Veterinárias de Jaboticabal, Brazil, pp. 56.

Peroni, E.F.C., E.S. Zanetti, L.D. Rola, M.S. Cursino and J.M.B. Duarte. 2012. Morfologia Espermática dos Veados-Cinza Brasileiros (*Mazama gouazoubira* e *Mazama nemorivaga*) Mantidos em Cativeiro: Resultados Preliminares. Proc. Congr ABRAVAS, Florianópolis, Santa Catarina, Brazil 15: 30.

Putman, R. 1988. The natural history of deer. Cornell University Press, New York, United States.

Remedi, S.V. 2014. Criopreservación de Semen de Venado de Campo (*Ozotoceros bezoarticus*): Comparación de la Efectividad de 1055. Dos Diluyentes Comerciales a la Descongelación. Thesis, Universidad de la Republica de Uruguay, Montevidéo, Uruguay.

Rola, L.D., E.S. Zanetti and J.M.B. Duarte. 2012. Avaliação de dois métodos para condicionamento e coleta de sêmen em quatro espécies do gênero *Mazama*. Pesq. Vet. Bras. 32(7): 658–662.

Rola, L.D., E.S. Zanetti and J.M.B. Duarte. 2013. Evaluation of semen characteristics of the species *Mazama americana* in captivity. Anim. Prod. Sci. 53(5): 472–477.

Rola, L.D., I.M. Tomazella, M.E. Buzanskas, J. Therrien, L.C. Smith and J.M.B. Duarte. 2020a. Isolamento, expansão e caracterização de três tipos de células-tronco Multipotentes de cervo-do-pantanal (*Blastocerus dichotomus*). VI Congresso Brasileiro de Recursos Genéticos, Brasília, Brazil.

Rola, L.D., F.F. Bressan, F. Meirelles, N.C.G. Pieri, J.M.B. Duarte and L.C. Smith. 2020b. Etapas iniciais para a produção *in vitro* de gametas a partir de linhagens celulares multipotentes de cervo-do-pantanal. VI Congresso Brasileiro de Recursos Genéticos, Brasília, Brazil.

Rola, L.D., M.E. Buzanskas, L.M. Melo, M.S. Chaves, V.J.F. Freitas and J.M.B. Duarte. 2021a. Assisted reproductive technology in neotropical deer: a model approach to preserving genetic diversity. Animals (Basel) 11: 1961.

Rola, L.D., E.S. Zanetti, M. Del Collado, E.F.C. Peroni and J.M.B. Duarte. 2021b. Collection and *in vitro* maturation of *Mazama gouazoubira* (brown brocket deer) oocytes obtained after ovarian stimulation. Zygote 29: 216–222.

Rola, L.D., F.F. Bressan, M.E. Buzanskas, L.C. Smith and J.M.B. Duarte. 2021c. Uso de diferentes técnicas para tentar derivar linhagens de células-tronco de pluripotência induzida em cervo-do-pantanal. XXIX Encontro e XXIII Congresso ABRAVAS.

Salviano, M.B., M.S. Cursino, E.D.S. Zanetti, V.V. Abril and J.M.B. Duarte. 2017. Intraspecific chromosome polymorphisms can lead to reproductive isolation and speciation: an example in red brocket deer (*Mazama americana*). Biol. Reprod. 96: 1279–1287.

Sandoval, E.D.P., L.D. Rola, J.A. Morales-Donoso, S. Gallina, R. Reyna-Hurtado and J.M.B. Duarte. 2022. Integrative analysis of *Mazama temama* (Artiodactyla; Cervidae) and designation of a neotype for the species. J. Mammal. 103(2): 447–458.

Santos, G.L., L. Ceravolo, S. Souza and J.M.B. Duarte. 2001. Sazonalidade reprodutiva e duração do ciclo estral e do cio de fêmeas de veado-catingueiro (*Mazama gouazoubira*) sob condições de cativeiro. Proceedings V Congreso Internacional Manejo del Fauna Silvestre en Amazonía y Latinoamerica. Cartagena, Colombia 232.

Schipper, J., J.S. Chanson, F. Chiozza, N.A. Cox, M. Hoffmann, V. Katariya, et al. 2008. The status of the world's land and marine mammals: diversity, threat, and knowledge. Science 322: 225–230.

Soto, A., P. Vallejos, M.I. Mezino and B.M. Carpiretti. 1995. Obtención de semen por medio de vagina artificial en venado de las pampas (*Ozotoceros bezoarticus* Linneaus 1756). Proc. JAM, La Plata, Argentina 10: 61.

Spindler, R.E. and D.E Wildt. 2010. Male reproduction: assessment, management, assisted breeding, and fertility control. pp. 429–446. *In*: D.G. Kleiman, K.V. Thompson and C.K. Baer (eds). Wild Mammals in Captivity, 2nd Ed. University of Chicago Press: Chicago, USA.

Stewart, J.L., C.F. Shipley, A.S. Katich, E. Po, R.E. Ellerbrock, F.S. Lima, et al. 2016. Cryopreservation of white-tailed deer (*Odocoileus virginianus*) semen using soybean-, liposome-, and egg yolk-based extenders. Anim. Reprod. Sci. 171: 7–16.

Strand, J., M.M. Ragborg, H.S. Pedersen, T.N. Kristensen, C. Pertoldi and H. Callesen. 2016. Effects of post-mortem storage conditions of bovine epididymides on sperm characteristics: investigating a tool for preservation of sperm from endangered species. Conserv. Physiol. 4(1): cow069.

Venegas, F., M. Guillomot, X. Vignon, J.L. Servely, C. Audouard, E. Montiel, et al. 2006. Obtaiment of Pudu (*Pudu pudu*) deer embryos by the somatic nuclear transfer technique. Int. J. Morphol. 24: 285–292.

Vidal, F., J.A.M. Smith-Flueck, W.T. Flueck and L. Bartoš. 2012. Variation in reproduction of a temperate deer, the southern pudu (*Pudu puda*). Anim. Prod. Sci. 52(8): 735–740.

Waldhalm, S.J., H.A. Jacobson, S.K. Dhungel and H.J. Bearden. 1989. Embryo transfer in the white-tailed deer: a reproductive model for endangered deer species of the world. Theriogenology 31(2): 437–449.

Weber, M. and S. Gonzalez. 2003. Latin American deer diversity and conservation: A review of status and distribution. Écoscience 10: 443–454.

Wemmer, C. 1998. Deer—Status survey and conservation action plan. IUCN, Gland, Switzerland.

Wildt, D.E. and C. Wemmer. 1999. Sex and wildlife: the role of reproductive science in conservation. Biodivers. Conserv. 8: 965–976.

Willard, S.T., R.G. Sasser, J.T. Jaques, D.R. White, D.A. Neuendorff and R.D. Randel. 1998. Early pregnancy detection and the hormonal characterization of embryonic-fetal mortality in fallow deer (*Dama dama*). Theriogenology 49: 861–869.

Zanetti, E.S., B.F. Polegato and J.M.B. Duarte. 2009. Primeiro relato de inseminação artificial com tempo-fixo em veado-mateiro (*Mazama americana*). XVIII Congresso Brasileiro de Reprodução Animal.

Zanetti, E.D.S. and J.M.B. Duarte. 2012. Comparison of three protocols for superovulation of brown brocket deer (*Mazama gouazoubira*). Zoo. Biol. 31: 642–655.

Zanetti, E.S., M.S. Munerato, M.S. Cursino and J.M.B. Duarte. 2014. Comparing two different superovulation protocols on ovarian activity and fecal glucocorticoid levels in the brown brocket deer (*Mazama gouazoubira*). Reprod. Biol. Endocrinol. 12: 1–9.

Chapter 3

The Tayassuids

Alexandre Rodrigues Silva[1]*, Luana Grasiele Pereira Bezerra[1],
Romário Parente dos Santos[1], Samara Sandy Jerônimo Moreira[1],
Gabriela Liberalino Lima[2] and Pierre Comizzoli[3]

[1]Laboratory of Animal Germplasm Conservation (LCGA),
Department of Animal Sciences, Universidade Federal Rural do Semi-Árido (UFERSA),
Francisco Mota Avenue, 572, Mossoró, Rio Grande do Norte, 59625-900, Brazil.
Email: alexrs@ufersa.edu.br, luana_grasielly@yahoo.com.br,
romario.parente@hotmail.com, samara.sandy@bol.com.br,

[2]Instituto Federal de Educação, Ciência e Tecnologia do Ceará (IFCE),
Campus Crato, CE – 292, S/N, Giselia Pinheiro, Crato, Ceará, 63115-500, Brazil
Email: gabriela.lima@ifce.edu.br

[3]Smithsonian's National Zoo and Conservation Biology Institute,
Veterinary Hospital, Washington, DC 20008, USA
Email: comizzolip@si.edu.br

INTRODUCTION

Tayassuidae are a part of a family belonging to the Cordata phylum, Mammalian class, and Cetardiodactyla order. It is composed of the collared peccary (*Pecari tajacu* – Figure 3.1A), the white-lipped peccary (*Tayassu pecari* – Figure 3.1B) and the Chacoan peccary (*Catagonus wagneri*) (Wetzel et al. 1975, Gongora et al. 2011). Tayassuidae live in the American continent, with the *Pecari tajacu* being the most widely distributed species (from the south of the United States to the north of Argentina), and currently classified as a least concern species (Gasparini 2013). On the other hand, the *Tayassu pecari* has a smaller distribution, ranging from southern Mexico to northern Argentina, being classified as a vulnerable species (Eisenberg and Redford 1999) while *Catagonus wagneri* is an endemic

*Corresponding author: alexrs@ufersa.edu.br

species that inhabits the forests of the Grande Chaco of Bolivia, Paraguay, and Argentina, being listed as endangered species (Altrichter et al. 2017).

Figure 3.1   (A) Collared peccaries (*Pecari tajacu*) bred in captivity at the Wild Animal Multiplication Center of the Federal University of the Semiarid Region, Mossoró, RN, Brazil; (B) White-lipped peccaries (Tayassu pecari) bred in captivity in wildlife center of the State University of Santa Cruz, Ilhéus, BA, Brazil.

Tayassuidae play important ecological roles in the balance and composition of food chains, including the maintenance of large carnivores that are their predators (Aranda 2002, Altrichter et al. 2015). They can spread some plant species through feces (Desbiez and Keuroghlian 2009) and even play the role of ecosystem engineer by creating and maintaining wallows that become habitats for many anuran species (Beck et al. 2010).

The *Pecari tajacu* is the smallest Tayassuidae (0.75 to 1.0 m long; 0.40 to 0.45 m high) and weigh between 14 to 30 kg (Sowls 1997). Its fur is made of long, rough, and usually black with white bristles, giving the animal a grayish coat. The erectile mane on the back is made of hairs that tend to be darker. In the neck region, there is a band of white hairs giving the necklace appearance (Sowls 1984). The species lives in groups of 5 to 50 individuals, occupying semi-desert habitats, savannas, tropical forests, and highland forests (Carrillo et al. 2002). It is an omnivore that mainly consumes food of plant origin and complements its diet with small invertebrates (Bodmer and Sowls 1996). In general, life expectancy is approximately 15 years in the wild and up to 24 years in captivity (Mena et al. 2000).

The *Tayassu peccari* is larger than the collared peccaries, measuring around 1.20 m in length and 0.40 to 0.50 m in height, and weighing between 25 and 40 kg (Sowls 1984). The fur varies from brown to black, having many white hairs in the jaw and snout region (Sowls 1997). The white-lipped peccaries are social ungulates that form large groups as those with up to 300 individuals in the Amazon rainforest (Fragoso 2004). They also live in grasslands, wet and dry forests, xerophytic areas, tropical dry forests, and coastal mangroves, with a life expectancy around 13 years (Keuroghlian et al. 2013). White-lipped peccaries are mainly frugivores but eat other plants, invertebrates, and fungi (Desbiez and Keuroghlian 2009, Keuroghlian et al. 2009).

The *Catagonus wagneri* is larger and heavier than the other Tayassuidae, measuring from 0.52 to 0.69 m in height, 0.96 to 1.17 m in length, and weighing from 29.5 to 40 kg. Dorsal bristles are longer, having a mixed coloration of various shades of grey-brown, black, and white (Altrichter et al. 2017). The Chacoan peccaries form small groups of 3 to 10 individuals, primarily inhabiting xerophytic deciduous forests with multiple layers, as the driest part of the Gran Chaco ecoregion (Altrichter et al. 2017). They feed on cacti, fruits, and flowers (Altrichter et al. 2017). In the wild, they live approximately nine years, but they can reach 18 years in captivity (Altrichter et al. 2017).

Over the last decades, various studies have characterized important traits of the reproductive physiology of male and female Tayassuidae. The collared peccaries reach puberty around 8–10 months of age. Females then have estrus cycle of 21 days and a pregnancy length of proximately 140 days (Mayor et al. 2007). White-lipped peccaries only reach puberty at 10–12 months of age, with females having estrus cycle lasting around 29 days and pregnancy length from 147 to 158 days. Finally, Chacoan peccaries reach puberty only after 14 months. While the duration of estrus cycle in this species is not reported yet, it is known that pregnancy length is 184 days (Yahnke et al. 1997).

Such information serves as a basis for the development of *in vivo* and *in vitro* strategies for conservation, with the collared peccary (the least concern species) serving as a model. This chapter is an overview of studies related to the development of assisted reproductive technologies that are used to monitor the fertility of these animals and to develop conservation strategies.

MONITORING BASIC INFORMATION RELATED TO MALE REPRODUCTION

Multiple studies describe important information related to male reproductive anatomy and physiology, mainly for collared and white-lipped peccaries. The reproductive tract consists of testes, spermatic cords, accessory glands (seminal vesicles, the prostate, and bulbourethral glands) in addition to the urethra, penis, and foreskin (Garcia et al. 2014). Both collared peccaries and white-lipped peccaries have oval testes, externally located close to the pelvic region (between the perineal and inguinal regions), with dorso-caudally inclined and spiral-shaped penis (Sowls 1997, Sonner et al. 2004).

Male collared peccaries reach sexual maturity at 11 months of age, marked by the presence of sperm in the lumen of the seminiferous tubules (Bellatoni 1991, Guimaraes et al. 2013), while white-lipped peccary maturity occurs at 10 months (Sowls 1984). Histological approaches show that sperm production/output in collared peccaries is very high compared to other mammals, with high volumetric dimensions of the seminiferous tubule (84% of the parenchyma) compared to small amounts of Sertoli cells per gram of testis (19×106) (Costa et al. 2004). The entire spermatogenic process lasts 55.1 ± 0.7 days (Costa et al. 2010, Guimaraes et al. 2013).

Ultrasonography has also allowed the evaluation of the testicular dimensions and tissue homogeneity, serving as an accurate non-invasive method for monitoring the gonadal function in collared peccaries. Among the various existing formulas for the evaluation of testicular biometry from ultrasound measurements, the formula proposed by Lambert (length × width × height × 0.71) was the one that provided the greatest reliability when related to the actual volume of the testis, as observed by the Archimedes principle (Peixoto et al. 2012).

COLLECTING AND PRESERVING PECCARY SPERM

The main method currently reported for sperm collection in all the Tayassuidae is electroejaculation under anesthesia. Training animals to be collected with an artificial vagina without sedation is impossible in that species. The anesthetic protocol adopted for collared peccaries consists of the use of 5 mg/kg propofol (Propovan®, Cristalia, Fortaleza, Brazil) in bolus (Souza et al. 2009). For the semen collection, an electroejaculator (Autojac®, Neovet, Campinas, SP, Brazil) connected to a 12 V source and a probe measuring 0.9 cm in diameter and 12.5 cm in length with two longitudinal electrodes is used (Fig. 3.2A and B). The electroejaculation protocol consists of 10 stimuli from 5 V to 12 V, increased by 1 V in each cycle, each stimulus having a duration of 3 s, with a pause of 2 s between stimuli (Castelo et al. 2010a). In 80% of the cases, semen is obtained in collared peccaries (Souza et al. 2009).

For the electroejaculation, however, there are marked differences related to the anesthetic protocols used for the restraint of other Tayassuidae. In white-lipped peccaries, the anesthetic of choice consists of a combination of acepromazine (0.2 mg/kg I.M., Acepran 0.2%®, Univet S.A., São Paulo-SP) followed by ketamine a (5.0 mg/kg I.M, Dopalen®, Vetbrands Saúde Animal, Jacareí-SP, Brazil). Also, an intravenous administration of 5 IU oxytocin before the procedure is suggested. The stimuli protocol, however, is adapted from the one used in the collared peccaries but with stimuli varying from 5 to 10 V but using a larger probe measuring 18 cm for length and 1.0 cm for diameter (Vieira et al. 2021). The same stimuli protocol is used for the Chacoan peccaries but following an anesthesia protocol based on the combination of intramuscular tiletamine/zolazepam (30 mg; Zoletil® 50, Virbac, Fort Worth, TX, USA), azaperone (10 mg; 50 mg/ml; ZooPharm, Windsor, CO, USA), and medetomidine (1 mg; 10 mg/ml; ZooPharm) (Goblet et al. 2018).

The collared peccary ejaculate is characterized by presenting 3 distinct fractions that are usually mixed during electroejaculation. The first fraction comprises secretions from accessory glands, having a light color and poor in sperm cells. The second fraction has a milky appearance and a yellowish color, characteristic of a fraction rich in sperm cells. Finally, the third fraction is composed of secretions from the bulbourethral glands, presenting a gel-like appearance and acting as a buffer (Hellgren et al. 1989). The existence of distinct fractions has not been reported in the other Tayassuidae.

After collection, peccary semen is usually subjected to the evaluation of conventional parameters as volume, concentration, motility, vigor, viability, membrane integrity, morphology (Figure 3.2C), and pH, as presented in Table 3.1.

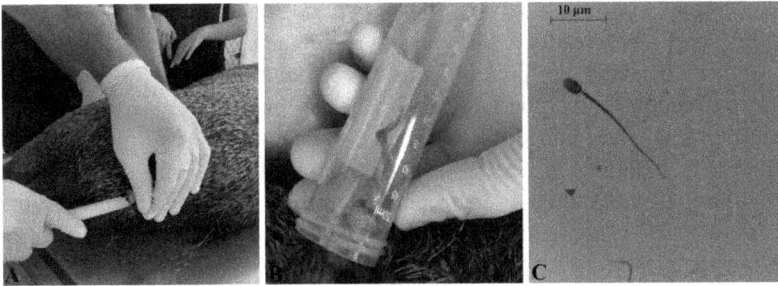

Figure 3.2 (A) Electroejaculation in collared peccaries; (B) Exposed penis during semen collection; (C) Normal sperm stained with Bengal Rose.

Table 3.1 Parameters (means ± SEM) of Tayassuid semen samples collected by electroejaculation

Semen parameters	Collared peccaries	White-lipped peccary	Chacoan peccaries
Volume (mL)	3.5 ± 3.8	0.53 ± 0.6	2.9 ± 0.7
Concentration (sperm/mL)	$0.8 ± 0.7 × 10^9$	$967.2 ± 947.3 × 10^6$	1.58 ± 1.01
Motility (%)	85.1 ± 10.7	75.5 ± 16.3	18.3 ± 8.5
Vigor (0–5)	4.2 ± 0.9	2.1 ± 0.8	0.6 ± 0.3
Sperm viability (%)	86.3 ± 13.3	62.4 ± 43.9	25.1 ± 5.6
Membrane integrity (%)	76.9 ± 13.2	85.3 ± 21.0	–
Normal morphology (%)	79.9 ± 12.8	86.5 ± 4.0	13.9 ± 2.4
pH	8.1 ± 0.4	–	7.7 ± 0.3
Reference	Peixoto et al. 2012	Vieira et al. 2021	Goblet et al. 2018

Additional techniques have been described for the analysis of semen, especially for the collared peccaries. Using a micro-morphometric technique with the aid of an image analyzer software (ImageJ software, Wayne Rasband—National Institute of Health, Maryland, United States), the morphometry of collared peccary sperm was 50.68 ± 0.121 μm in length, of which 6.34 ± 0.018 μm correspond to the head and 32.25 ± 0.076 μm to the tail (Sousa et al. 2013). This information was essential for the configuration of a computer added sperm analyzer system (CASA – IVOS 7.4G; Hamilton-Thorne Research, Beverly, MA, USA) according to the following settings validated for the peccaries: temperature 37°C; 60 frames/s; minimum contrast, 45; straightness threshold, 30%; low-velocity average pathway (VAP) cutoff, 10 m/s; and medium VAP cutoff, 30 m/s (Souza et al. 2016). In addition, functional tests using heterologous substrates to evaluate the sperm binding ability of collared peccary sperm have also been developed. These tests were based on the interactions among the peccary sperm and perivitelline membrane of the egg yolk or porcine oocytes. They are important to test the functional viability of the sperm, especially after preservation procedures, mainly because we do not have enough females available for conduction of artificial insemination trials (Campos et al. 2017).

Besides the sperm, the seminal plasma of collared peccaries was investigated through a proteomic approach using two-dimensional electrophoresis followed by

Coomassie blue staining, analysis of polypeptide maps with PDQuest Software (Bio-Rad), and identification of proteins by tandem mass spectrometry (LC–MS/MS). Authors detected 179 protein spots per gel and 98 spots were identified by mass spectrometry, corresponding to 23 different proteins, from which the main protein identified were clusterin, spermadhesin porcine seminal plasma protein 1 and bodhesin 2 (Santos et al. 2014). Additionally, the use of an enzyme linked sorbent assay (ELISA) and the spectrophotometry were proposed for the evaluation of the activity of superoxide dismutase (SOD) and catalase in peccary seminal plasma, respectively; however, only traces of SOD (0.033 ± 0.049 AU/mgP) were identified (Santos et al. 2018). Recently, the biochemistry of collared peccary seminal plasma was investigated by using commercial kits evaluated through spectrophotometry. Among all the organic and inorganic compounds described, fructose and calcium were highlighted for presenting higher concentrations during rainy season when compared to those observed during dry season. Moreover, fructose is positively correlated with various sperm kinetic parameters of peccary seminal plasma (Moreira et al. 2019).

Another option for obtaining male gametes from the collared peccary is the recovery of epidydimal sperm, which is mainly indicated for genetically valuable individuals that suddenly die. For this purpose, epididymis and vas deferens are dissected and the sperm can be collected for both retrograde washing and flotation. Both techniques provide viable sperm, but a high number of debris and blood cells is observed at the use of flotation, which could impair further preservation procedures, the washing retrograde technique being the most appropriate for this purpose (Bezerra et al. 2014).

There are two main options for the peccary sperm preservation, the chilling and the freezing, even if research focused on semen chilling is usually scarce. At this time, most prominent results for collared peccary semen were achieved at the use of a Tris-based extender plus egg yolk (20%) and gentamicin (70 μg/mL) that provided efficient preservation of sperm parameters (for example > 40% motile sperm) from 10 individuals for 36 h at 5°C in a biological incubator (Santos et al. 2021). For white-lipped peccaries, a study conducted with only two individuals reported the maintenance of >80% motile sperm for 24 h at 15°C in a commercial thermal box for semen transportation (Botuflex®, Botupharma, Botucatu, SP, Brazil), using Beltsville Thawing Solution or powdered coconut water extenders (ACP®, ACP-Biotecnologia, Fortaleza, Brazil) (Barros et al. 2019).

Regarding semen cryopreservation, the first attempt of freezing collared peccary semen was reported in 2010 by Castelo et al. (2010a, b). Since then, the protocol was improved and nowadays the semen is diluted in a Tris-fructose-citric acid extender supplemented with egg yolk (20%) and glycerol (3%), and is subjected to a fast-freezing curve in which the samples are refrigerated at 15°C for 40 min in isothermal boxes and stabilized at 5°C for another 30 min in a biological incubator (Quimis, Diadema, SP, Brazil). Then, they are filled in 0.25 mL plastic that were placed in contact with the nitrogen vapor (5 cm) for 5 min and finally stored in a cryobiological container at −196°C. After thawing, samples usually present values higher than 40% motile sperm (Moreira et al. 2022). In addition, epididymal sperm obtained by retrograde washing method can be efficiently cryopreserved using the same methodologies described for ejaculated sperm (Bezerra et al. 2019). Interestingly, the sperm resilience to freezing in peccaries seems to be influenced by environmental variables, since ejaculates present better post-thawing results when frozen during the rainy season in comparison to those frozen during dry season in Caatinga biome (Maia et al. 2019).

TESTICULAR TISSUE CRYOPRESERVATION AND CULTURE

Testicular tissue cryopreservation seems to be a promising tool to aid the assisted reproduction of Tayassuidae, as it allows the storage of fragments containing many germ cells at various stages of development, including undifferentiated spermatogonia that can be cultured *in vitro* or through xenografts. The technique may promote promote restoration of both gametogenic and endocrine function after cryopreservation (Silva et al. 2019b). Testicular tissue can be collected from sexually mature and immature animals, as well as from animals that have just died. In addition, the techniques can vary from slow freezing to vitrification, associating different internal and external cryoprotectants at different concentrations. Table 3.2 shows the main results related to the application of testicular tissue cryopreservation techniques in collared peccaries (Figure 3.3A, B and C).

After cryopreservation, the next step is to provide adequate conditions for the tissues to resume spermatogenesis. This can be achieved either in *in vivo* or *in vitro* culture. In collared peccary, tissue fragments and testicular cell suspension from immature animals were xenografted in SCID mice (nZ48) (Campos-Junior et al. 2014). After six and eight months of tissue xenograft, spermatogenesis was observed; however, a delay in spermatogenesis progression was observed in testicular cell suspension xenografts, with production of fertile spermatozoa only eight months after engraftment. Furthermore, sperm collected from xenografts resulted in diploid embryos that expressed the paternal imprint gene NNAT after ICSI (Campos-Junior et al. 2014).

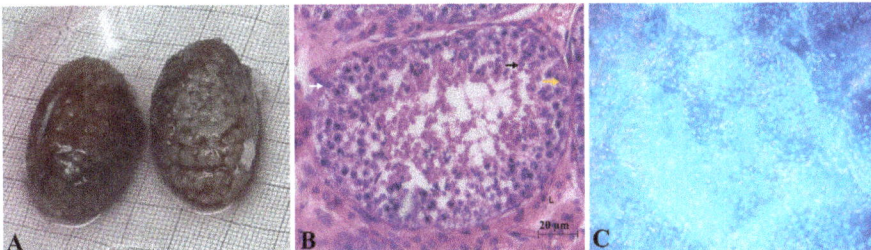

Figure 3.3 (A) Pair of collared peccaries' testicles; (B) Micrographs of testicular tissue from collared peccaries stained with eosin-nigrosine evidencing the seminiferous tubules, Leydig cells (L) around the tubules, spermatogonia (black arrow), spermatids (white arrows) and Sertoli cells (yellow arrows). (C) Tissue labeled with fluorescent probes to assess cell viability. Cells labeled in blue (Hoecht 33342) are viable cells, cells labeled in light pink are non-viable cells.

Table 3.2 Main characteristics of testicular tissues frozen with different techniques in collared peccaries

Cryopreservation method	Cryoprotectant	Main results				References
		Cell viability	Histomorphology	Cell proliferation	DNA integrity	
Fresh Control	—	~85%	Normal testicular tissue histomorphology.	Normal potential proliferative capacity.	—	Silva et al. 2019a
Solid-Surface Vitrification	1.5 M DMSO + 1.5 M EG	~40%	Seminiferous tubule lumen and tubular organization conserved. Better basement membrane preservation than DMSO (P < 0.05).	Maintenance of the potential proliferative capacity of spermatogonia and Sertoli cells.	—	
Fresh Control	—		Normal testicular tissue histomorphology.	Normal potential proliferative capacity.	~97.9%	Silva et al. 2021
Slow Freezing	0.25 M sucrose + 0.75 M DMSO + 0.75 M EG	~67%		Efficient preservation of spermatogonia proliferative potential.	~90.2&	
	0.25 M sucrose + 0.75 M DMSO + 0.75 M Glycerol	~60%		Efficient preservation of spermatogonia proliferative potential.	~90%	
Conventional vitrification	0.25 M sucrose + 1.5 M DMSO + 1.5 M EG	~61%		Efficient preservation of spermatogonia proliferative potential.	~88%	
	0.25 M sucrose + 1.5 M EG + 1.5 M Glycerol	~52.8%		Efficient preservation of spermatogonia proliferative potential.	~74.6%	
Solid-Surface Vitrification	0.25 M sucrose + 1.5 M EG + 1.5 M Glycerol	~58%		Efficient preservation of spermatogonia proliferative potential.	~73.8%	

Monitoring and Control of the Female Reproduction

Most of the studies conducted for monitoring estrous cycle in peccaries are based on serum steroids analysis using ELISA assays. In collared peccaries, mean values for the estrogen peak were 55.6 ± 20.5 pg/mL and 53.4 ± 8.1 pg/mL for animals raised in Amazon or Caatinga biomes, respectively. Moreover, values for progesterone during luteal phase are also similar for animals from both Amazon (30.8 ± 4.9 ng/mL) or Caatinga (35.3 ± 4.4 ng/mL) biomes (Mayor et al. 2006, 2007, Maia et al. 2014a). On the other hand, the correspondence between vaginal cytology and estradiol profile reported for the Amazonian individuals is not seen in animals from the Caatinga. For the first ones, a positive correlation among estrous serum, vaginal cytology, and vulvar appearance was verified, with 62.5% of coincidence among the methods employed. However, the variations between individuals or even between different groups could not be discarded (Mayor et al. 2007).

The ovarian monitoring by ultrasonography is another possible method for verifying structural changes along the estrous cycle. Although it is a noninvasive way of identifying the ovarian structures, an accurate monitoring of the ovarian dynamics may be difficult due to the presence of a *bursa ovarica* surrounding ovary. Also, the need for a repetitive physical or chemical restraint for sequential examinations limits its use (Peixoto et al. 2019).

Even if it is known as a reliable method for estrous detection, the blood collection requires continuous animal restraint, as it is also necessary for ultrasound and vaginal cytology, being stressful and potentially unsafe for both, the animal, and the handling. Thus, non-invasive approaches are required, for example, the fecal metabolites analysis. This is a simple methodology to detect the ovarian activity in many species. In collared peccary, a high correlation between serum progesterone and fecal progesterone metabolites levels suggests that it can be used to monitor their reproductive function (Mayor et al. 2019). Through this analysis, estrous cycle length of 27.9 ± 4.5 days (ranging from 21 to 36 days) was comparable to the average length days reported using serum evaluation (27.8– Mauget et al. 1997; 21.0 ± 5.7 days – Maia et al. 2014a).

Regarding the pregnancy monitoring, different methods have already been described for carrying out the gestational diagnosis in peccaries; however, the main limiting factor is the difficulty in animal restraint, which can cause stress and trauma. Early pregnancy detection in collared peccaries can be performed by ultrasonography, based on the embryo size and its structures and characteristics. It can be carried out as early as 18 to 26 days after mating. The accuracy of this method on days 22, 26 and 28 after fertilization is 56%, 93% and 100%, respectively (Mayor et al. 2005). Additionally, both serum progesterone and fecal progesterone metabolite concentrations are not accurate for early pregnancy diagnosis in the collared peccary, with accuracy increasing during the late stage of pregnancy. The use of serum 17β—estradiol and progesterone for this purpose is better (73%) when the threshold was established at 20 ng/mL of progesterone and 20 pg/mL of serum estradiol. This accuracy increased for pregnancy diagnosis in mid (50–100 days) and late (>100 days) pregnancy, being 78% and 95%, respectively (Mayor et al. 2012).

To control ovarian cycles, estrous synchronization was first attempted by using 60 µg D-cloprostenol in two doses 9 days apart, which promoted estrus synchronization in 80% of the females after 9.5 ± 0.5 days (Maia et al. 2014b). The association of 400 IU of equine chorionic gonadotropin (eCG) with 200 IU of human chorionic gonadotropin (hCG) also has 100% of efficiency, with synchronized estrus 6 days after drug administration. The first effort of an artificial insemination was performed with fresh semen. Despite the absence of fetuses (ultrasonography examination), an increase of progesterone values was observed in 70% of females for 60 days after the treatments (Peixoto et al. 2019).

Female Germplasm Assessment and Preservation

Initial studies performed in *P. tajacu* allowed to estimate the preantral follicle (PF) population of 33.273,45 ± 3.019,30 follicles per ovary (Figure 3.4A and B), of which most belong to the primordial follicle category (91.56%), followed by primary (6.29%) and secondary (2.15%) and most of them is morphologically normal (94.4 ± 1.54%). The mean diameter of primordial, primary, and secondary follicles is 31.82 ± 1.0; 40.90 ± 2.14 and 196.21 ± 17.06 µm, respectively. The presence of polyovular follicles (group of oocytes enclosed in a single follicle) was reported and it was more frequently present in primordial follicles. The oocyte cytoplasm presented amounts of lipid droplets, shown as vacuoles at light microscopy, which was confirmed through ultrastructural analysis. It is probably involved in energy storage and is also an important building material for the cytoplasmic membranes of future embryos (Lima et al. 2012).

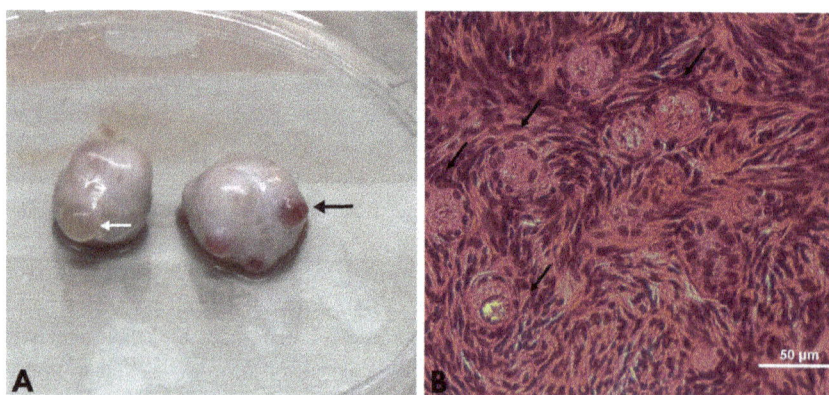

Figure 3.4 (A) Collared peccary ovary showing an antral follicle (white arrow) and a corpus luteum (black arrow). (B) Histological section of the ovary showing the preantral ovarian follicles of the collared peccary (black arrow).

Regarding antral follicles, the number range from 60.0 ± 26.1 to 103.5 ± 45.6 in luteal and follicular phase, respectively (Guimaraes et al. 2012), higher than that observed in white lipped peccary, in which the number ranged from 11.62 ± 5.10 to 26.67 ± 8.67, in pregnancy and follicular phase, respectively (Mayor et al. 2009).

To improve the female gametes reserve, protocols aiming the *in vitro* culture (IVC) and development of PFs have been tested. In collared peccaries, the use of Tissue Culture Medium 199 (TCM-199+) supplemented with 50 ng/ml of recombinant FSH maintained the proportion of intact PFs like day 1 (63.2%) and promoted PFs granulosa cell proliferation and extracellular matrix preservation through 7 days (Lima et al. 2018).

To improve IVC method, 25 ng/mL of bone morphogenetic protein-15 (BMP-15) was added to the TCM-199+ medium of PFs from the same species. It maintained 79.67% $\pm$ 0.69% of morphologically healthy follicles, stimulated the activation of primordial follicles and did not affect oocyte and follicular growth after 6 days in culture (Gomes et al. 2020). This represents an important step for further *in vitro* follicle development and maturation.

The female germplasm conservation was performed using distinct methods in *P. tajacu* (Lima et al. 2014, 2019, Campos et al. 2019). The main results are presented in Table 3.3. In general, the solid surface vitrification (SSV) and the use of an ovarian tissue cryosystem (OTC) are efficient tools for collared peccaries' female germplasm preservation, but OTC offers the advantage of being a practice closed system, representing a promising tool for other Tayassuidae conservation.

Table 3.3 Ovarian tissue frozen with different methods in the collared peccary

Methods	Main results	References
Fresh control	Samples presenting 94.4% normal morphology and 86.7% viability	Lima et al. 2014
Short-term preservation (~4 to 8°C)	Powdered coconut water (ACP®) based medium provided preantral follicles morphology (66.7%) preservation for up 36 h and viability preservation of more than 30% for 24 h.	
Fresh control	Samples presenting 92.2% normal morphology and 97% viability	Lima et al. 2019
Solid surface vitrification (SSV)	Conservation of 74.2 $\pm$ 7.3% of morphology and 97% of viability using 3M ethylene glycol.	
Fresh control	Samples presenting 75.6% normal morphology and 84% cell viability, with a normal proliferative capacity and only 36% activated caspase indicating apoptosis.	Campos et al. 2019
Ovarian tissue cryosystem vitrification (OTC)	3M ethylene glycol provides cell viability (79.0 $\pm$ 4.3%) and morphology (67.8 $\pm$ 6.8%), maintains the cell proliferation capacity and prevents apoptosis with only 46% activated caspase.	

In vitro Embryo Production

In Tayassuidae, embryo technology is in its initial stages only in collared peccaries. The first steps for embryo production in Tayassuidae were conducted using a collared peccary model by Borges et al. (2018a). Authors proved that *in vitro* maturation (IVM) during 48 h of incubation would be the ideal time for collared peccary oocytes to acquire meiotic competence from the expansion of cumulus cells (100.0 ± 0.0), emergence of the first polar body (90.5 ± 2.0) and nuclear state in second metaphase (76.2 ± 1.3).

Recently, Borges et al. (2020a) evaluated various conditions of *in vitro* maturation (IVM) of oocytes and chemical activation in collared peccaries. Initially, IVM were evaluated in the absence or presence of epidermal growth factor (EGF). There was no significant difference between cumulus-oocyte complexes (COCs) matured with or without EGF, except for the thickness of the zona pellucida that was reduced in the presence of this factor. Oocyte activation then was artificially triggered by the addition of ionomycin and four combinations of secondary activators (6-dimethylaminopurine (6D), 6D and cytochalasin B (6D + CB), cycloheximide (CHX), and CHX and CB (CHX + CB)). The highest rates of blastocysts/good grade COCs were obtained in the group (6D 27.6% ± 0.3); in addition, only this treatment triggered the production of embryos from poor grade COCs (25.0% ± 0.2). Thus, the association of ionomycin and 6D could produce collared peccary parthenogenic embryos by activating COCS, leading to the implementation of assisted reproduction techniques for the conservation of this species.

Somatic Cell Banks and Cloning

The preservation of somatic samples has become critical for the conservation of biodiversity, as well as in the reproductive improvement of these species (Costa et al. 2016). In Tayassuidae, but specifically in collared peccaries, somatic tissues can be acquired through management-related procedures, such as identification management adopted in authorized breeding sites (Silva et al. 2017). Basically, the process consists of cleaning and fragmentation of the tissues, which can easily be used for the formation of biobanks (Borges et al. 2017a), as well as immediately processed and cultured to obtain somatic cells (Santos et al. 2016). In the extrapolation of this technique to the collared legs, it is important to highlight the pioneering work of Santos et al. (2016) who verified the composition of the most suitable culture medium for the culture of somatic cells derived from the ear tissue of collared peccaries, aiming at obtaining cells. It was evidenced that 20% fetal bovine serum (FBS) are suitable for the recovery of somatic cells in modified essential medium (DMEM), presenting viability results around 98.2%.

Other information extracted from the *in vitro* culture of these fragments and cells is the need to elucidate the components of the architecture of the peripheral auricular tissue of collared peccaries. This knowledge would help to establish more specific protocols in the processing and culture of tissues and cells somatic (Silva et al. 2017). For instance, Borges et al. (2017b) were able to

Table 3.4 Current strategies of the use of somatic cells and fibroblasts in the conservation of collared peccaries (*Pecari tajacu* Linnaeus 1758)

Sample source	*In vitro* culture conditions	Cryopreservation conditions	Results	Authors
Ear tissue	DMEM supplemented with 2.2 g/L (w/v) sodium bicarbonate solution, 2% (v/v) antibiotic-antimycotic solution, and 10% (v/v) FBS at 38.5° C, 5% CO_2 in air and 95% relative humidity.	**Solid surface vitrification (SSV):** DMEM supplemented with 20% (v/v, 3.58 M) ethylene glycol (EG), 20% (v/v, 2.82 M) dimethyl sulfoxide (DMSO), 0.25 M (w/v) sucrose and serum fetal bovine (FBS) at 10% (v/v).	SSV proved to be a more efficient method for vitrification of the skin tissue of collared peccaries when compared to DVC. Morphological integrity of the cytoplasm (25.0%).	(Borges et al. 2017a)
Ear tissue	Silver solution prepared in 1 part 2% gelatin in 1% aqueous formic acid and 2 parts 50% aqueous silver nitrate solution, and the slides were exposed in a dark room for 30 min. Slides were washed in 5% thiosulfate solution for 10 min.	**Solid surface vitrification (SSV):** DMEM composed of 2.2 g/L sodium bicarbonate and 10% FBS (DMEM+), supplemented with sucrose, EG-SUC (DMEM+ +3.0 M EG +0.25 M sucrose +10% FBS).	**Solid surface vitrification (SSV):** DMEM composed of 2.2 g/L sodium bicarbonate and 10% FBS (DMEM+), supplemented with sucrose, EG-SUC (DMEM+ +3.0 M EG +0.25 M sucrose +10% FBS).	(Borges et al. 2018b)
Ear tissue	Tissue fragments were cultured in DMEM plus sodium bicarbonate, penicillin G, streptomycin, amphotericin B and FBS and/or EGF at 38.5°C, 5% CO_2 in air and 95% relative humidity.	**Slow freezing:** Freezing medium (DMEM supplemented with 10% DMSO as cryoprotectant permeant and 10% FBS and 0.2 M sucrose as non-permeable cryoprotectants). Cells at a concentration of $5.0 \times 10\ 4$ cells/mL were first exposed to the DMSO-FBS solution for 15 min at 4°C, then the sucrose solution was added followed by an additional incubation for 15 min at 4°C.	Cryopreservation did not affect viability as assessed by trypan blue staining (87.4 ± 0.3% vs. 74.0 ± 5.9%, P = 0.11). Furthermore, after two passages of the thawed cells, the viability was 86.4 ± 3.2%. In addition, no difference (P = 0.77) was observed for metabolic activity between cryopreserved cells (85.2 ± 10.0%) and non-cryopreserved cells (100.0 ± 36.4%).	(Borges et al. 2020b)
Ear tissue – fibroblasts	DMEM plus 10% FBS and 2% antibiotic-antimycotic solution. The skin was cultured at 38.5°C in a controlled environment with 5% CO_2 and 95% air, according to the method described.	**Slow freezing:** DMEM supplemented with 10% DMSO as cryoprotectant permeant and 10% FBS and 0.2 M sucrose as non-permeable cryoprotectants. First exposed to the DMSO-FBS solution for 15 min at 4°C, then the sucrose solution was added followed by an additional incubation for 15 min at 4°C.	No significant difference in cell viability (74.5–84.4%) after passages. However, metabolic activity was reduced in the tenth. Quality of fibroblast lines showed viability of (87.4 ± 0.3% vs. 74.0 ± 5.9%). The cryopreserved cells showed higher levels of intracellular ROS.	(Lira et al. 2020)

histomorphologically characterize the integumentary tissue of collared peccaries, observing some peculiarities regarding other domestic mammals, such as the number of epidermal cells, number of layers and thickness of the epidermis, melanocytes, and proliferative parameters. Based on these initial studies, it was possible to implement conservationist techniques to efficiently preserve the tissue and somatic cells of collared peccaries (Table 3.4).

Silva et al. (2017) mention that, in summary, the development of adequate protocols for the storage and processing of somatic fragments of collared peccaries, as well as the culture conditions for the recovery of cells are already established. Through these procedures, it will be possible to use these cells properly as karyoplasts for future cloning in this species (Silva et al. 2017).

FINAL CONSIDERATIONS

Major progress has been made in Tayassuidae regarding gametes characterization and preservation, estrous cycle monitoring and manipulation, and *in vitro* culture conditions. These encouraging studies can increase reproductive performance, with practical implications for the management and conservation of the species. Although some assisted reproductive technologies have been already described for *Pecari tajacu* and *Tayassu tajacu*, scarce information about reproductive morphology and physiology is available for *Catagonus wagneri* which currently limits the development of biotechnologies. Enhancing the use of biotechnologies in Tayassuidae will not only enable the preservation of these species itself, but also of predators depending on peccaries, such as jaguars. Thus, reproductive biotechnologies are valuable tools for improving reproductive performance of those species, also allowing the creation of germplasm banks.

ACKNOWLEDGMENTS

Authors thank the Coordenação de Aperfeiçoamento de Pessoal de Nível Superior– Brazil (CAPES; Financial Code–001) and the National Council for the Scientific Development (CNPq; Grants no. 303929/2018-9) for financial support.

REFERENCES

Altrichter, M., A. Taber, A. Noss, L. Maffei and J. Campos. 2015. *Catagonus wagneri*. The IUCN Red List of Threatened Species e.T4015A72587993.

Altrichter, M., S.S. Ballesai, J. Decarre, M. Camino, A. Yanosky, J.M.C. Krauer, et al. 2017. Situación de conservación del pecarí del Chaco o tagua (*Catagonus wagneri*): distribución, aptitud de hábitat y viabilidad poblacional. Paraquaria Nat. 4(2): 30–39.

Aranda, M. 2002. Importancia de los pecaríes para la conservación del jaguar en México. Em El Jaguar en el Nuevo Mileno. pp. 101–106. *In*: R.A. Medellín, C. Equihua, C.L.B. Chetkiewicz, P.G.Jr. Crawshaw, A. Rainowitz, K.H. Redford, et al.

(eds). Fondo de Cultura Económica, Ediciones Científicas Universitarias, Universidad Nacional Autónoma de México, and Wildlife Conservation Society, México.

Barros, C.H.S.C., W.M. Machado, R.L.A. Vieira, I.B. Allaman, S.L.G. Nogueira-Filho, R.F. Bittencourt, et al. 2019. Use of the ACP® and BTS extenders for cooling at 15°C white-lipped peccary (*Tayassu pecari*) semen. Pesq. Vet. Bras. 39: 332–341.

Beck, H., P. Thebpanya and M. Filiaggi. 2010. Do Neotropical peccary species (*Tayassuidae*) function as ecosystem engineers for anurans? J. Trop. Ecol. 26: 407–414.

Bellatoni, E. 1991. Habitat use by mule deer and collared peccaries in an urban environment. Technical report, Tucson: University of Arizona.

Bezerra, J.A.B., A.M. Silva, G.C.X. Peixoto, M.A. Silva, M.F. Oliveira and A.R. Silva. 2014. Influence of recovery method and centrifugation on epididymal sperm from collared peccaries (*Pecari tajacu* Linnaeus, 1758) Zool. Sci. 31: 238–42.

Bezerra, L.G.P., A.L.P. Souza, A.E.A. Lago, L.B. Campos, T.L. Nunes, V.V. Paula et al. 2019. Addition of equex STM to Extender improves post-thawing longevity of collared peccaries sperm. Biopreserv. Biobank. 17: 143–147.

Bodmer, R.E. and L.K. Sowls. 1996. El pecary de colar. pp. 5–15. *In*: W.L.R. Oliver. (ed.). *Pecaries*. Quito, Ecuador: IUCN.

Borges, A.A., G.L. Lima, L.B. Queiroz Neta, M.V.O. Santos, M.F.O, A.R. Silva, et al. 2017a. Conservation of somatic tissue derived from collared peccaries (*Pecari tajacu* Linnaeus, 1758) using direct or solid-surface vitrification techniques. Cytotechnology. 69: 643–654.

Borges, A.A., F.V.F. Bezerra, F.N. Costa, L.B. Queiroz Neta, M.V.O. Santos, M.F. Oliveira, et al. 2017b. Histomorphological characterization of collared peccary (*Pecari tajacu* Linnaeus, 1758) ear integumentary system. Arq. Bras. Med. Vet. Zootec. 69: 948–954.

Borges, A.A., M.V.O. Santos, L.B.Q. Neta, M.F. Oliveira, A.R. Silva and A.F. Pereira. 2018a. *In vitro* maturation of collared peccary (*Pecari tajacu*) oocytes after different incubation times. Pesq. Vet. Bras. 38: 1863–1868.

Borges, A.A., L.B. Quiroz Neta, M.V.O. Santos, M.F. Oliveira, A.R Silva and A.F. Pereira. 2018b. Combination of ethylene glycol with sucrose increases survival rate after vitrification of somatic tissue of collared peccaries (*Pecari tajacu* Linnaeus, 1758). Pes. Vet. Bras. 38: 1678–5150.

Borges, A.A., M.V.O. Santos, L.E. Nascimento, G.P.O. Lira, E.A. Praxedes, M.F. Oliveira, et al. 2020a. Production of collared peccary (*Pecari tajacu* Linnaeus, 1758) parthenogenic embryos following different oocyte chemical activation and *in vitro* maturation conditions. Theriogenology 142: 320–327.

Borges, A.A., G.P.O. Lira, L.E. Nascimento, M.V.O Santos, M.F. Oiveira, A.R Silva et al. 2020b. Isolation, characterization, and cryopreservation of collared peccary skin-derived fibroblast cell lines. PeerJ. 8: e9136.

Campos, L.B., G.C.X. Peixoto, A.M. Silva, A.L.P. Souza, T.S. Castelo, K.M. Maia, et al. 2017. Estimating the binding ability of collared peccary (Pecari tajacu Linnaeus, 1758) sperm using heterologous substrates. Theriogenology 92: 57–62.

Campos, L.B., A.M. Silva, E.C.G. Praxedes, L.G.P. Bezerra, T.L.B.G. Lins, V.G. Menezes, et al. 2019. Vitrification of collared peccary ovarian tissue using open or closed systems and different intracellular cryoprotectants. Cryobiology 91: 77–83.

Campos-Junior, P.H., G.M. Costa, G.F. Avelar, S.M. Lacerda, N.N. da Costa, O.M. Ohashi, et al. 2014. Derivation of sperm from xenografted testis cells and tissues of the peccary (*Tayassu tajacu*). Reproduction 147(3): 291–299.

Carrillo, E., J.C. Saenz and T.K. Fuller. 2002. Movements and activities of white-lipped peccaries in Corcovado National Park, Costa Rica. Biol. Conserv. 108: 317–324.

Castelo, T.S., F.S.B. Bezerra, G.L. Lima, H.M. Alves, I.R.S. Oliveira, E.A.A. Santos, et al. 2010a. Effect of centrifugation and sugar supplementation on the semen cryopreservation of captive collared peccaries (*Tayassu tajacu*)?. Cryobiology 61: 275–279.

Castelo, T.S., F.S.B. Bezerra, A.L.P. Souza, M.A.P. Moreira, V.V. Paula, M.F. Oliveira et al. 2010b. Influence of the thawing rate on the cryopreservation of semen from collared peccaries (*Tayassu tajacu*) using Tris-based extenders. Theriogenology 74: 1060–1065.

Costa, C.A.S., A.A. Borges, M.V.O. Santos, L.B. Queiroz Neta and A.F. Pereira. 2016. Tools for the evaluation of somatic cells and tissues after cryopreservation in mammals. A review. Rev. Bras. Higien. Sanid. Anim. 10: 820–829.

Costa, D.S., M. Henry and T.A.R. Paula. 2004. Spermatogenesis of collared peccaries (*Tayassu tajacu*). Arq. Bras. Med. Vet. Zootec. 56: 46–51.

Costa, G.M.J., M.C. Leal, J.V. Silva, A.N.S. Ferreira, D.A. Gimarães and L.R. França. 2010. Spermatogenic Cycle Length and Sperm Production in a Feral Pig Species (Collared Peccary, *Tayassu tajacu*). J. Androl. 31: 221–230.

Desbiez, A.L.J and A. Keuroghlian. 2009. Can bite force be used as a basis for niche separation between native peccaries and introduced feral pigs in the Brazilian Pantanal? Mammalia 73(4): 369–372.

Eisenberg, J.F. and K.H. Redford. 1999. Mammals of the Neotropics: The Central Neotropics: Ecuador, Peru, Bolivia, Brazil. Chicago, Illinois, USA: The University of Chicago Press.

Fragoso, J.M.V. 2004. A long-term study of White-lipped peccary (*Tayassu pecari*) population fluctuations in northern Amazonia—Anthropogenic versus "natural" causes. pp. 286–296. *In*: K.M. Silvius, R.E. Bodmer and J.M.V. Fragoso (eds). People and nature: Wildlife Conservation in South and Central America. Columbia Univerity Press, New York, USA.

García, M.V., E.Q. Sánchez, U. Aguilera-Reyes, O. Monroy-Vilchis, J.M.V Mora, A.L. Blasio, et al. 2014. El aparato urogenital del pecarí de collar (*Pecari tajacu* Chordata: Artiodactyla): un estudio anatómico. Cienc. Ergo-sum. 22: 54–62.

Gasparini, G.M. 2013. Records and stratigraphical ranges of South American Tayassuidae (Mammalia, Artiodactyla). J. Mammal. Evol. 20: 57–68.

Goblet, C., G. West, J.M. Campos-Krauer and A.E. Newell-Fugate. 2018. Semen analysis parameters from a captive population of the endangered Chacoan peccary (*Catagonus wagneri*) in Paraguay. Anim. Reprod. Sci. 195: 162–167.

Gomes, H.A.N., L.B., Campos, É.C.G. Praxedes, M.F. Oliveira, A.F. Pereira, A.R. Silva et al. 2020. BMP-15 activity on *in vitro* development of collared peccary (*Pecari tajacu* Linnaeus, 1758) preantral follicles. Reprod. Domest. Anim. 55: 958–964.

Gongora, J., R. Reyna-Hurtado, H. Beck, A. Taber, M. Altrichter and A. Keuroghlian. 2011. *Pecari tajacu*. The IUCN Red List of Threatened Species e.T41777A10562361.

Guimaraes, D.A., S.C.G. Garcia, M.A.P. Ferreira, S.S.B. Silva, N.I. Albuquerque and Y. Le Pendu. 2012. Ovarian folliculogenesis in collared peccary: *Pecari tajacu*. (Artiodactyla: Tayassuidae). Rev. Biol. Trop. 60: 437–455.

Guimaraes, D.A., D.L. Cardoso, M.A.P. Ferreira and N.I. Albuquerque. 2013. Puberty in male collared peccary (*Pecari tajacu*) determined by quantitative analysis of spermatogenic cells. Acta Amazon. 43: 99–103.

Hellgren, E.C., R.L. Lochmiller, M.S. Amoss, S.W.J. Seager, S.J. Magyar, K.P. Coscarelli et al. 1989. Seasonal variation in serum testosterone, testicular measurements and semen characteristics in the collared peccary (*Tayassu tajacu*). J. Reprod. Fert. 85: 677–686.

Keuroghlian, A., D.P. Eaton and A.L.J. Desbiez. 2009. The response of a landscape species, white-lipped peccaries, to seasonal resource fluctuations in a tropical wetland, the Brazilian Pantanal. Int. J. Biodivers. Conserv. 1(4): 87–97.

Keuroghlian, A., A. Desbiez, R. Reyna-Hurtado, M. Altrichter, H. Beck, A. Taber, et al. 2013. Tayassu pecari. The IUCN Red List of Threatened Species e.T41778A44051115.

Lima, G.L., E.A.A. Santos, V.B. Luz, A.R. Silva and A.P.R. Rodrigues. 2012. Morphological Characterization of the Ovarian Preantral Follicle Population of Collared Peccaries (Tayassu tajacu Linnaeus, 1758). Anat. Histol. Embryol. 42: 304–311.

Lima, G.L., E.A.A. Santos, L.F. Lima, V.B. Luz, A.P.R. Rodrigues and A.R. Silva. 2014. Short-term preservation of *Pecari tajacu* ovarian preantral follicles using phosphate buffered saline (PBS) or powdered coconut water (ACP®) media. Arq. Bras. Med. Vet. Zootec. 66: 1623–1630.

Lima, G.L., V.B. Luz, L.F. Lima, R.M.P. Rocha, S.V. Castro, T.S. Castelo, et al. 2018. Interactions between different media and follicle-stimulating hormone supplementation on in vitro culture of preantral follicles enclosed in ovarian tissue derived from collared peccaries (*Pecari tajacu* Linneaus, 1758). Reprod. Domest. Anim. 53: 880–888.

Lima, G.L., V.B. Luz, F.O. Lunardi, A.L.P. Souza, G.C.X. Peixoto, A.P.R. Rodrigues, et al. 2019. Effect of cryoprotectant type and concentration on the vitrification of collared peccary (Pecari tajacu) ovarian tissue. Anim. Reprod. Sci. 205: 126–133.

Lira, G.P.O., A.A. Borges, M.B. Nascimento, L.V.C. Aquino, M.F. Oliveira, A.R. Silva et al. 2020. Cryopreservation of collared peccary (*Pecari tajacu* Linnaeus, 1758) somatic cells is improved by sucrose and high concentrations of fetal bovine serum. Cryoletters 41: 271–279.

Maia, K.M., G.C.X. Peixoto, L.B. Campos, J.A.B. Bezerra, A.R.F. Ricarte, N. Moreira, et al. 2014a. Estrus cycle monitoring of captive collared peccaries (*Pecari tajacu*) in semiarid conditions. Pesq. Vet. Bras. 34: 1115–1120.

Maia, K.M., G.C. Peixoto, L.B. Campos, A.M. Silva, T.S. Castelo, A.R. Ricarte, et al. 2014b. Estrous synchronization in captive collared peccaries (*Pecari tajacu*) using a prostaglandin F2α analog. Zoolog Sci. 31: 836–839.

Maia, K.M., A.L.P. Souza, A.M. Silva, J.B.F. Souza-Jr, L.L.M. Costa, F.Z. Brandao, et al. 2019. Environmental effects on collared peccaries (*Pecari tajacu*) serum testosterone, testicular morphology, and semen quality in the Caatinga biome. Theriogenology 126: 286–294.

Mauget, R., F. Feer, O. Henry and G. Dubos. 1997. Hormonal and behavioural monitoring of ovarian cycles in peccaries. Proc. first international symposium on physiology and ethology of wild and zoo animals suppl. II; 1997. p. 145e9. Berlin, Germany.

Mayor, P., F. Lopez-Gatius and M. Lopez-Bejar. 2005. Ultrasonography within the reproductive management of collared peccary (*Tayassu tajacu*). Theriogenology 63: 1832–1843.

Mayor, P., D.A. Guimaraes, F. Lopez-Gatius and M. Lopez-Bejar. 2006. First postpartum estrus and pregnancy in the female collared peccary (*Tayassu tajacu*) from the amazon. Theriogenology 66: 2001–2007.

Mayor, P., H. Gálvez, D.A. Guimarães, F. López-Gatius and M. López-Béjar. 2007. Serum estradiol-17β, vaginal cytology and vulval appearance as predictors of estrus cyclicity in the female collared peccary (*Tayassu tajacu*) from the eastern Amazon region. Anim. Reprod. Sci. 97: 165–174.

Mayor, P., R.E. Bodmer, L. Schettini, Mariño and G.O. López-Béjar M. 2009. Anatomicohistological characteristics of the female genital organs of the white-lipped peccary (*Tayassu pecari*) in the Peruvian Amazon. Anat. Histol. Embryol. 38: 467–74.

Mayor, P., D.A. Guimaraes and M. López-Béjar. 2012. Progesterone and estradiol-17β as a potential method for pregnancy diagnosis in the collared peccary (*Pecari tajacu*). Res. Vet. Sci. 93:1413–1417.

Mayor, P., D.A. Guimaraes, J. Silva, F. Jori and M. Lopez-Bejar. 2019. Reproductive monitoring of collared peccary females (*Pecari tajacu*) by analysis of fecal progesterone metabolites. Theriogenology. 134: 11–17.

Mena, P.V., J.R. Stallings, J.B. Regalado and R.L. Cueva. 2000. The sustainability of current hunting practices by the Huaorani. pp. 57–78. *In*: J.G. Robinson and E.L. Bennett (eds). Hunting for Sustainability in Tropical Forests. Columbia University Press, New York.

Moreira, S.S.J., A.M. Silva, E.C.G. Praxedes, L.B. Campos, C.S. Santos, A.L.P. Souza et al. 2019. Composition of collared peccary seminal plasma and sperm motility kinetics in semen obtained during dry and rainy periods in a semiarid biome. Anim. Reprod. Sci. 211: 106229.

Moreira, S.S.J., C.S. Santos, T.S. Castelo, L.G.P. Bezerra, E.C.G. Praxedes, T.M. Matos et al. 2022. Investigating the need for antibiotic supplementation to the extender used for semen cryopreservation in collared peccaries. Front. Vet. Sci. 9:954921.

Peixoto, G.C.X., M.A. Silva, T.S. Castelo, A.M. Silva, J.A.B. Bezerra, A.L.P. Souza, et al. 2012. Individual variation related to testicular biometry and semen characteristics in collared peccaries (*Tayassu Tajacu* Linnaeus, 1758). Anim. Reprod. Sci. 134: 191–196.

Peixoto, G.C.X., G.L. Lima, K.M. Maia, A.L.P. Souza, T.S. Castelo, A.L.C. Paiva, et al. 2019. Single injection of eCG/hCG leads to successful estrous synchronization in the collared peccary (*Pecari tajacu* Linnaeus, 1758). Anim. Reprod. Sci. 208: 106–112.

Santos, E.A.A., P.C. Sousa, J.A.M. Martins, R.A. Moreira, A.C.O. Monteiro-Moreira, F.B.M.B. Moreno, et al. 2014. Protein profile of the seminal plasma of collared peccaries (*Pecari tajacu* Linnaeus, 1758). Reproduction 147: 753–764.

Santos, M.L.T., A.A. Borges, L.B. Queiroz Neta, M.V.O Santos, M.F. Oliveira, A.R. Silva, et al. 2016. *In vitro* culture of somatic cells derived from ear tissue of collared peccary (*Pecari tajacu* Linnaeus, 1758) in medium with different requirements. Pesq. Vet. Bras. 36: 1194–1202.

Santos, E.A.A., P.C. Sousa, A.M. Silva, A.L.C. Paiva, A.E.A. Lago, V.V. Paula, et al. 2018. Superoxide dismutase and catalase activity in collared peccary (Pecari tajacu) seminal plasma and their relation to sperm quality. Semin. Cienc. Agrar. 39: 787–796.

Santos, C.S., L.B. Campos, É.C.G. Praxedes, S.S.J. Moreira, J.B.F. Souza-Júnior, P. Comizzoli, et al. 2021. Influence of antibiotics on bacterial load and sperm parameters during short-term preservation of collared peccary semen. Anim. Reprod. 18(3): e20210021. doi: 10.1590/1984-3143-AR2021-0021.

Silva, A.R., L.B. Campos, K.M. Maia and A.A. Borges. 2017. Reproductive biotechniques applied to collared peccaries (*Pecari tajacu* Linnaeus, 1758) – an experimental strategy. Ver. Bras. Reprod. Anim. 41: 110–115.

Silva, A.M., L. Bezerra, E. Praxedes, S. Moreira, C. de Souza, M. de Oliveira, et al. 2019a. Combination of intracellular cryoprotectants preserves the structure and the cells proliferative capacity potential of adult collared peccary testicular tissue subjected to solid surface vitrification. Cryobiology 91: 53–60.

Silva, A.M., A.F. Pereira and A.R. Silva. 2019b. Testicular tissue cryopreservation and culture as a tool for the conservation of wild mammals. R. bras. Reprod. Anim. 43(2): 229–234.

Silva, M.A., A.G. Pereira, A.V. Brasil, L.B. Macedo, J. Souza-Junior, C.E.M. Bezerra, et al. 2021. Influence of freezing techniques and glycerol-based cryoprotectant combinations on the survival of testicular tissues from adult collared peccaries. Theriogenology. 167: 111–119.

Sonner, J.B., M.A. Miglino, T.C. Santos, R. Carvalhal, A.C. Assis Neto, A.C. Moura, et al. 2004. Macroscopic and morphometric aspects of the testes in collared peccaries and peccaries. Biota. Neotrop. 4: 1–12.

Sousa, P.C., E.A.A. Santos, A.L.P. Souza, G.L. Lima, F.F.P.C Barros, M.F. Oliveira, et al. 2013. Sperm morphological and morphometric evaluation in captive collared peccaries (Pecari tajacu). Pesq. Vet. Brasil. 33: 924–930.

Souza, A.L.P., G.L. Lima, C.G.X. Peixoto, A.M. Silva, M.F. Oliveira and A.R. Silva. 2009. Evaluation of anesthetic protocol for the collection of semen from captive collared peccaries (*Tayassu tajacu*) by electroejaculation. Anim. Reprod. Sci. 116: 370–375.

Souza, A.L. Paz., G.L. Lima, G.C.X. Peixoto, A.M. Silva, M.F. Oliveira and A.R. Silva. 2016. Use of Aloe vera-based extender for chilling and freezing collared peccary (*Pecari tajacu*) semen. Theriogenology. 85: 1432–1438.

Sowls, L.K. 1984. The Peccaries. Tucson, University of Arizona Press.

Sowls, L.K. 1997. Javelines and Other Peccaries: Their Biology, Management and Use. 2.Ed. Tucson: Texas A&M University Press.

Vieira, R.L.A., M.A.R. Feliciano, R.H.R. Moreira and C.M. Costa. 2021. Seminal characterization of white-lipped peccary (*Tayassu peccary*). Reprod. Domest. Anim. 00: 1–6.

Wetzel, R.M., R.E. Dubos, R.L. Martin and P. Myers. 1975. Catagonus, an "extinct" peccary, alive in Paraguay. Science 189: 379–381.

Yahnke, C., J. Unger, B. Lohr, D. Meritt and W. Heuschele. 1997. Age specific fecundity, litter size and sex ratio in the Quimilero (*Catagonus wagneri*). Zoo Biol. 16: 301–302.

Part II
Carnivora

Part Co-Editor: Regina Celia R. Paz
Laboratório de Pesquisa em Animais de Zoológico
Universidade Federal de Mato Grosso
Cuiabá, Brazil
reginacrpaz@gmail.com

Chapter 4

The Canids

Thyara de Deco-Souza[1]* and Gediendson Ribeiro de Araujo[2]

[1]Faculdade de Medicina Veterinária e Zootecnia,
Universidade Federal de Mato Grosso do Sul – UFMS Av. Sen. Filinto Müler,
2443 – Pioneiros, Campo Grande – MS, Brazil, 79070-900.
Email: thyara.araujo@ufms.br

[2]Laboratório Reprobiote, Biotério Central, UFMS Av. Senador Filinto Muller,
1555 Vila Ipiranga, Campo Grande – MS, Brazil, 79070-900.
Email: gediendson.araujo@ufms.br

INTRODUCTION

The South American canids are not a monophyletic group and emerged from three canid invasions, now represented by the bush dog (*Speothos venaticus*), the maned wolf (*Chrysocyon brachyurus*), and the South America foxes (Sillero-Zubiri et al. 2013). Only the genus Pseudalopex, Lycalopex, and perhaps Atelocynus, appear to have a South American origin. The morphologic and genetic similarity of Pseudalopex and Lycalopex species suggests that they should be grouped as a single genus where the term Lycalopex has priority (see review in Sillero-Zubiri et al. 2013). Thus, six extant canids genus can be found in South America: *Atelocynus* (1 species); *Cerdocyon* (1 species); *Chrysocyon* (1 species); *Lycalopex* (6 species); *Speothos* (1 species); and *Urocyon* (1 species). The most endangered species is *Lycalopex fulvipes*, an endemic canid that occurs in Chile (Silva-Rodriguez et al. 2016). The other species are near threatened (N=5) or Least concerned (N=5) (IUCN 2021).

Conservation efforts aim to maintain a genetically viable population, reducing the loss of individuals and increasing the size and genetic variability of the population. In this context, the *ex situ* specimens play a key role as genetic

---

*Corresponding author: thyara.araujo@ufms.br

backups to be accessed when *in situ* genetics is threatened. Assisted reproductive technologies (ART) can improve genetic exchange and reproductive efficiency, overcoming complications resulting from animal translocation and optimizing access to individuals' genetics. The challenge is modulating the ART for each species, considering physiological and environmental specificities (Deco-Souza et al. 2021). In this chapter, we discuss the basic aspects of the reproductive physiology and the application of Assisted Reproduction technologies in South American canids.

CANID REPRODUCTIVE PHYSIOLOGY

The canine species have unique reproductive characteristics such as a pre-ovulatory progesterone rise, ovulation of immature oocytes, and a prolonged anestrus period (Concannon 2011). Although the reproductive physiology of most wild canid species is similar to the domestic dog, considerable variations in terms of seasonality, ovulation mechanism, and characteristics of the estrus cycle have been demonstrated among some of the species studied (DeMatteo et al. 2006, Johnson et al. 2014b, Silva et al. 2022). Therefore, even though the domestic dog serves as a starting point, the particularities of each species should not be ignored when developing assisted reproduction programs. Among the South American canids, reproductive information is extremely scarce and is concentrated mostly in the maned wolf and crab-eating fox (*Cerdocyon thous*).

Although domestic dogs do not show reproductive seasonality, most South American canids are seasonal breeders (Concannon 2011). The natural breeding season for the maned wolf range from May to September with most births occurring in June, and copulation starting in April (Maia and Gouveia 2002). It has been shown that photoperiodism influences reproductive physiology in this species, as captive animals housed in North America breed from October to March (Maia and Gouveia 2002). In Southern Brazil, crab-eating foxes give birth during spring (Faria-Corrêa et al. 2009); the same season was found in a captive population, in the Midwest region, where pups are born from August to November (mating from June to September) (Souza et al. 2012). Captive animals at the National Zoological Park, Washington could give birth twice a year – around eight months of birth interval (Brady 1978). Besides this, reduced birth interval has not been found in another *ex situ* populations (Souza et al. 2012). Faria-Corrêa et al. (Faria-Corrêa et al. 2009) reported other authors who also found a biannual birth interval in *C. thous* (Nowak 1999, Eisenberg and Redford 1999, Parera 2002). Chilla (*Lycalopex griseus*), Culpeo (*L. culpaeus*), and hoary fox (*L. vetulus*) also are reported as seasonal breeders, reproducing from June to September (Jiménez and Novaro 2004a, b, Courtenay et al. 2006, Candeias et al. 2020). The gray fox (*Urocyon cinereoargenteus*) is also reported as a seasonal breeder; however, this information is only valuable within the North American population (Follman 1978, Fritzell 1987, Wood 1958). A known exception among South American canids is the bush dog, which reproduces throughout the year (Porton et al. 1987, DeMatteo et al. 2006). Several factors can interfere with the reproductive periodicity in canids such as photoperiod, nutrition (food availability), and rainfall (Concannon 2012, McNutt et al. 2019). The variety of environments where South American canids can be found may explain the difference in evolutionary strategies and local adaptations presented in its reproductive strategies; thus, there is still much to be studied to understand how the environment influences each species' seasonality and how it can be managed to improve ART.

In some species, seasonality is known to affect even sperm production. In maned wolf there is a reduction, but not complete cessation, of sperm production. The volume, the number of spermatozoa per ejaculate, and the percentage of live sperm and testicular volume reduce during the non-breeding season (Teodoro et al. 2012, Nagashima and Songsasen 2021). The same path was observed in free-living crab-eat foxes, with a reduction in the mean total sperm collected during the non-breeding season (Silva et al. 2022). Knowledge about reproductive seasonality and the factors that stimulate reproductive activity is essential for planning assisted reproduction programs actions, such as couples pairing, gamete collection and storage, artificial insemination, and embryo transfer.

Canids have a long interestrus interval, that can last as long as 12 months, which is why they are classified as monoestrous – referring to a single estrus per estrous cycle (Johnston et al. 2001, Songsasen et al. 2006, Concannon 2011). However, it has been suggested that peripubertal bush dog has multiple estrous cycles, with interestrus intervals ranging from 15 to 44 days (Porton et al. 1987). It seems that this polyestrous behavior is transient, as mature females have an obligatory pseudo-pregnancy if they ovulate and do not get pregnant, just like the other canines (DeMatteo et al. 2006). Bush dog estrous cycles are also somehow affected by the male presence. Despite being a spontaneous ovulator, the interestrus intervals of mature females reduce when the male is present (DeMatteo et al. 2006). On the other hand, peripubertal females do not show any sign of estrus until paired with a male (Porton et al. 1987). Male presence also influences estrous cycles in the maned wolf. This species is a well known induced ovulator, and olfactive stimuli from the male are required to complete the estrous cycle (Velloso et al. 1998, Songsasen et al. 2006, Johnson et al. 2014b).

Canids' estrus cycle consists of four phases: proestrus, estrus, diestrus, and anestrus. The proestrus is marked by a progressive increase in serum estradiol and female attractivity. The LH and FSH pulses become less detectable until their pre-ovulatory surge (Concannon 2009, 2011). In late proestrus, follicle luteinization begins, and progesterone concentrations increase (Concannon 1977). Estrus is marked by female receptivity to copulation in response to the decrease in the estrogen: progesterone ratio (Concannon et al. 1979). Canine oocytes ovulate around 36–50 h after LH pick, in Prophase I, and complete maturation in the oviduct up to 48–72 h after being released (Reynaud et al. 2005, Chastant-Maillard et al. 2011). After maturation, oocyte needs an additional 12 to 24 h to achieve full competence to be fertilized (Nagashima et al. 2015). The luteal phase length is similar in pregnant and non-pregnant cycles, lasting around two months (Concannon 2009, 2011). During anestrus, hormonal levels are basal until some stimuli increase GnRH pulsatility, frequency of LH pulse and follicular growth (Tani et al. 1996). The long luteal phase and anestrus reflect the long interestrus interval even in non-seasonal canids. Endocrinology of the estrous cycle in wild canids can be accessed by longitudinal non-invasive hormonal monitoring of fecal estrogen and progestagen metabolites and has been presented as similar to that of the bitch for the crab-eating fox, maned wolf, and hoary fox (Velloso et al. 1998, Songsasen et al. 2006, Souza et al. 2012, Candeias et al. 2020).

SPERM COLLECTION

Several techniques are available for male gamete recovery, allowing sperm collection in live animals and post-mortem. Choosing the ideal method depends on several factors such as equipment and logistics available, as well as animal age, the experience of the team, and the temperament of the animal. Sperm can be used immediately after collection or can be stored for a few hours (maintained at 5°C) or several years (in liquid nitrogen). However, the process of reducing cellular activity (either by refrigeration or cryopreservation) stresses the sperm and reduces its fertilizing capacity. Thus, the refinement of protocols that meet the species-specific characteristics is crucial to minimize these damages and increase sperm fertility.

Epididymal sperm collection consists in recovering sperm from epididymis and vas deferens in living animals or post-mortem. Sperm can be collected by flushing, squeezing, slicing, or percutaneous epididymal sperm aspiration (PESA) (Martins et al. 2009, Varesi et al. 2013, Muñoz-Fuentes et al. 2014, Ali Hassan et al. 2021, Nagashima and Songsasen 2021, Galarza et al. 2022). The first three methods are performed in dead animals or after castration. PESA is indicated for living vasectomized males, and to diagnosticate azoospermia; however, repeated puncture sessions may lead to an inflammatory reaction in the epididymis (Ali Hassan et al. 2021). After being removed from the animal, the testes can be stored at 4°C in sterile 0.9% saline or ringer lactate solutions for about a week until processed (Yu and Leibo 2002, Galarza et al. 2022). In the laboratory, cauda epididymis and vas deferens (CE-Vas) are isolated from the testicle, dissected to remove blood vessels, and rinsed with 0.9% saline solution or DPBS (Dulbecco's Phosphate-Buffered Saline) to reduce blood cells contamination (Yu and Leibo 2002, Zmudzinska et al. 2022).

Removing the diploid cells, like cellular debris and erythrocytes, may improve cryo survival as they can be a source of reactive oxygen species (Dorado et al. 2013, Muñoz-Fuentes et al. 2014). After sperm collection, those cells can be removed by several methods which also can select functional spermatozoa, improving sample quality. The colloidal centrifugation (Density-gradient centrifugation – DGC and single layer centrifugation – SLC) involves centrifuging sperm on a (double or single) column of colloidal solution, where the sperms form a pellet (Dorado et al. 2013, Muñoz-Fuentes et al. 2014, Galarza et al. 2022). Based on the ability of motile spermatozoa to move from one suspension to another, the Swim-up method consists of placing the sperm under a layer of a medium and co-incubate it to recover the supernatant containing the spermatozoa (Dorado et al. 2016, Galarza et al. 2022).

Unfortunately, there are no robust studies collecting epididymis sperm from South American canids. This technique is essential to recovering the genetics from animals that were killed in road accidents, died due to environmental or anthropic catastrophes, or were slaughtered or castrated for population control, reducing the impact of the loss of this individual. In maned wolfs, epididymal sperm collection has been described in six animals (with sperm cells in only three of them). Interestingly, in one animal 23.5 $\times$ 10^6 sperm cells, with 60–65% of sperm motility, could be collected after more than 24 h following animal death/ necropsy (Nagashima and Songsasen 2021).

Digital manipulation of the penis is the standard technique for collecting dog semen, and also has been used in South American canids (Table 4.1) (Carvalho et al. 2020, Teodoro et al. 2012, 2021). Five captive crab-eating foxes were conditioned for semen collection by digital manipulation (Carvalho et al. 2020). After 30 days of training, all five animals allowed physical touch, but after 50 days of training, semen was collected only from two of them. From those two animals, the researchers could obtain 13 semen samples. Captive maned wolves (N=3) were conditioned for being restrained using a yoke and muzzle for semen collection. Semen was collected in all attempts (N=70) during the breeding season; however, only 80.0% of them contained sperm cells (Teodoro et al. 2012). Semen collection by digital manipulation requires a great effort for animal conditioning and is restricted to individuals kept in captivity. However, it has the advantage of being performed without anesthesia and has positive benefits on animal welfare; thus, it should be considered in centers that aim to develop assisted reproduction.

The most used method for semen collection in wild animals is the electroejaculation (EE), as it can be performed under anesthesia and does not require animal training. The main disadvantage of electroejaculation is the high incidence of urine contamination, which is toxic to spermatozoa (Johnston et al. 2007, Nagashima and Songsasen 2021). Some factors may favor the contamination of samples with urine, such as cranial insertion of the rectal probe and high electrical stimulation. To reduce or even eliminate urine contamination, several authors wash the bladder with 0.9% saline solution before EE (Johnston et al. 2007, Deco et al. 2010, Johnson et al. 2014a). The stimulation protocol may vary according to the species (Johnston et al. 2007, Johnson et al. 2014a, Assumpção et al. 2017). In maned wolves and crab-eating foxes, the protocol was similar to that described for other carnivores (Howard 1993) and consists of 60–90 stimuli with 2–5 V. The stimuli were delivered during three series (3 to 7 min intervals) in sets of around 30 each, with increasing intensity (2–3–4V; 3–4–5V and 4–5 V) (Souza et al. 2011, Johnson et al. 2014a, Assumpção et al. 2017). Good-quality semen samples could be collected from maned wolves using EE (Table 4.1). However, it did not succeed in the crab-eating foxes, as in all attempts there was urine contamination (Souza et al. 2011, Assumpção et al. 2017).

In recent years, pharmacological semen collection has been developed in wild and domestic carnivores. The technique is based on the ability of some drugs, particularly alpha-2 agonists, to stimulate the emission of sperm in the animal's urethra (Chaturapanich et al. 2002, Sanbe et al. 2007). Sperms are collected by urethral catheterization (CT). Among carnivores, the technique is successful

Table 4.1 Sperm collection in South American Canids

Species	Method	Semen volume (μL)	Total sperm $\times 10^6$	Total motility (%)	Progressive motility (%)	References
	Epididymis		24.8	61.7		(Nagashima and Songsasen 2021)
	Digital manipulation	1300	73.9	76.1		(Teodoro et al. 2012)
Maned wolf	Digital manipulation	2500	214.2	88.3		(Teodoro et al. 2021)
	Electroejaculation	2000	78.1	59.8		(Johnson et al. 2014a)
	Urethral catheterization	100	1	40	30	(Lueders et al. 2013)
Crab-eating fox	Digital manipulation	393.2	217.4	68	48.8	(Carvalho et al. 2020)
	Urethral catheterization	54	10.15	62.5		(Silva et al. 2022)

within the felines, and highly concentrated, good-quality semen samples are collected (Zambelli et al. 2008, Lueders et al. 2012, Araujo et al. 2018). However, the results in canids are still controversial. The first CT collection in canids was described in African wild dogs (*Lycaon pictus*) and one maned wolf, using Medetomidine (1 mg)/Zoletil® (30 mg) and Medetomidine (0.5 mg)/Ketamine (50 mg)/Butorphanol (5 mg), respectively (Lueders et al. 2013). With the dosage used in the maned wolf (0.025 mg/kg, considering a 20 kg male), fewer sperms were collected compared to digital manipulation and electroejaculation (Table 4.1). Fewer sperms were also collected in dogs using dexmedetomidine (0.015 mg/kg)/ketamine (3 mg/kg), with 278.6×10^6 total sperm collected, in contrast to an average of 978×10^6 total sperm collected by digital manipulation (Kuczmarski et al. 2020, Hermansson et al. 2021). It is well established that medetomidine dosage influences the number of sperm collected by CT (Cunto et al. 2015). In felines, the medetomidine at 0.1 mg/kg (associated with ketamine at 5 mg/kg) is often used (Zambelli et al. 2008, Lueders et al. 2012, Araujo et al. 2018), and this same dosage was tested in free-living crab-eating foxes (Figure 4.1) (Silva et al. 2022). Even so, the total spermatozoa collected was lower than by digital manipulation (Table 4.1). While digital manipulation and electroejaculation are likely to allow the collection of more sperm, they demand great commitment and have issues with urine contamination, which also limits their efficiency. Using more accurate procedures, such as intrauterine insemination, IVF, and ICSI may enable the use of CT-collected sperm in assisted reproduction programs.

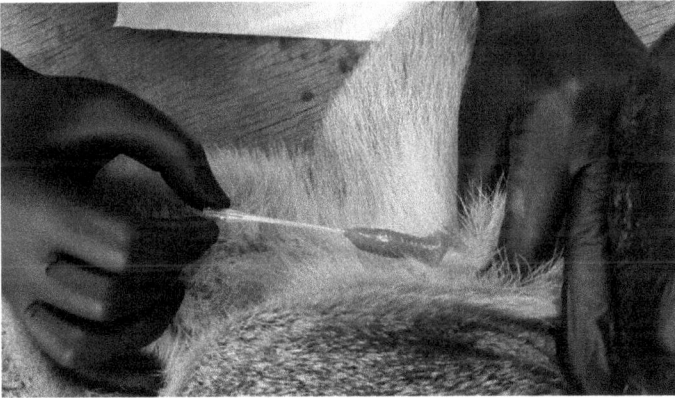

Figure 4.1 Pharmacological semen collection using the association of medetomidine and ketamine in free-living crab-eating fox (Silva et al. 2022).

SPERM PRESERVATION

The reduction of sperm metabolism, by refrigeration or cryopreservation, enables the genetic exchange and allows the application of other reproductive technologies in individuals who are in different locations, avoiding the transport of animals. Despite allowing storage for a shorter period, refrigeration is less stressful to the sperm and can be applied in situations where the fertilization procedure can be performed within a few days after semen collection. The chilled semen viability/ fertility depends on several factors such as sperm quality, extender used, storage temperature, and the insemination dose (Tsutsui et al. 2003, Lojkić et al. 2022).

In dogs, an 83.3% conception rate was observed after artificial insemination (AI) of 400×10^6 sperms stored at 4°C for 4 days in an egg yolk tris-fructose citrate extender (Tsutsui et al. 2003). In transport conditions, the temperature may vary up to 4°C, but it has been demonstrated that sperm motility did not decrease until the first 48 h even when kept at 12°C (Hori et al. 2014). Despite being neglected in wild canids, sperm refrigeration must be considered to achieve better conception rates in situations with easy animal access (dos Santos and Silva 2022).

The cryopreservation of spermatozoa begins with dilution in an extender containing reagents that protect the cells from the stresses caused by the reduction in temperature. Subsequently, they are subjected to a period of refrigeration to stabilize the membrane, followed by freezing and storage in liquid nitrogen. Cholesterol proportion as well as the rate of saturated X unsaturated fatty acids influences the susceptibility of sperm plasma and acrosomal membranes to the changes in temperature during cryopreservation (Amann and Pickett 1987, Hammerstedt et al. 1990, Bailey et al. 2000, Holt 2000). As membrane composition varies between species, species-specific extenders must be considered to achieve better conception rates. Thus, it is necessary to know the function of each reagent used in sperm cryopreservation extender and their potential damage to cell structure.

Egg yolk (EY) still is the most used reagent in refrigeration and cryopreservation extender in canids. The Low-density lipoproteins (LDL) and their phospholipids are the main EY components that protect sperm cells from injuries, by stabilizing the sperm membrane phospholipids and by forming a protective film on its surface (see review in Manjunath 2012, Bustani and Baiee 2021). Besides its well-known benefits, EY also has some disadvantages, such as the variability in the composition, the risk of bacterial contamination, and the international commercial restriction on animal-based extenders.

Soy lecithin, polyvinyl alcohol, and LDL are common EY substitutes used in dogs (Bencharif et al. 2008, Dalmazzo et al. 2018, Nabeel et al. 2019). However, in South American canids the EY remains the principal option (Johnson et al. 2014a, Teodoro et al. 2021). Egg yolk replacement is of particular importance to enable the cryopreservation of semen in free-ranging animals since in most cases the capture sites do not have the structure to keep the extender frozen. The LDL, for example, can be stored at room temperature and are therefore better suited for this purpose.

Another key point in the extender is the cryoprotectant, whose toxicity varies according to the species. In dogs, glycerol is recommended as the cryoprotectant whereas dimethyl sulfoxide shows toxicity at 0.6 M (Songsasen et al. 2002, Martins-Bessa et al. 2006, Rota et al. 2006). By contrast, in maned wolves, semen cryopreserved in 1 M dimethyl sulfoxide showed higher post-thaw motility and membrane integrity, when compared to glycerol (Johnson et al. 2014a). However, good post-thaw quality was found in semen cryopreserved with 7% glycerol from one maned wolf (Teodoro et al. 2021).

MALE GERM CELL CULTURE

Aiming to access genetic material from prepubertal animals or even to take advantage of the potential of the spermatogonia reserve to obtain more gametes, researchers have been dedicated to developing methodologies for *in vitro* production of sperm from testicular tissue. In dogs, progression of spermatogenesis from testis fragments has been studied since 2010, with full spermatogenesis being completed in 11% of the seminiferous tube after 13 months of xenografting incubation (Abrishami et al. 2010). From then on, some studies have been developed to evaluate testicular vitrification methods and culture medium (see review in Silva 2022).

The optimal testicular transportation conditions (extender, temperature, and length) have not been studied in dogs yet. Until now, tests were shipped in DPBS on ice within 1 hour or processed immediately (Abrishami et al. 2010, Shiraz et al. 2014). In grey wolf (*Canis lupus*), testicles were shipped overnight in supplemented Leibovitz-15 medium (L15) on ice (Andrae et al. 2021). Defining long-term transport conditions is essential for the application of the method in wild animals since in most cases the material will not be available near reference laboratories for processing. Once in the lab, testicles can be freshly cultivated or vitrified. In dogs, 13 months of *in vivo* cultivation (by xenografting) of fresh testicle tissue alow male pronuclear formation after Intracytoplasmic sperm injection (ICSI) (Shiraz et al. 2014). Xenotransplantation was also made in vitrified dog testicles, allowing the formation of spermatogonial stem cells (Lee et al. 2016).

Although until now there are no publications on South American Canids, the development of methods that allow transportation, cryopreservation, and cultivation are essential to maximizing the utilization of genetic material from endangered species. The main challenge is to access a sufficient amount of testicles with good manipulation conditions for a robust design of experimental protocols.

ARTIFICIAL INSEMINATION

The success of artificial insemination (AI) depends on several factors in addition to sperm quality, such as the correct synchronization of insemination with ovulation, localization of sperm deposition, the amount of sperm inseminated, and the animal's welfare condition. It demands deep acknowledgment of female anatomy and physiology, to establish the hormonal path during the estrous cycle, ovulation mechanisms, and responsiveness to hormonal protocols.

As discussed before, in dogs, ovulation occurs around 36–50 h after LH pick. Considering that: 1. bitch ovulate as primary oocytes, and needs additional two to five days to complete maturation; 2. oocytes remain viable for two to four more days, and 3. fresh dog spermatozoa can survive in the uterus for up to six days; theoretically, the AI can be made from about two days before and eight days after ovulation (Linde-Forsberg 1995). The timing of ovulation in relation to the LH pick is usually determined by measuring serum progesterone or LH. Also, clinical (vaginal smear morphology, vulva swelling, and color of vaginal discharge) and behavior (tail flexing and male receptivity) signs may be monitored (Thomassen and Farstad 2009). This method may have limited application in wild species, as sequential restrain procedures will be required for the serum hormonal monitoring and vaginal smear (unless the animal is conditioned for this purpose), and the inherent stress of this procedure may interfere with reproduction. However, in farmed blue foxes (*Alonex lagonus*) estrous cycle was monitored for AI with a modified ohm meter, that measures electrical resistance in the vaginal mucus (Farstad et al. 1992). Females inseminated on the second day from vaginal electrical resistance peak had the highest conception rate.

Artificial insemination may also be performed after the stimulation of ovarian function. Many protocols have been used for estrous induction in bitches (see review in Kutzler 2005); however, the studies are still limited in South American Canids. Deslorelin, a GnRH agonist, was used to stimulate ovarian function in captive maned wolves. Females were implanted in the vestibular mucosa of the vulva or in the ear, for 7 to 11 days. All of the animals responded to the treatment and exhibited a rise in estrogen metabolites; however, ovulation only occurred when females were paired or treated with LH on the day of deslorelin removal (Johnson et al. 2014b). Thus, as maned wolfs are induced ovulators, the GnRH agonists may induce ovarian function but not ovulation in this species.

Surgical and non-surgical techniques are available for AI. Vaginal and intrauterine insemination may be performed using disposable plastic and steel catheter, respectively (Linde-Forsberg 1995). However, intrauterine AI yielded significantly higher whelping rates than vaginal AI, especially when frozen-thawed semen is used (Farstad 1984, Linde-Forsberg et al. 1999). The intra-uterine AI may also be performed non-surgically using an endoscope, that allows visualization of the entire vaginal lumen, and the vaginal portion of the cervix. The main advantage of non-surgical techniques is the possibility to inseminate several times in a short interval, as it does not require sedation. However, for most wild canids sedation may be necessary even for these non-surgical procedures, as they may be aggressive and the physical restrain may cause additional stress. Laparotomy or laparoscopy can be used to ensure that the semen is properly deposited into the uterus, near the physiological fertilization site, with the latter having the advantage of being a minimally invasive technique. However, as those surgical methods are performed only once, they demand an accurate estimate of ovulation (see review in Thomassen and Farstad 2009).

EMBRYO TECHNOLOGY

In most mammal species, oocytes may be collected for *in vitro* embryo production throughout *in situ* methodologies (oocyte pick-up and laparoscopic oocyte pick-up) or fowling ovarian removal (after animal death or castration) (Prentice and Anzar 2011). In canids, due to the ovarian bursa, the usual *in situ* methodologies are not feasible (Nagashima and Songsasen 2021). However, ovulated oocytes can be collected by retrograde flushing of the oviduct (Hossein et al. 2007), and mature (metaphase II) oocytes can be flushed on day 6 after LH surge (Nagashima et al. 2015). Oocyte recovery after ovarian removal may be a feasible choice; however, they are still immature and metaphase I oocyte cultivation is not reliable yet (see review in Nagashima et al. 2019). Therefore, retrograde flushing of the oviduct, after the period of oocyte maturation, is still the method of choice for collecting oocytes in canids.

The *in vitro* embryo production (either by classic *in vitro* fertilization or by ICSI) depends on cultivation in proper media, which varies between species. In dogs, male and female gametes were co-incubated for 14 h in NCSU-23 medium, achieving high hate (78.8%) of embryo development and the birth of live cubs after embryo deposition in the cranial uterine horn. In this study, they also found that the addition of magnesium in the extender was fundamental for proper sperm capacitation before IVF (Nagashima et al. 2015). The site of embryo deposition depends on its stage, with an 8-cells embryo being transferred into the oviduct, and the uterine horn being the preferred site for oocyte deposition for later stage embryos – >8-cells (see review in Nagashima et al. 2019). However, two puppies were born after the intrauterine transfer of three early embryos matured *in vivo*, suggesting that this may be a feasible choice for both early and late-stage embryos (Tsutsui et al. 2006).

The preservation of oocytes and embryos has also been studied in the domestic dog. Because of its lipid-richness, which increases oocyte sensitivity to chilling injuries, few papers described oocyte vitrification in canids (Boutelle et al. 2011, Abe et al. 2008, Turathum et al. 2010). On the other hand, cryopreserved domestic dog embryos have resulted in live births. Dog embryos were vitrified either using a commercially available Vit Kit (Irvine Scientific, Santa Ana, CA) or using ethylene glycol and sucrose (Nagashima et al. 2015, Abe et al. 2011). Corroborating the latter studies, the Cryotop vitrification method (with ethylene glycol, DMSO, and sucrose) showed a higher pregnancy rate than the slow-freezing method in dog embryos produced *in vivo* (Hori et al. 2016).

PREANTRAL FOLLICLES MANIPULATION

Preantral follicles (PAF), i.e. primordial to secondary follicles, consists of the major follicular population within the ovarian cortex. However, most of them will be lost during follicular waves and will never become available for fertilization. Aiming to take advantage of this potential population, which can be accessed even from prepubertal females, several methodologies have been developed to access the ovarian cortex for *in vitro* embryo production. Whole ovary can be collected after castration or after animal death, aiming to save its genetic potential, but small ovarian fragments can be obtained through ovarian biopsy pick-up in live animals. The latter is of special application in seasonal species (as some South American canids) enabling the production of mature gametes regardless of the time of year. Recent literature reviews are evaluable with deep discussion on this field (Araújo et al. 2014, Mohammed 2022, Green and Shikanov 2016).

Despite the potential as a tool for wild animal conservation, research developing PAF technologies in South American Canids is incipient. However, some advances have been made in dogs.

FINAL CONSIDERATIONS

The few valuable data in South American Canids Assisted Reproduction Technologies are concentrated in male technologies, especially on sperm collection and cryopreservation. Considering the unique aspects of reproductive biology and the variety of species, we still have a long way to go for the effective application of these technologies in assisted reproduction programs. What can be concluded is that semen collection by digital manipulation may be an alternative when dealing with conditioned animals. Electroejaculation is a feasible choice; however, its effectiveness is influenced by the technician's experience and urine contamination may impair its application. Urethral catheterization after pharmacological induction of sperm emission still is not as successful as in feline species but can be used in more precise fertilization methodologies such as IVF and ICSI. Also, the seasonal aspects of reproduction must be checked and considered when studying this species.

REFERENCES

Abe, Y., D.-S. Lee, S.-K. Kim and H. Suzuki. 2008. Vitrification of canine oocytes. J. Mamm. Ova Res. 25(1): 32–36. https://doi.org/10.1274/jmor.25.32.

Abe, Y., Y. Suwa, T. Asano, Y.Y. Ueta, N. Kobayashi, N. Ohshima, et al. 2011. Cryopreservation of canine embryos. Bio. Reprod. 84(2): 363–368. https://doi.org/10.1095/biolreprod.110.087312.

Abrishami, M., S. Abbasi and A. Honaramooz. 2010. The effect of donor age on progression of spermatogenesis in canine testicular tissue after xenografting into immunodeficient mice. Theriogenology 73(4): 512–22. https://doi.org/10.1016/j.theriogenology.2009.09.035.

Ali Hassan, H., G. Domain, G.C. Luvoni, R. Chaaya, A. Van Soom, E. Wydooghe, et al. 2021. Canine and feline epididymal semen—a plentiful source of gametes. Animals 11(10): 1–10. https://doi.org/10.3390/ani11102961.

Amann, R.P. and B.W. Pickett. 1987. Principles of cryopreservation and a review of cryopreservation of stallion spermatozoa. J. Equine. Vet. Sci. 7(3): 145–173. https://doi.org/10.1016/S0737-0806(87)80025-4.

Andrae, C.S., E.C.S. Oliveira, M.A.M.M. Ferraz and J. B. Nagashima. 2021. Cryopreservation of grey wolf (Canis Lupus) testicular tissue. Cryobiology 100(June 2020): 173–179. https://doi.org/10.1016/j.cryobiol.2021.01.010.

Araujo, G.R., T.A.R. Paula, T. Deco-Souza, R.G. Morato, L.C.F. Bergo, L.C. Silva, et al. 2018. Comparison of semen samples collected from wild and captive jaguars (*Panthera Onca*) by urethral catheterization after pharmacological induction. Animal Reprod. Sci. 195(July 2017): 1–7. https://doi.org/10.1016/j.anireprosci.2017.12.019.

Araújo, V.R., M.O. Gastal, J.R. Figueiredo and E.L. Gastal. 2014. *In vitro* culture of bovine preantral follicles: a review. Reprod. Biol. Endocrinol 12: 78. https://doi.org/10.1186/1477-7827-12-78.

Assumpção, T.I., A.L.Q. Santos and E.A. Canelo. 2017. Biometria Testicular e Características Morfológicas Dos Espermatozoides de Cachorros-Do-Mato Cerdocyon Thous Linnaeus, 1766 (Carnivora, Canidae). Rev. Bras. Ciência Vet. 24(2): 99–103. https://doi.org/10.4322/rbcv.2017.020.

Bailey, J.L., J.F Bilodeau and N. Cormier. 2000. Semen cryopreservation in domestic animals: a damaging and capacitating phenomenon. J. Androl. 21(1): 1–7. https://doi. org/10.1002/j.1939-4640.2000.tb03268.x.

Bencharif, D., L. Amirat, M. Anton, E. Schmitt, S. Desherces, G. Delhomme, et al. 2008. The advantages of LDL (Low Density Lipoproteins) in the cryopreservation of canine semen. Theriogenology 70(9): 1478–1488. https://doi.org/10.1016/j.theriogenology.2008. 06.095.

Boutelle, S., K. Lenahan, R. Krisher, K.L. Bauman, C.S. Asa and S. Silber. 2011. Vitrification of oocytes from endangered mexican gray wolves (Canis Lupus Baileyi). Theriogenology 75(4): 647–654. https://doi.org/10.1016/j.theriogenology.2010.10.004.

Brady, C.A. 1978. Reproduction, growth and parental care in crab-eating foxes cerdocyon thous at the national zoological park, Washington. Int. Zoo Yearb. 18(1): 130–134. https://doi.org/10.1111/j.1748-1090.1978.tb00243.x.

Bustani, G.S. and F.H. Baiee. 2021. Semen extenders: an evaluative overview of preservative mechanisms of semen and semen extenders. Vet. World 14(5): 1220–1233. https://doi. org/10.14202/vetworld.2021.1220-1233.

Candeias, Í.Z., C.F. da Motta Lima, F.G. Lemos, K.M. Spercoski, C. Al. de Oliveira, N. Songsasen, et al. 2020. First assessment of hoary fox (*Lycalopex Vetulus*) seasonal ovarian cyclicity by non-invasive hormonal monitoring technique. Conserv. Physiol. 8(1): 1–12. https://doi.org/10.1093/conphys/coaa039.

Carvalho, J.C., F.E. Silva, G. Rizzoto, C.R. Dadalto, L.S. Rolim, M.J. Mamprim, et al. 2020. Semen collection, sperm characteristics and ultrasonographic features of reproductive tissues in crab-eating fox (Cerdocyon Thous). Theriogenology 155 (October): 60–69. https://doi.org/10.1016/j.theriogenology.2020.06.016.

Chastant-Maillard, S., C. Viaris de Lesegno, M. Chebrout, S. Thoumire, T. Meylheuc, A. Fontbonne, et al. 2011. The canine oocyte: uncommon features of *in vivo* and *in vitro* maturation. Reprod. Fertil. Dev. 23(3): 391. https://doi.org/10.1071/RD10064.

Chaturapanich, G., S. Maythaarttaphong, V. Verawatnapakul and C. Pholpramool. 2002. Mediation of contraction in rat cauda epididymidis by alpha-adrenoceptors. Reproduction 887–892. https://doi.org/10.1530/rep.0.1240887.

Concannon, P.W. 1977. Changes in LH, Progesterone and sexual behavior associated with preovulatory luteinization in the bitch. Biol. Reprod. 17(4): 604–613. https://doi.org/10. 1095/biolreprod17.4.604.

Concannon, P.W., N. Weigand, S. Wilson and W. Hansel. 1979. Sexual behavior in ovariectomized bitches in response to estrogen and progesterone treatments. Biol. Reprod. 20(4): 799–809. https://doi.org/10.1095/biolreprod20.4.799.

Concannon, P.W. 2009. Endocrinologic control of normal canine ovarian function. Reprod. Domest. Anim. 44 (SUPPL. 2): 3–15. https://doi.org/10.1111/j.1439-0531.2009.01414.x.

Concannon, P.W. 2011. Reproductive cycles of the domestic bitch. Animal Reprod. Sci. 124(3–4): 200–210. https://doi.org/10.1016/j.anireprosci.2010.08.028.

Concannon, P.W. 2012. Research challenges in endocrine aspects of canine ovarian cycles. Reprod. Domest. Anim. 47 (SUPPL. 6): 6–12. https://doi.org/10.1111/rda.12121.

Courtenay, O., D.W. Macdonald, S. Gillingham, G. Almeida and R. Dias. 2006. First observations on South America's largely insectivorous canid: the hoary fox (*Pseudalopex Vetulus*). J. Zool. 268(1): 45–54. https://doi.org/10.1111/j.1469-7998.2005.00021.x.

Cunto, M., D.G. Küster, C. Bini, C. Cartolano, M. Pietra and D. Zambelli. 2015. Influence of different protocols of urethral catheterization after pharmacological induction (Ur.

Ca.P.I.) on semen quality in the domestic cat. Reprod. Domest. Anim. 50(6): 999–1002. https://doi.org/10.1111/rda.12626.

Dalmazzo, A., J.D.A. Losano, C.C. Rocha, R.H. Tsunoda, D.S.R. Angrimani, C.M. Mendes, et al. 2018. Effects of soy lecithin extender on dog sperm cryopreservation. Anim. Biotechnol. 29(3): 174–182. https://doi.org/10.1080/10495398.2017.1334662.

Deco, T.S., T.A.R. Paula, D.S. Costa, G.R. Araujo, R.M. Garay, G.S.C. Vasconcelos, et al. 2010. Coleta e Avaliação de Sêmen de Pumas (Puma Concolor Linnaeus, 1771) Adultos Mantidos Em Cativeiro. Rev. Bras. Reprod. Anim. 34(4): 252–259.

Deco-Souza, T., C.S. Pizzutto, L.S.B. Souza, P.N. Jorge-Neto, A.C. Csermak-Jr and G.R. Araújo. 2021. Desafios Para a Reprodução Assistida Em Animais de Vida Livre. Rev. Bras. Reprod. Anim. 45(4): 253–258. https://doi.org/10.21451/1809-3000. RBRA2021.033.

DeMatteo, K.E., I.J. Porton, D.G. Kleiman and C.S. Asa. 2006. The effect of the male bush dog (*Speothos Venaticus*) on the female reproductive cycle. J. Mammal. 87(4): 723–732. https://doi.org/10.1644/05-MAMM-A-342R1.1.

Dorado, J., M.J. Gálvez, J.M. Morrell, L. Alcaráz and M. Hidalgo. 2013. Use of single-layer centrifugation with androcoll-c to enhance sperm quality in frozen-thawed dog semen. Theriogenology 80(8): 955–962. https://doi.org/10.1016/j.theriogenology.2013.07.027.

Dorado, J., M.J. Gálvez, S. Demyda-Peyrás, I. Ortiz, J.M. Morrell, F.Crespo, et al. 2016. Differences in preservation of canine chilled semen using simple sperm washing, single-layer centrifugation and modified swim-up preparation techniques. Reprod. Fertil. Dev. 28(10): 1545. https://doi.org/10.1071/RD15071.

Eisenberg, J.F. and K.H. Redford. 1999. Mammals of the Neotropics—The Central Neotropics. University of Chicago Press, Chicago, USA.

Faria-Corrêa, M., R.A. Balbueno, E.M. Vieira and T.R.O. de Freitas. 2009. Activity, habitat use, density, and reproductive biology of the crab-eating fox (*Cerdocyon thous*) and comparison with the pampas fox (*Lycalopex gymnocercus*) in a Restinga area in the southern Brazilian Atlantic Forest. Mamm. Biol. 74(3): 220–229. https://doi.org/10.1016 /j.mambio.2008.12.005.

Farstad, W. 1984. Bitch fertility after natural mating and after artificial insemination with fresh or frozen semen. J. Small Anim. Pract. 25(9): 561–565. https://doi.org/10.1111/ j.1748-5827.1984.tb03429.x.

Farstad, W., J.A. Fougner and C.G. Torres. 1992. The optimum time for single artificial insemination of blue fox vixens (Alopex lagopus) with Frozen-thawed semen from silver foxes (Vulpesvulpes). Theriogenology 38(5): 853–865. https://doi.org/10.1016/0093-691X(92)90161-J.

Follman, E.H. 1978. Annual Reproductive Cycle of the Male Gray Fox. Trans. Ill. State Acad. Sci. 71: 304–311.

Fritzell, E.K. 1987. Gray fox and island gray fox. pp. 408– 420. *In:* M. Novak, J.A. Baker, M.E. Obbard and D. Malloch. Wild Furbearer Management and Conservation in North America. Ontario Trappers Association and Ontario Ministry of Natural Resources, Toronto, Canada.

Galarza, D.A., D.I. Jara, E.B. Paredes, J.X. Samaniego, M.S. Méndez, M.E. Soria, et al. 2022. BoviPure® Density-gradient centrifugation procedure enhances the quality of fresh and cryopreserved dog epididymal spermatozoa. Animal Reprod. Sci. 242(May): 1–9. https://doi.org/10.1016/j.anireprosci.2022.107003.

Green, Lisa J. and Ariella Shikanov. 2016. *In vitro* culture methods of preantral follicles. Theriogenology 86(1): 229–238. https://doi.org/10.1016/j.theriogenology.2016.04.036.

Hammerstedt, R.H., J.K. Graham and J.P. Nolan. 1990. Cryopreservation of mammalian sperm: what we ask them to survive. J. Androl. 11(1): 73–88. https://doi.org/https://doi.org/10.1002/j.1939-4640.1990.tb01583.x.

Hermansson, U., A. Johannisson and E. Axnér. 2021. Cryopreservation of dog semen in a tris extender with two different 1% soybean preparations compared with a tris egg yolk extender. Vet. Med. Sci. 7(3): 812–819. https://doi.org/10.1002/vms3.445.

Holt, W.V. 2000. Basic aspects of frozen storage of semen. Animal Reprod. Sci. 62(1–3): 3–22. https://doi.org/10.1016/S0378-4320(00)00152-4.

Hori, T., R. Yoshikuni, M. Kobayashi and E. Kawakami. 2014. Effects of storage temperature and semen extender on stored canine semen. J. Vet. Med. Sci. 76(2): 259–263. https://doi.org/10.1292/jvms.13-0303.

Hori, T., H. Ushijima, T. Kimura, M. Kobayashi, E. Kawakami and T. Tsutsui. 2016. Intrauterine embryo transfer with canine embryos cryopreserved by the slow freezing and the cryotop method. J. Vet. Med. Sci. 78(7): 1137–1143. https://doi.org/10.1292/jvms.16-0037.

Hossein, M.S., M.K. Kim, G. Jang, H.Y. Fibrianto, H.J. Oh, H.J. Kim, et al. 2007. Influence of season and parity on the recovery of *in vivo* canine oocytes by flushing fallopian tubes. Animal Reprod. Sci. 99(3–4): 330–341. https://doi.org/10.1016/j.anireprosci.2006.05.016.

Howard, J.G. 1993. Semen collection and analysis in carnivores. pp. 390–399. *In:* M.E. Fowler and R.E. Miller (eds). Zoo and Wildlife Animal Medicine: Current Therapy III. W.B. Saunders, Philadelphia, USA.

IUCN. 2021. The IUCN Red List of Threatened Species. Version 2021-3. 2021. https://www.iucnredlist.org.

Jiménez, J.E. and A.J. Novaro. 2004a. 3.4 Culpeo *Pseudalopex Culpaeus*. pp. 44–49. *In:* C. Sillero-Zubiri, M. Hoffmann and D.W. Macdonald (eds). Canids: Foxes, Wolves, Jackals and Dogs. Status Survey and Conservation Action Plan. IUCN/SSC Canid Specialist Group, Gland, Switzerland and Cambridge, UK.

Jiménez, J.E. and A.J. Novaro. 2004b. 3.6 Chilla *Pseudalopex Griseus*. pp. 56–63. *In*: C. Sillero-Zubiri, M. Hoffmann and D.W. Macdonald (eds). Canids: Foxes, Wolves, Jackals and Dogs. Status Survey and Conservation Action Plan. IUCN/SSC Canid Specialist Group, Gland, Switzerland and Cambridge, UK.

Johnston, S.D., M.V.R. Kustritz and P.S. Olson. 2001. Canine and Feline Theriogenology. Saunders, Philadelphia, EUA.

Johnston, S.D., D. Ward, J. Lemon, I. Gunn, C.A. MacCallum, T. Keeley, et al. 2007. Studies of male reproduction in captive african wild dogs (*Lycaon Pictus*). Animal Reprod. Sci. 100(3–4): 338–355. https://doi.org/10.1016/j.anireprosci.2006.08.017.

Johnson, A.E.M., E.W. Freeman, D.E. Wildt and N. Songsasen. 2014a. Spermatozoa from the maned wolf (*Chrysocyon Brachyurus*) display typical canid hyper-sensitivity to osmotic and freezing-induced injury, but respond favorably to dimethyl sulfoxide. Cryobiology 68(3): 361–370. https://doi.org/10.1016/j.cryobiol.2014.04.004.

Johnson, A.E.M., E.W. Freeman, M. Colgin, C. McDonough and N. Songsasen. 2014b. Induction of ovarian activity and ovulation in an induced ovulator, the maned wolf (*Chrysocyon Brachyurus*), using GnRH agonist and recombinant LH. Theriogenology 82(1): 71–79. https://doi.org/10.1016/j.theriogenology.2014.03.009.

Kuczmarski, A.H., M. Alves de Barros, L.F. Souza de Lima, T.F. Motheo, H.J. Bento, G.A. Iglesias, et al. 2020. Urethral catheterization after pharmacological induction for semen collection in dog. Theriogenology 153(September): 34–38. https://doi.org/10.1016/j.theriogenology.2020.04.035.

Kutzler, M.A. 2005. Induction and synchronization of estrus in dogs. Theriogenology 64(3): 766–775. https://doi.org/10.1016/j.theriogenology.2005.05.025.

Lee, K.H., W.Y. Lee, D.H. Kim, S.H. Lee, J.T. Do, C. Park, et al. 2016. Vitrified canine testicular cells allow the formation of spermatogonial stem cells and seminiferous tubules following their xenotransplantation into nude mice. Sci. Rep. 6(November 2015): 1–11. https://doi.org/10.1038/srep21919.

Linde-Forsberg, C. 1995. Artificial insemination with fresh, chilled extended and frozen-thawed semen in the dog. Semin. Vet. Med. Surg. Small Anim. 10(1): 48–58.

Linde-Forsberg, C., B. Ström Holst and G. Govette. 1999. Comparison of fertility data from vaginal vs intrauterine insemination of frozen-thawed dog semen: a retrospective study. Theriogenology 52(1): 11–23. https://doi.org/10.1016/S0093-691X(99)00106-5.

Lojkić, M., N. Maćešić, T. Karadjole, B. Špoljarić, G.B. Silvijo Vince, I. Getz, et al. 2022. A retrospective study of the relationship between canine age, semen quality, chilled semen transit time and season and whelping rate and litter size. *Vet. Arh.* 92(3): 301–310. https://doi.org/10.24099/vet.arhiv.1733.

Lueders, I., I. Luther, G. Scheepers and G. van der Horst. 2012. Improved semen collection method for wild felids: urethral catheterization yields high sperm quality in african lions (*Panthera leo*). Theriogenology 78(3): 696–701. https://doi.org/10.1016/j.theriogenology.2012.02.026.

Lueders, I., I. Luther, K. Müller, G. Scheepers, A. Tordiffe and G. Horst. 2013. Semen collection via urethral catheter in exotic feline and canine species: a simple alternative to electroejaculation. Proc. Int. Conf. Dis. Zoo Wild Animals. Austria: 161.

Maia, O.B. and A.M.G. Gouveia. 2002. Birth and mortality of maned wolves chrysocyon brachyurus (Illiger, 1811) in captivity. Braz. J. Biol. 62(1): 25–32. https://doi.org/10.1590/S1519-69842002000100004.

Manjunath, P. 2012. New insights into the understanding of the mechanism of sperm protection by extender components. Anim. Reprod. 9(4): 809–815. https://www.animal-reproduction.org/article/5b5a6053f7783717068b46cc.

Martins, M.I.M., L.C. Padilha, F.F. Souza and M.D. Lopes. 2009. Fertilizing capacity of frozen epididymal sperm collected from dogs. Reprod. Domest. Anim. 44 (SUPPL. 2): 342–344. https://doi.org/10.1111/j.1439-0531.2009.01431.x.

Martins-Bessa, A., A. Rocha and A. Mayenco-Aguirre. 2006. Comparing ethylene glycol with glycerol for cryopreservation of canine semen in egg-yolk TRIS extenders. Theriogenology 66(9): 2047–2055. https://doi.org/10.1016/j.theriogenology.2006.06.004.

McNutt, J.W., R. Groom and R. Woodroffe. 2019. Ambient temperature provides an adaptive explanation for seasonal reproduction in a tropical mammal. J. Zool. 309(3): 153–160. https://doi.org/10.1111/jzo.12712.

Mohammed, A.E.A. 2022. Developmental potential of ovarian follicles in mammals: involvement in assisted reproductive techniques. Pak. J. Zool. 1–11. https://doi.org/10.17582/journal.pjz/20220128140127.

Muñoz-Fuentes, V., C. Linde Forsberg, C. Vilà and J. M. Morrell. 2014. Single-layer centrifugation separates spermatozoa from diploid cells in epididymal samples from gray wolves, *Canis lupus* (L.). Theriogenology 82(5): 773–776. https://doi.org/10.1016/j.theriogenology.2014.04.029.

Nabeel, A.H.T., Y. Jeon and I.J. Yu. 2019. Use of polyvinyl alcohol as a chemically defined compound in egg yolk-free extender for dog sperm cryopreservation. Reprod. Domest. Anim. 54(11): 1449–1458. https://doi.org/10.1111/rda.13547.

Nagashima, J.B., S.R. Sylvester, J.L. Nelson, S.H. Cheong, C. Mukai, C. Lambo, et al. 2015. Live births from domestic dog (*canis familiaris*) embryos produced by *in vitro* fertilization. Plos One 10(12): e0143930. https://doi.org/10.1371/journal.pone.0143930.

Nagashima, J.B., A.J. Travis and N. Songsasen. 2019. The domestic dog embryo: *in vitro* fertilization, culture and transfer. Methods Mol. Biol. 247–267. https://doi.org/10.1007/978-1-4939-9566-0_18.

Nagashima, J.B. and N. Songsasen. 2021. Canid reproductive biology: norm and unique aspects in strategies and mechanisms. Animals 11(3): 653. https://doi.org/10.3390/ani11030653.

Nowak, R.M. 1999. Walker's Mammals of the World. Johns Hopkins University Press, Baltimore.

Parera, A. 2002. Los Mamíferos de La Argentina y La Región Austral de Sudamérica. El Ateneu, Buenos Aires.

Porton, I.J., D.G. Kleiman and M. Rodden. 1987. Aseasonality of bush dog reproduction and the influence of social factors on the estrous cycle. J. Mammal. 68(4): 867–871. https://doi.org/10.2307/1381569.

Prentice, J.R. and M. Anzar. 2011. Cryopreservation of mammalian oocyte for conservation of animal genetics. Vet. Med. Int. 2011: 146405. https://doi.org/10.4061/2011/146405.

Reynaud, K., A. Fontbonne, N. Marseloo, S. Thoumire, M. Chebrout, C. Viaris de Lesegno et al. 2005. *In vivo* meiotic resumption, fertilization and early embryonic development in the bitch. Reproduction 130(2): 193–201. https://doi.org/10.1530/rep.1.00500.

Rota, A., C. Milani, G. Cabianca and M. Martini. 2006. Comparison between glycerol and ethylene glycol for dog semen cryopreservation. Theriogenology 65(9): 1848–1858. https://doi.org/10.1016/j.theriogenology.2005.10.015.

Sanbe, A., Y. Tanaka, Y. Fujiwara, H. Tsumura, J. Yamauchi, S. Cotecchia, et al. 2007. α_1–Adrenoceptors are required for normal male sexual function. Br. J. Pharmacol. 152(3): 332–340. https://doi.org/10.1038/sj.bjp.0707366.

Santos, R.P. dos and A.R. Silva. 2022. Sperm cooling as an assisted reproduction tool for wildlife: an underrated technology. Biopreserv. Biobank 00(00). https://doi.org/10.1089/bio.2022.0025.

Shiraz, A., A. Khadivi and N. Shams-Esfandabadi. 2014. Male pronuclear formation using dog sperm derived from ectopic testicular xenografts, testis, and epididymis. Avicenna J. Med. Biotechnol. 6(3): 140–146.

Sillero-Zubiri, C., M. Hoffman and D.W. Mcdonald. 2013. Black-backed jackal. pp. 56–56. *In:* C. Sillero-Zubiri, M. Hoffman and D.W. Mcdonald (eds). Animals of the Masai Mara. Princeton University Press, Gland, Switzerland. https://doi.org/10.1515/9781400844913.56.

Silva, L.D.M. 2022. Canine and feline testicular preservation. Animals 12(1): 124. https://doi.org/10.3390/ani12010124.

Silva, M.C.C., P.N. Jorge-Neto, G.M. Miranda, A.C. Csermak-Jr, R. Zanella, C.S. Pizzutto et al. 2022. Reproductive parameters of male crab-eating foxes (Cerdocyon thous) subjected to pharmacological semen collection by urethral catheterization. Theriogenology Wild 1: 100004. https://doi.org/10.1016/j.therwi.2022.100004.

Silva-Rodriguez, E., A. Farias, D. Moreira-Arce, J. Cabello, E. Hidalgo-Hermoso, M. Lucherini, et al. 2016. *Lycalopex fulvipes*, Darwin's Fox. The IUCN red list of threatened species 2016: E.T41586A107263066, 17. https://doi.org/10.2305/IUCN.UK. 2016-1.RLTS.T41586A85370871.en.

Songsasen, N., I. Yu, S. Murton, D.L. Paccamonti, B.E. Eilts, R.A. Godke, et al. 2002. Osmotic sensitivity of canine spermatozoa. Cryobiology 44(1): 79–90. https://doi. org/10.1016/S0011-2240(02)00009-3.

Songsasen, N., M. Rodden, J.L. Brown and D.E. Wildt. 2006. Patterns of fecal gonadal hormone metabolites in the maned wolf (*Chrysocyon brachyurus*). Theriogenology 66(6–7): 1743–1750. https://doi.org/10.1016/j.theriogenology.2006.01.044.

Souza, N.P, L.D'A Guimarães and R.C.R Paz. 2011. Dosagem Hormonal e Avaliação Testicular Em Cachorro-Do-Mato (*Cerdocyun thous*) Utilizando Diferentes Protocolos Anestésicos. Arq. Bras. Med. Vet. Zootec. 63(5): 1224–1228. https://doi.org/10.1590/ S0102-09352011000500025.

Souza, N.P., P.V. Furtado and R.C. Rodrigues da Paz. 2012. Non-invasive monitoring of the estrous cycle in captive crab-eating foxes (*Cerdocyon thous*). Theriogenology 77(2): 233–239. https://doi.org/10.1016/j.theriogenology.2011.08.021.

Tani, H., T. Inaba, H. Tamada, T. Sawada, J. Mori and R. Torii. 1996. Increasing gonadotropin-releasing hormone release by perifused hypothalamus from early to late anestrus in the beagle bitch. Neurosci. Lett. 207(1): 1–4. https://doi.org/10.1016/0304-3940(96)12471-X.

Teodoro, L.O., A.A. Melo-Junior, K.M. Spercoski, R.N. Morais and F.F. Souza. 2012. Seasonal aspects of reproductive physiology in captive male maned wolves (*Chrysocyon brachyurus*, Illiger 1815). Reprod. Domest. Anim. 47 (SUPPL. 6): 250–255. https://doi. org/10.1111/rda.12071.

Teodoro, L.O., L.S. Camargo, V.F.C. Scheeren, C.P. Freitas-Dell'Aqua, F.O. Papa, C.S. Honsho, et al. 2021. First successful frozen semen of the maned wolf (*Chrysocyon brachyurus*). Reprod. Domest. Anim. 56(11): 1464–1469. https://doi.org/10.1111/rda. 13999.

Thomassen, R. and W. Farstad. 2009. Artificial insemination in canids: a useful tool in breeding and conservation. Theriogenology 71(1): 190–199. https://doi.org/10.1016/j. theriogenology.2008.09.007.

Tsutsui, T., T. Tezuka, Y. Mikasa, H. Sugisawa, N. Kirihara, T. Hori, et al. 2003. Artificial insemination with canine semen stored at a low temperature. J. Vet. Med. Sci. 65(3): 307–312. https://doi.org/10.1292/jvms.65.307.

Tsutsui, T., T. Hori, S. Endo, A. Hayama and E. Kawakami. 2006. Intrauterine transfer of early canine embryos. Theriogenology 66(6–7): 1703–1705. https://doi.org/10.1016/j. theriogenology.2006.01.009.

Turathum, B., K. Saikhun, P. Sangsuwan and Y. Kitiyanant. 2010. Effects of vitrification on nuclear maturation, ultrastructural changes and gene expression of canine oocytes. Reprod. Biol. Endocrinol. 8: 1–9. https://doi.org/10.1186/1477-7827-8-70.

Varesi, S., V. Vernocchi, M. Faustini and G.C. Luvoni. 2013. Quality of canine spermatozoa retrieved by percutaneous epididymal sperm aspiration. J. Small Anim. Pract. 54(2): 87–91. https://doi.org/10.1111/jsap.12020.

Velloso, A.L., S.K. Wasser, S.L. Monfort and J.M. Dietz. 1998. Longitudinal fecal steroid excretion in maned wolves (*Chrysocyon brachyurus*). Gen. Comp. Endocrinol. 112(1): 96–107. https://doi.org/10.1006/gcen.1998.7147.

Wood, J.E. 1958. Age structure and productivity of a gray fox population. J. Mammal. 39: 74–86.

Yu, I. and S.P. Leibo. 2002. Recovery of motile, membrane-intact spermatozoa from canine epididymides stored for 8 days at 4°C. Theriogenology 57(3): 1179–1190. https://doi. org/10.1016/S0093-691X(01)00711-7.

Zambelli, D., F. Prati, M. Cunto, E. Iacono and B. Merlo. 2008. Quality and *in vitro* fertilizing ability of cryopreserved cat spermatozoa obtained by urethral catheterization after medetomidine administration. Theriogenology 69(4): 485–490. https://doi.org/10. 1016/j.theriogenology.2007.10.019.

Zmudzinska, A., M.A. Bromke, R. Strzezek, M. Zielinska, B. Olejnik and M. Mogielnicka-Brzozowska. 2022. Proteomic analysis of intracellular and membrane-associated fractions of canine (*canis lupus familiaris*) epididymal spermatozoa and sperm structure separation. Animals 12(6): 772. https://doi.org/10.3390/ani12060772.

Chapter 5

The Jaguar: *Panthera onca*

Regina Celia Rodrigues da Paz[1]*, Lindsey M. Vansandt[2],
William F. Swanson[3] and Cristina Harumi Adania[4]

[1]Laboratório de Pesquisa em Animais de Zoológico – Universidade Federal de Mato Grosso.
Postal address: Av. Fernando Correa da Costa, 2367,
Boa Esperança, Cuiabá/MT, Brazil – 78060-900.
Email: reginacrpaz@gmail.com

[2]Center for Conservation and Research of Endangered Wildlife – Cincinnati Zoo and
Botanical Garden, 3400 Vine St, Cincinnati, OH 45220, USA – 45220.
Email: Lindsey.Vansandt@cincinnatizoo.org

[3]Center for Conservation and Research of Endangered Wildlife – Cincinnati Zoo and
Botanical Garden, 3400 Vine St, Cincinnati, OH 45220, USA – 45220.
Email: bill.swanson@cincinnatizoo.org

[4]Centro Brasileiro para Conservação de Felinos Neotropicais Associação Mata Ciliar,
Avenida Emílio Antonon, 1000, Chácara Aeroporto,
Jundiaí/SP, Brazil – 13212-010.
Email: cristina.adania@mataciliar.org.br

INTRODUCTION

The jaguar (*Panthera onca*) is the largest feline in the Americas, weighing 35–130 kg, with a robust, compact, muscular body and color ranging from light yellow to yellowish-brown. The body is completely covered by black spots, which form rosettes of various sizes with black dots inside. Melanistic individuals also occur. They are solitary and terrestrial, with nocturnal and daytime activity. Their diet is generally comprised of medium-sized mammals and reptiles.

Jaguars are found from the coastal plains of Mexico to northern Argentina. Originally occurring throughout Brazil, they are currently restricted to the North region, parts of Central Brazil, and some isolated areas in the South and Southeast

---

*Corresponding author: reginacrpaz@gmail.com

regions. Jaguar habitat consists of areas of dense vegetation, including Tropical and Subtropical Forest, Cerrado, Caatinga and Pantanal (Oliveira and Cassaro 1999).

Due to poaching and fragmentation/loss of habitat, the number of jaguars *in situ* has decreased substantially. The species is classified as Near Threatened on the IUCN Red List, with a decreasing population trend in Latin America (Quigley et al. 2017). The most robust wild populations are found in the Amazon and Pantanal; however, even in these areas they are affected by restricted gene flow, increasing the risk of inbreeding, reduced genetic variability, and threats of extinction.

There is a large jaguar population (>500 individuals) housed under human care in Brazil and other Latin American countries, as well as in North American and European zoos. These felids collectively can serve as an assurance population that can protect against species extinction and potentially allow reintroduction of zoo-born animals back into the wild. There is a critical need to maintain genetic diversity within these populations, which is in turn dependent on successful reproduction; however, several factors may interfere with *ex situ* natural breeding success: inadequate enclosures, stress, diseases, nutritional deficiencies, physical impairments and behavioral incompatibilities. Hence, reproductive biotechnology has emerged as a potentially important tool to help preserve genetic diversity, either through improving the success of natural breeding or developing techniques for assisted reproduction such as artificial insemination (AI) or *in vitro* fertilization (IVF) and embryo transfer (ET). Increased knowledge about basic reproductive physiology and new adaptations of existing methodologies will be crucial for improving reproductive management of jaguars *in situ* and *ex situ*.

REPRODUCTIVE MANAGEMENT

Reproductive management plans are essential to ensure that imperiled populations maintain adequate genetic and demographic variability and remain representative of the species as a whole. When necessary, these plans may require introduction of new individuals or their genetic material (e.g. semen) to reinvigorate or expand the managed population.

Studbook data typically form the basis for developing reproductive management plans. A studbook is basically a family tree for the managed population, comprising a record of all living and dead animals descended from a group of ancestors (i.e. founders), often born in the wild. Each registered animal is assigned a studbook number and relevant information such as date/location of birth, history of institutional transfers, birth of offspring, and date/location of death is included. Studbooks are designed to help manage captive populations by allowing detailed genetic and demographic analyses to optimize genetic diversity, make reproductive recommendations, and decide on specific animal pairings within and between institutions. In Brazil, a large felid studbook, which includes the jaguar, has been compiled and maintained for many years but was last published in 2005 (Adania et al. 2005).

Reproductive management plans historically were based on natural breeding alone, but the need for animal transport among zoos as well as numerous pairing failures due to behavioral incompatibility resulted in limited offspring production for some species and this approach was not always cost effective. Beginning in the 1990s, reproductive biotechnologies have been explored as a means to enhance the conservation of endangered species, focused on maintaining genetic diversity through enhanced animal propagation. Initial births in a few wildlife species globally from assisted reproduction created high levels of optimism and support for its expanded use. However, thirty years later, these efforts are still only resulting in limited success—but hope springs eternal. Recently, a technical cooperation agreement was signed between the Brazilian Association of Aquariums and Zoos (AZAB), Chico Mendes Institute for Biodiversity Conservation (ICMBIO) and Ministry of the Environment (MMA) for the establishment of "The Jaguar Management Plan," which includes development and application of the reproductive biotechnologies discussed below (BRASIL 2018).

According to AZAB records, the Brazilian jaguar population census for the last 50 years recorded a total 351 animals. In 2021, the total population consisted of 82 jaguars under human care in 34 institutions (AZAB 2022).

The population decline observed since 2000 is largely related to aging; however, over the past 20 years we have observed that the population has stabilized, not because of births but due to the arrival of free-living animals, mainly cubs from the Amazon biome. Jaguars from the Amazon biome have substantial genetic representation in the *ex situ* population, totaling 51 individuals (19 males, 32 females) maintained in 26 institutions.

We also observed an imbalance between the number of males and females (32/50), with the number of females superior to that of males over the last 20 years. Currently, there are 51 (62%) jaguars that were wild-born, 21 (26%) jaguars born under human care, and 10 (12%) jaguars with an unknown birth status.

Despite adapting to *ex situ* conditions and reaching advanced ages, jaguars in Brazilian zoological institutions were not incorporated into reproductive management programs when the situation was favorable, and today, a large number of institutions have animals that have failed to reproduce. In addition, many zoos have difficulty keeping more than one breeding pair in their care and must wait for the animals to die or be transferred to receive new individuals with reproductive profile.

Faced with this situation, several breeding facilities were created to house jaguars that needed to be removed from the wild but could not be transferred to zoos already at maximum holding capacity. These breeding facilities have become an important tool for changing the critical management situation of the *ex situ* population in Brazil.

Suggestions for new pairings and other recommendations are regularly made by the AZAB Endangered Species Program—ICMBio. However, given the jaguar's large size and aggressiveness, financial, logistical, and/or safety concerns are often prohibitive for animal transport and prevent recommended pairings from being established.

Collectively, natural breeding efforts for Brazilian jaguars in zoos and breeding facilities alone does not produce enough births to create a stable population. These animals represent an important conservation tool, serving as a genetic bank and assurance population for their wild-living counterparts. Implementation of assisted reproduction programs may be imperative to augment natural breeding efforts and secure a future for this species.

REPRODUCTIVE BIOLOGY

Measurement of hormone metabolites in fecal samples, using validated radio immunoassays and enzyme immunoassays, has proven invaluable for assessing endocrine traits in felids and other wildlife species. In jaguars, analysis of fecal estrogen metabolites determined that females are polyestrous with an average estrous period of 6.5 ± 0.3 days (Barnes et al. 2016). Although jaguars are classically described as induced ovulators (Wildt et al. 1979, 2010), they also may be capable of spontaneous ovulation. In one study (Barnes et al. 2016), three of five females, housed in the presence of males, appeared to spontaneously ovulate (defined by an elevation in fecal progestin metabolites without observed breeding activity or other exogenous stimuli). However, spontaneous ovulation was not detected in the four control females that were housed in the absence of males. Jorge-Neto et al. (2020) reported spontaneous ovulation in jaguars stimulated with equine chorionic gonadotropin (eCG) and then evaluated 3–5 days later with laparoscopy; two of the four females in visual, olfactory and/or auditory contact with males ovulated, whereas both females housed in complete isolation of males failed to ovulate. This suggests that sensory contact with a male is required, or at least enhances the likelihood for spontaneous ovulation during natural or induced estrous periods.

Observations of jaguar reproductive behavior indicate that several copulations usually occur during a normal estrous phase. However, a recent study (Jorge-Neto et al. 2018) reported that in 42% of observed mating attempts, despite the female's receptivity and the beginning of pelvic movements typical of copulation, the male quickly dismounted the female, suggesting that there was no penile penetration. In the remaining 58% of observations, feline copulatory behavior was observed, in which the male vocalized and bit the nape of the female, resulting in a lateral-dorsal rolling movement of the female suggestive of penile penetration. It is hypothesized that this repeated mounting behavior, whether actual penetration occurs or not, is necessary to induce multiple ovulations.

If pregnancy occurs following mating, the gestation period for jaguars ranges from 90 to 111 days, with two cubs born in the average litter. The cubs open their eyes around day 13 post-partum and begin to eat solid food at approximately 75 days, although they may continue to nurse until five months of age (Oliveira 1994). According to AZAB/ICMbio records (note that ages are estimated for wild-born individuals), the youngest male and female to reproduce were 1.3 and 1.9 years old, respectively. The oldest male and female to reproduce were 18.6 and 13.7 years old, respectively.

SEMEN COLLECTION

Electroejaculation is the most commonly used technique for semen collection in wild cats, including jaguars (Howard 1993). The first detailed paper describing the use of electroejaculation as a semen collection method in jaguars was published in 1998 (Morato et al. 1998) with numerous other studies also reporting the effectiveness of this approach (Morato et al. 2000, Paz et al. 2000, Swanson et al. 2003, Silva et al. 2019a, b).

A standardized protocol has been developed for electroejaculation in felids, in which a total of 80 electrical stimuli are divided into 3 series: (1) 10 stimulations each at 2, 3, and 4 V, (2) 10 stimulations each at 3, 4, and 5 V and (3) 10 stimulations each at 4 and 5 V, with a 5 to 10-minute rest between each series (Howard 1993). With jaguars, electroejaculation requires a bipolar rectal probe, measuring 2.3 cm in diameter and containing three longitudinal copper electrodes (5 cm long and 0.2 cm high, with 0.4 cm separation between adjacent electrodes; Paz et al. 2000).

An alternative semen collection method has been described in the domestic cat (Zambelli et al. 2007, 2008), using pharmacological treatment with an α-adrenergic receptor agonist to stimulate adrenoreceptors and cause contraction of smooth muscle in the vas deferens (MacDonald et al. 1980). This in turn induces release of high concentration spermatozoa into the urethra, allowing recovery by insertion of a urethral catheter. Semen collection by urethral catheterization after pharmacological induction results in low semen volume but at much higher sperm concentrations compared to electroejaculation (Zambelli et al. 2008). This pharmacological-based semen collection approach subsequently has been applied to multiple wild cat species, beginning with excellent results in the lion (*Panthera leo*) a decade ago (Lueders et al. 2012).

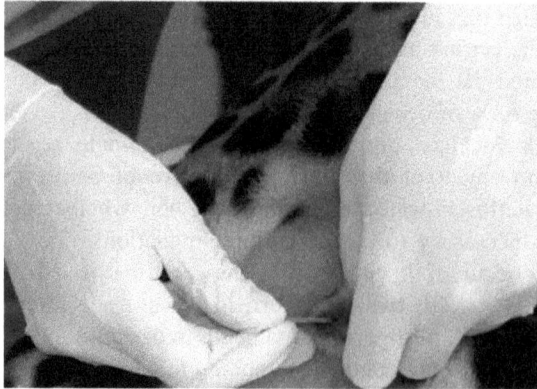

Figure 5.1   Pharmacological semen collection by urethral catheterization in *Panthera onca.*

In the jaguar, several different α-adrenergic receptor agonists have been investigated for their semen recovery capacity with urethral catheterization (Figure 5.1). In one protocol, detomidine (mean dose 50 µg/kg, range 38–92) was combined with ketamine (mean dose 8.5 mg/kg, range 5.9–10.8) to anesthetize males for semen collection (Vansandt et al. 2017). Catheterization of the penile urethra using a 5-Fr catheter 20–30 minutes post-injection yielded sperm in 14 of 16 attempts. An alternative drug protocol, using the α-adrenergic receptor agonist medetomidine (0.08–0.1 mg/kg) combined with ketamine (5 mg/kg) also produced good results (Araujo et al. 2017). These authors also reported that transrectal prostate massage promoted sperm release into the urethra and enhanced total ejaculate volume.

To maximize semen recovery, it is possible to combine the pharmacological collection method with electroejaculation. In ocelots and other felids, additional sperm often can be recovered if electroejaculation is conducted immediately after urethral catheterization, indicating that residual sperm remains in the vas deferens. This combination approach could be most relevant for free-living animals, in which the opportunity for semen collection may occur only on a single occasion and maximal sperm recovery is desirable.

SEMEN ANALYSIS AND CRYOPRESERVATION

Semen analyses include assessment of sperm motility, morphology, and functionality. Sperm motility may be evaluated by subjectively estimating the percentage of motile sperm and rate of progressive movement (on scale of 0–5) under light microscopy or, more objectively, by using Computer Assisted Sperm Analysis (CASA). Sperm morphology may be characterized by using phase contrast microscopy (1000x) to assess a wet semen preparation after formaldehyde salt or glutaraldehyde fixation.

Morphological sperm traits are classified as primary defects (i.e. those that occur during the process of spermatogenesis) or secondary defects (i.e. a consequence of maturation and transport in the epididymis).

Plasma membrane functionality can be assessed by the hypo-osmotic swelling test, in which sperm with functional membranes respond with tail coiling when incubated in low osmolarity solution (Comercio et al. 2013). Eosin-nigrosin and rose bengal/fast green stains can also be used to analyze plasma membrane viability (Bjorndahl 2003) and acrosome integrity (Pope et al. 1991), respectively.

Sperm cryopreservation allows the transfer of genetic material, rather than living animals, between geographically distant populations and ensures that genetic diversity is not lost if the animals die without reproducing. However, poor survivability of frozen-thawed sperm has been a persistent challenge in felids, despite extensive efforts to improve sperm cryopreservation methods. Paz et al. (2007) evaluated the potential fertility of jaguar sperm frozen in a lactose-based medium containing 20 percent egg yolk and 8 percent glycerol. Frozen-thawed sperm showed a significant reduction in motility (mean: 70 ± 3.3 percent motile pre-freeze; 26.7 ± 4.4 percent motile post-freeze). Using the hamster zona-free oocyte penetration assay as a measure of capacitation, only 15.4 percent of oocytes were penetrated. The low post-thaw sperm quality in this study may have been affected by the high percentage of pre-freeze morphological defects (mean: 72.4 ± 4.3 percent abnormal). Silva et al. (2020) reported that a tris-hydroxymethyl-aminomethane (TRIS)-based extender provided a higher post-thaw motility (48.8 ± 4.8 percent motile) and number of sperm bound to the egg yolk perivitelline membrane (29.5 ± 3.3) compared to a coconut water-based extender (25.0 ± 4.6 percent motile; 18.6 ± 1.5 bound sperm).

While the aforementioned endpoints are correlated with fertility, the capacity to fertilize an oocyte is a more direct indicator of sperm competence. However, it is not feasible to collect large numbers of oocytes from jaguars and there are a limited number of reports on the collection and maturation of jaguar oocytes (Johnston et al. 1991, Morato et al. 2002). Spermatozoa from non-domestic cats can readily fertilize domestic cat oocytes; therefore, heterologous *in vitro* fertilization (IVF) using domestic cat oocytes represents a more practical option to evaluate sperm function than homologous IVF in the jaguar. Santos et al. (2022) utilized heterologous IVF to compare jaguar sperm processed in TRIS and coconut water semen extenders and found similar rates of cleavage at 48 hours post-insemination (TRIS: 16.7 ± 3.6 percent; coconut water: 22.9 ± 3.2 percent). However, only embryos produced with TRIS-treated sperm were able to progress to the blastocyst stage during culture (2 blastocysts/12 cleaved oocytes; 16.7 ± 1.5 percent).

OVARIAN STIMULATION

Induction of ovarian activity and ovulation in felids requires intramuscular injection of exogenous gonadotropins, which are comprised of large glycoproteins. Equine chorionic gonadotropin (eCG) and human chorionic gonadotropin (hCG) are often used due to their long half-lives in circulation (24–48 h). Other gonadotropin options include porcine follicle hormone (pFSH) and porcine luteinizing hormone (pLH), which have much shorter half-lives (~2 h) requiring multiple injections (for pFSH) to produce similar efficacy (Wildt et al. 1981, Dresser et al. 1988, Crichton et al. 2003).

The repeated administration of exogenous gonadotropins at short intervals may produce an immunologically mediated decrease in ovarian responsiveness (Swanson et al. 1995). Repeated gonadotropin treatment can induce generation of neutralizing immunoglobulins that attenuate gonadotropin stimulation of the ovaries. However, alternating gonadotropin regimens (e.g. eCG/hCG; pFSH/pLH) in sequential treatments can help to mitigate immunological interference due to the variable affinity of immunoglobulins to the different exogenous gonadotropins (Maurer et al. 1968, Swanson et al. 1996, Paz et al. 2005).

In jaguars, initial attempts at AI and IVF used two different gonadotropin regimens for ovarian stimulation. For attempted laparoscopic AI, an eCG/hCG combination was first assessed in three females: one failed to respond to treatment, one had an ovarian tumor (observed at time of AI) and one exhibited ovulatory follicles and was inseminated; however, no pregnancy resulted (Jimenez et al. 1999). For IVF in jaguars, a pFSH/pLH combination was assessed for ovarian stimulation with attempted laparoscopic oocyte collection and insemination (Morato et al. 2000), but oocyte recovery/IVF success was limited and no pregnancies or offspring have been reported following ET procedures.

More recently, an alternative exogenous gonadotropin regimen, using eCG and pLH, has been evaluated for laparoscopic AI in domestic cats. This protocol resulted in consistent ovulation and high pregnancy success without excessive formation of secondary follicles or corpora lutea (CL) that are typical of eCG/hCG treatment (Conforti et al. 2013). The same protocol (eCG/pLH) has proven effective for producing pregnancies in ocelots (*Leopardus pardalis*), tigers (*Panthera tigris*) (Lambo et al. 2014) and jaguars (*Panthera onca*) (Vansandt et al. 2019).

However, the timing of exogenous gonadotropin administration relative to the female's natural estrous cycle can produce highly variable ovarian responses, including excessive follicle formation and high estrogen concentrations, premature ovulation, and aberrant progesterone levels. For this reason, pre-treatment of female cats with oral progestin (altrenogest) to suppress endogenous ovarian activity prior to stimulation with exogenous gonadotropins has improved the success of fixed-time artificial insemination (FTAI) procedures, resulting in pregnancies in several species of wild felids (Swanson 2019).

The ovarian quiescence engendered by synthetic progestin treatment prior to gonadotropin injection results in a more consistent ovulatory response and better oocyte quality in domestic cats (Pelican et al. 2010). Progestins primarily induce ovarian quiescence by suppressing the release of endogenous pituitary gonadotropins (FSH, LH) via a negative feedback loop on the pituitary and hypothalamus. Acting directly on the ovary, progestins also promote more efficient recruitment and growth of high-quality oocytes and reduce atresia of antral follicles to increase the total number of gonadotropin-responsive follicles (Peluso 2006, Pelican et al. 2010). Initial studies in jaguars assessing the effectiveness of altrenogest for inducing ovarian quiescence revealed a general insensitivity relative to domestic cats, requiring much higher dosages to be used on a per body weight basis (Vansandt et al. 2019). Presumably, progestins suppressed endogenous gonadotropin release in jaguars in a similar manner as in domestic cats but the direct effects of progestins on the jaguar ovary still need to be defined. However, preliminary results suggest that altrenogest pre-treatment is effective for reducing ovarian hyperstimulation and increasing ovulatory responses after eCG/pLH treatment in this wild cat species.

ARTIFICIAL INSEMINATION

In felids, the success of AI is strongly affected by the semen deposition site and associated AI method. AI using non-surgical methods (e.g. vaginal or transcervical insemination) often produces a lower pregnancy rate when compared to surgical methods, which allow insemination deep into the uterine horns or the oviducts. The development of less invasive surgical techniques, such as laparoscopy, have helped to minimize trauma to the AI recipient while facilitating deposition of low sperm numbers directly into the uterus (LU-AI) or the oviduct (LO-AI).

In jaguars, the reproductive tract can be easily accessed and examined using laparoscopy, allowing detailed evaluation of the uterine horns for diameter, firmness, and other morphological traits. Similarly, the ovaries can be visualized to characterize the presence, number, dimensions and morphology of pre-ovulatory follicles and corpora lutea.

In jaguars, the LU-AI technique after eCG/hCG administration was first attempted by Jimenez et al. (1999) using freshly-collected semen. A total of three females were subjected to gonadotropin treatment and laparoscopy, but only one female presented with ovulatory follicles and was inseminated. However, no pregnancy resulted, likely due in part to insemination with low sperm numbers. Typically, LU-AI procedures are more likely to succeed when higher (> 10 million motile) sperm numbers are used. With lower sperm numbers, LO-AI procedures are preferred over LU-AI in domestic cats to increase pregnancy percentages, especially following insemination with frozen-thawed semen (Swanson 2012, Conforti et al. 2013).

In wild felids, LO-AI has become the preferred technique, particularly in smaller-sized species in which average sperm numbers per ejaculate are frequently low (Lambo et al. 2014, Swanson 2019). With larger wild cats, laparoscopy can be more challenging to perform due to the greater body size of the animal and difficulty in handling. However, if few sperm are recovered with semen collection or if frozen semen is to be used, LO-AI still represents the best option for large cat species. For example, the first tiger pregnancy produced using the LO-AI technique occurred following insemination with only 120,000 motile spermatozoa which represented the entire ejaculate recovered from the male tiger (Lambo et al. 2014).

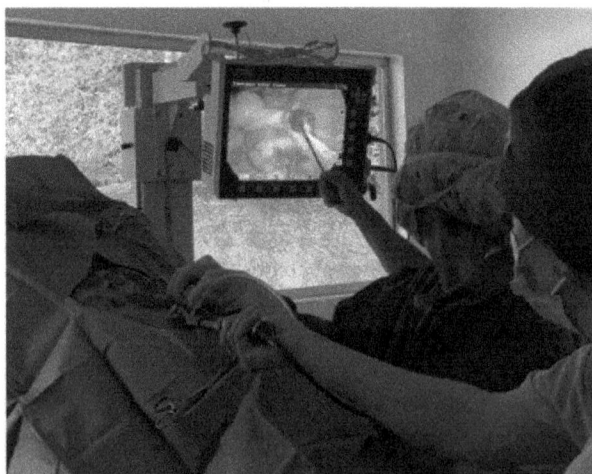

Figure 5.2 Video laparoscopy for LO-AI in *Panthera onca*.

In jaguars, the feasibility of fixed time LO-AI using low numbers of viable spermatozoa was conclusively demonstrated with the recent conception and birth of a jaguar cub (Figure 5.2; Vansandt et al. 2019). This pregnancy success using freshly-collected semen was attributed to the surgical technique used, but also was facilitated by systematic refinement of the exogenous gonadotropin regimen, including altrenogest pre-treatment. By adjusting altrenogest and gonadotropin dosages, the duration of the progestin withdrawal period and the gonadotropin injection interval, a more optimal ovarian response was obtained with multiple freshly-ovulated CL. These findings in jaguars also may be relevant for refining and improving AI protocols for other large felid species in which AI has been rarely successful.

Although robust at birth, the jaguar cub born from the first successful fixed time LO-AI survived for only two days before apparent infanticide and consumption by the dam. It is unknown if the cub's health was compromised prior to death, but reports of infanticide are frequent not only in captivity, but also in free-living animals (Tortato et al. 2017). However, the unfortunate loss of this cub doesn't negate the scientific advances achieved in the development and application of these assisted reproductive techniques for jaguars and other large felid species. With this strong scientific basis, it is expected that additional fixed time LO-AI procedures will prove successful in other jaguars in the near future.

EMBRYO TECHNOLOGY

Embryo technology, as applied to jaguars, has received very little scientific attention. With the exception of heterologous IVF with jaguar sperm (as a means to assess sperm function) and preliminary attempts at oocyte collection and IVF with *in vivo* and *in vitro* matured oocytes (discussed previously), there have been no in-depth studies focused on embryo production in this species. To our knowledge, there also have been no reports of any attempted embryo transfers in jaguars.

FINAL CONSIDERATIONS

There is an increasing need for effective management of jaguars under human care as this neotropical cat species becomes more imperiled in the wild. Our knowledge of their basic reproductive characteristics has grown in recent years, but we still lack a broad understanding of their reproductive biology. This information is essential not only for improving natural breeding within zoos but also for developing effective assisted reproductive technologies. Although semen can be easily collected from jaguars, semen cryopreservation is still sub-optimal and, to date, no jaguars have been produced with frozen semen by IVF/ET or AI. However, the recent birth of a jaguar cub by LO-AI with fresh semen provides some optimism that LO-AI with frozen sperm also may prove feasible. If consistently successful, LO-AI of jaguars with frozen semen could greatly improve management of zoo-housed jaguar populations as well as free-living animals isolated in fragmented habitats throughout Latin America.

ACKNOWLEDGEMENTS

The assistance of animal care staff working with jaguars at zoos and wildlife facilities in Brazil and the United States is greatly appreciated. Funding support for jaguar studies has been provided by the Cincinnati Zoo & Botanical Garden, Associação Mata Ciliar, and Foundation for International Aid to Animals.

REFERENCES

Adania, C.H., J.C.R. Silva, C.Y. Hashimoto and E.F. Santos. 2005. Studbook dos Grandes Felinos Brasileiros. Associação Mata Ciliar, Jundiai, SP. pp. 80.

Araujo, G.R., T.A.R. Paula, T. Deco-Souza, R.G. Morato, L.C.F. Bergo, L.C. Silva, et al. 2017. Comparison of semen samples collected from wild and captive jaguars (*Panthera onca*) by urethral catheterization after pharmacological induction. Anim. Reprod. Sci. 195: 1–7.

AZAB. 2022. Associação de Zoológicos e Aquários do Brasil. Plano de manejo das onças pintadas (Panthera onca) [Internet]. [cited 2022 Sep 09]. Available from: https://wwwazab.org.br/more/19/programa-de-manejo-ex-situ-de-especies-ameacadas.

Barnes, S.A., J.A. Teare, S. Staaden, L. Metrione and L.M. Penfold. 2016. Characterization and manipulation of reproductive cycles in the jaguar (*Panthera onca*). Gen. Comp. Endocrinol. 225: 95–103

Bjorndahl, L. 2003. Evaluation of the one-step eosin-nigrosin staining technique for human sperm vitality assessment. Hum. Reprod. 18(4): 813–816.

BRASIL. 2018. Acordo de cooperação técnica de 05 de junho de 2018. Diário Oficial da União, Edição: 106; Seção: 3; Página: 108, Brasília, DF.

Comercio, E.A., N.E. Monachesi, M.E. Loza, M. Gambarotta and M.M. Wanke. 2013. Hypo-osmotic test in cat spermatozoa. Andrologia. 45(5): 310–314.

Conforti, V.A., H.L. Bateman, M.W. Schook, J. Newsom, L.A. Lyons, R.A. Grahn, et al. 2013. Laparoscopic oviductal artificial insemination improves pregnancy success in exogenous gonadotropin-treated domestic cats as a model for endangered felids. Biol. of Reprod. 89: 1–9.

Crichton, E.G.E. Bedows, A.K. Miller-Lindholm, D.M. Baldwin, D.L. Armstrong and H.G. Laura, et al. 2003. Efficacy of porcine gonadotropins for repeated stimulation of ovarian activity for oocyte retrievel and in vitro embryo production and cryopreservation in siberian tigers (*Panthera tigris altaica*). Biol. Reprod. 68: 105–113.

Dresser, B.L., E.J. Gelwicks, K.B. Wachs and G.L. Keller. 1988. First successful transfer of cryopreserved feline (*Felis catus*) embryos resulting in live offspring. J. Exp. Zool. 246: 180–186.

Howard, J.G. 1993. Semen collection and analysis in carnivores. *In:* M. Fowler (ed.). Zoo and Wild Animal Medicine Current Therapy, III. W.B. Saunders, Philadelphia, USA.

Jiminez, G.T., R.M. Zuge, R.C.R. Paz, J. Lopez and G.A. Crudeli. 1999. Sincronización de celo e inseminación artificial por vídeo-laparoscopia en yaguareté (*Panthera onca*) en cautiverio. Comunicaciones Cientificas y Tecnologicas. Argentina. 4: 67–70.

Johnston, L.A., A.M. Donoghue, S.J. O'Brien and D.E. Wildt. 1991. Rescue and maturation *in vitro* of follicular oocytes collected from nondomestic felid species. Biol. Reprod. 45(6): 898–906.

Jorge-Neto, P.N., C.S. Pizzutto, G.R.D. Araújo, T.D. Deco-Souza, J.A., L.C. Silva, J. Salomão Júnior, et al. 2018. Copulatory behavior of the Jaguar *Panthera onca* (Mammalia: Carnivora: Felidae). Journal of Threatened Taxa. 10(15): 12933–12939.

Jorge-Neto, P.N., T.C. Luczinski, G.R.D. Araújo, J.A. Salomão Júnior, A.D.S. Traldi, J.A.M.D. Santos, et al. 2020. Can jaguar (*Panthera onca*) ovulate without copulation? Theriogenology 147(57): 57–61.

Lambo, C.A., H.L. Bateman and W.F. Swanson. 2014. Application of laparoscopic oviductal artificial insemination for conservation management of Brazilian ocelots and Amur tigers. Reproduction, Fertility and Development. 26: 116 (abstract).

Lueders, I., G. Scheepers and G. Van Der Horst. 2012. Improved semen collection method for wild felids: urethral catheterization yields high sperm quality in African lions (*Panthera leo*). Theriogenology 78: 696–701.

Maurer, R.R., W.L. Hunt and R.H. Foote. 1968. Repeated superovulation following administration of exogenous gonadotrophins in Dutch-belted rabbits. J. Reprod. Fert. 15: 93–102.

MacDonald, A. and J.C. Mcgrat H. 1980. The distribution of adrenoreceptors and other drug receptors between the two ends of the rat vas deferens as revealed by selective agonists and antagonists. Br. J. Pharmacol. 71: 445–458.

Morato, R.G., M.A.B.V. Guimarães, A.L.V. Nunes, A.C. Carciofi, F. Ferreira, V.H. Barnabe, et al. 1998. Colheita e avaliação do sêmen em onça pintada (*Panthera onca*). Braz. J. Vet. Ani. Sci. 35(4): 178–181.

Morato, R.G., E.G. Crichton, R.C.R. Paz, R.M. Zuge, C.A. Moura, A.L.V. Nunes, et al. 2000. Ovarian stimulation and successful in vitro fertilization in the jaguar (*Panthera onca*). Theriogenology 53: 339 (abstract).

Morato, R.G., E. Crichton, R.C.R. Paz, R.M. Zuge, C.A. Moura, A.L.V. Nunes, et al. 2002. Ovarian stimulation, oocyte recovery and *in vitro* fertilization in the jaguar (*Panthera onca*). Revista Brasileira de Reprodução Animal 26: 317–324.

Oliveira, T.G. 1994. Neotropical Cats Ecology and Conservation. EDUFMA, São Luis:, Brasil.

Oliveira, T.G. and K. Cassaro. 1999. Guia de Identificação dos Felinos Brasileiros. Fundação Parque Zoológico de São Paulo, São Paulo, Brazil.

Paz, R.C.R., R.M. Zuge, V.H. Barnabe, R.G. Morato, P.A.N. Felippe, R.C. Barnabe, et al. 2000. Capacidade de penetração de semen congelado de onça pintada (*Panthera onca*) em oócitos heterólogos. Braz. J. Vet. Res. Anim. Sci. 37(6): 462–466.

Paz, R.C.R., W.F. Swanson, E.A. Dias, C.H. Adania, V.H. Barnabe and R.C. Barnabe. 2005. Ovarian and immunological responses to alternating exogenous gonadotropin regimens in the ocelot (*Leopardus pardalis*) and tigrina (*Leopardus tigrinus*). Zoo. Biol. 24(3): 247–260.

Paz, R.C.R., R.M. Züge and V.H. Barnabe. 2007. Frozen Jaguar (*Panthera onca*) sperm capacitation and ability to penetrate zona free hamster oocytes. Braz. J. Vet. Res. Anim. Sci. 44: 337–344.

Pelican, K.M., R.E. Spindler, B.S. Pukazhenthi, D. Wildt, M.A. Ottinger and J.G. Howard. 2010. Progestin exposure before gonadotropin stimulation improves embryo development after in vitro fertilization in the domestic cat. Biol. Reprod. 83: 558–567.

Peluso, J.J. 2006. Multiplicity of progesterone's actions and receptors in the mammalian ovary. Biol. Reprod. 75: 2–8.

Pope, C.E., Y.Z. Zhang and B.L. Dresser, 1991. A simple staining method for quantifying the acrosomal status of cat spermatozoa. J. Zoo and Wilidl. Med. 22(1): 87–95.

Quigley, H., R. Foster, L. Petracca, E. Payan, R. Salom and B. Harmsen. 2017. *Panthera onca* (errata version published in 2018). The IUCN Red List of Threatened Species, 2017: e.T15953A123791436. https://dx.doi.org/10.2305/IUCN.UK.2017-3.RLTS. T15953A50658693.en. Accessed on 09 June 2022.

Santos, M.V.O., H.V.R. Silva, L.G.P. Bezerra, L.R.M. Oliveira, M.F. Oliveira, N.D. Alves, et al. 2022. Heterologous in vitro fertilization and embryo production for assessment of jaguar (Panthera onca Linnaeus, 1758) frozen-thawed semen in different extenders. Anim. Reprod. 19: 1–11.

Silva, H.V., T.G.P. Nunes, L.R. Ribeiro, L.A. Freitas, M.F. Oliveira, A.C. Assis-Neto, et al. 2019a. Morphology, morphometry, ultrastructure, and mitochondrial activity of jaguar (*Panthera onca*) sperm. Anim. Reprod. Sci. 203: 84–93.

Silva, H.V., A.R. Silva, L.D.M. Silva and P. Comizzoli. 2019b. Semen cryopreservation and banking for the conservation of Neotropical carnivores. Biopreserv. Biobanking 17(2): 183–188.

Silva, H.V.R., T.G.P. Nunes, B.F. Brito, L.B. Campos, A.M. da Silva, A.R. Silva, et al. 2020. Influence of different extenders on morphological and functional parameters of frozen-thawed spermatozoa of jaguar (*Panthera onca*). Cryobiology 92: 53–61.

Swanson, W.F., D.W. Horohov and R.A. Godke. 1995. Production of exogenous gonado-trophin-neutralizing immunoglobulins in cats after repeated eCG-hCG treatment and relevance for assisted reproduction in felids. J. Reprod. Fertili. 105: 35–41.

Swanson, W.F., J.G. Howard, T. Roth, J.L. Brown, T. Alvarado, M. Burton, et al. 1996. Responsiveness of ovaries to exogenous gonadotrophins and laparoscopic artificial insemination with frozen-thawed spermatozoa in ocelots (*Felis pardalis*). J. Reprod. Fertili. 106: 87–94.

Swanson, W.F., W.E. Johnson, R.C. Cambre, S.B. Citino, K.B. Quigley, D.M. Brousset, et al. 2003. Reproductive status of endemic felid species in Latin American zoos and implications for *ex situ* conservation. Zoo Biol. 22: 421–441.

Swanson, W.F. 2012. Laparoscopic oviductal embryo transfer and artificial insemination in felids: Challenges, strategies and successes. Reprod. Domest. Anim. 47(6): 136–140.

Swanson, W.F., H.L. Bateman and L.M. Vansandt. 2016. Urethral catheterization and sperm vitrification for simplified semen banking in felids. Reprod. Dom. Ani. 51(3): 1–6.

Swanson, W.F. 2019. Practical application of laparoscopic oviductal artificial insemination for the propagation of domestic cats and wild felids. Reprod. Fertili. and Devel. 31(1): 27–39.

Tortato, F.R., A.L. Devlen, J.A. May-Junior, P. Crawshaw, T. Izzo, H. Quigley, et al. 2017. Infanticide in a jaguar (*Panthera onca*) population—does the provision of livestock carcasses increase the risk? Acta Ethol. 20: 69–73.

Vansandt, L.M., C.H. Adania, P.R. Yanai, P.E. Kunze, M.E. Iwaniuk and W.F. Swanson. 2017. Improving and Simplifying Semen Banking in the Brazilian Jaguar (*Panthera onca*). 50th Annual Meeting of the Society for the Study of Reproduction. Washington DC. Flash talk and poster presentation.

Vansandt, L.M., C.H. Adania, P.R. Yanai, J.S. Paulino, R.C.R. Paz, H.L. Bateman, et al. 2019. Ovarian Synchronization, Ovulation Induction, and Successful Artificial Insemination in the Jaguar (*Panthera Onca*). 51st Annual Conference of the American Association of Zoo Veterinarians. St. Louis, Missouri. Poster presentation.

Wildt, D.E., C.C. Platz, P.K. Chakraborty and S.W. Seager. 1979. Oestrous and ovarian activity in a female jaguar (*Panthera onca*). J. of Reprod. and Fert.. 56: 555–558.

Wildt, D.E., C.C. Platz, S.W. Seager and M. Bush. 1981. Induction of ovarian activity in the cheetah *(Acinonyx jubatus)*. Biol. Reprod. 24: 217–222.

Wildt, D.E., W. Swanson, J. Brown, A. Sliwa and A. Vargas. 2010. Felids *ex-situ*: managed programs, research, and species recovery. *In:* D.W. MacDonald and A.J. Loveridge (eds). Biology and Conservation of Wild Felids. Oxford University Press, Oxford.

Zambelli, D., M. Cunto, F. Prati and B. Merlo, 2007. Effects of ketamine or medetomidine administration on quality of electroejaculated sperm and on sperm flow in the domestic cat. Theriogenology. 68: 796–803.

Zambelli, D., F. Prati, M. Cunto, E. Lacono and B. Merlo. 2008. Quality and *in vitro* fertilizing ability of cryopreserved cat spermatozoa obtained by urethral catheterization after medetomidine administration. Theriogenology 69: 485–90.

Chapter 6

The Procyonids

Herlon Victor Rodrigues Silva[1]* and Regina Célia Rodrigues da Paz[2]

[1]Universidade Estadual do Ceará. Postal address: Av. Dr. Silas Munguba, 1700, Campus do Itaperi, Fortaleza, CE, Brazil – 60.714.903. Email: herlonvrs@hotmail.com

[2]Laboratório de Pesquisa em Animais de Zoológico – Universidade Federal de Mato Grosso. Postal address: Av. Fernando Correa da Costa, 2367, Boa Esperança, Cuiabá/MT, Brazil – 78060-900. Email: reginacrpaz@gmail.com

INTRODUCTION

Procyonids are animals belonging to the family Procyonidae and are placed in the order Carnivora. They are divided into 13 endemic species of the American continent, including kinkajous and olingos; however, the most popular are raccoons and coatis. Procyonids are small animals with short legs and dense fur, and they are plantigrade with five well-developed digits, which resemble the hands of primates, thus facilitating climbing to the top of trees (Eisenberg and Redford 1999, Beisiegel and Mantovani 2006).

Procyonids are called new world carnivores or neotropical carnivores because they are found exclusively on the American continent and inhabit different types of landscapes, from dense forests and flooded regions to semi-arid environments. They have a varied omnivorous diet, feeding on fruits, small vertebrates, and plant matter. In South America, it is possible to find the South American raccoon (*Procyon cancrivorus*), also known as the crab-eating raccoon, ring-tailed coati (*Nasua nasua*), kinkajou (*Potos flavus*), olingo (*Bassaricyon gabbii*), and the most recent member discovered in 2013, the olinguito (*Bassaricyon neblina*), about whom basic information is still lacking (Koepfli et al. 2007, Helgen et al. 2020).

*Corresponding author: herlonvrs@hotmail.com

GENERAL CHARACTERISTICS OF THE PROCYONIDS

Ring-tailed coati (*Nasua nasua*): Among the procyonid species, the coati can be considered the most famous in South America and the easiest to find, mainly in the south of the continent. In coatis, the color is variable in shades of reddish-brown throughout the dorsal part and yellowish-brown in the ventral part (Figure 6.1). The tail is long with hair that forms rings by interspersing the dark color with light. It has a long and thin nasal region with a flexible end (Gompper and Decker 1998). Coatis are found mainly in forested areas and exhibit diurnal activity. They have a generalist omnivorous diet, feeding on small animals, and even fruits, and may be an important seed disperser and may be an important seed disperser, thus contributing to the forests regeneration (Alves-Costa et al. 2004, Alves-Costa and Eterovick 2007). The species *N. nasua* is considered by the International Union for Conservation of Nature (IUCN) as a species of "Least Concern." However, populations of this species are in decline and are sensitive to the loss of forest habitats (Emmons and Helgen 2016).

Figure 6.1 Coati (*Nasua nasua*). *Source:* Personal archives.

Coatis present a highly complex social behavior, in which they form large groups of thirty individuals on average; females are the leaders of the group, with a dominant female. In these groups, male pups and young are permanently accepted but are expelled from the group when they reach sexual maturity. Adult males tend to be solitary but are always close to groups of females, being temporarily accepted into the group only during the mating period. There are very rare cases of solitary females. Occasionally, when there are no disputes over food and these resources are abundant in the environments where the groups are located, females accept the introduction of some adult males; however, they remain totally submissive (Hirsch 2007, 2009, 2011).

The reproductive period usually occurs in such a manner that parturition ensues when there is increased availability of food for future lactating females and their offspring. Owing to seasonal reproductive characteristics, the preference for the reproductive period occurs mainly from October to January (Gompper and Decker 1998). Females have an estrous cycle of approximately 5 to 7 days, and similar to the domestic dog, there is a prolonged anestrus immediately after the fertile phase. Pregnant individuals tend to separate from the group and build their nests in trees or on rocky substrates protected from predators. After a gestation period of approximately 74 to 77 days, they give birth to litters of three to seven young. The pups are altricial, that is, they are totally dependent on the mother in this initial phase of life. They weigh an average of 80 g at 5 days of life, and after completing 10 days of life, they begin to open their eyes. At 25 days, they can walk perfectly and climb tree trunks. After approximately 6 weeks of birth, the females and their young return to the original group. Females become sexually mature at 2 years of age, whereas males acquire sexual maturity at 3 years of age (Gompper and Decker 1998).

Crab-eating raccoon (*Procyon cancrivorus*: Raccoons are the largest procyonids, weighing up to 10 kg, and their fur is dense, especially in the tail region (Figure 6.2). The most predominant coloration in the raccoon is grayish, but sometimes appears mixed with shades of brown. The animal has a characteristic marking on the face that resembles a black mask around the eyes. The presence of long fingers makes them capable of manipulating objects similar to that observed in primates, and their tail, although not as long as that of the coatis, also has ring-shaped markings (Nowak 1999).

Figure 6.2   Crab-eating raccoon (*Procyon cancrivorus*). *Source:* Personal archives.

Crab-eating raccoons can be found from Costa Rica to most of South America, preferring habitats in swampy and jungle areas. The crab-eating raccoon has solitary habits and a preference for the nocturnal period; it can stay in the upper part of the trees, but mainly prefers to live on the ground. They are almost always found near streams, lakes, and rivers. According to the IUCN, its conservation status is classified as a species of "Least Concern" (Reid et al. 2016).

In the raccoon reproduction system, males are usually polygenic, that is, they mate with several females during the reproductive season, which is the only period in which females accept the proximity of a male. However, recently pubescent young males tend not to have sexual partners, because during the competition for females, these males are at a disadvantage against the more experienced males (Nowak 1999).

The reproduction of these species tends to occur between July and September. The complete estrous cycle is estimated to last from 80 to 140 days, with a gestation period lasting between 60 and 73 days, giving birth to litters of up to 7 pups, with the average being 3–4 pups per litter. If the female loses her litter, it is possible that she will restart her estrous cycle and become fertile. Females prefer to give birth in isolated places, such as crevices in rocks, trees, or abandoned burrows of other animals.

Raccoon pups are born without teeth and with eyes closed; however, after 21 days, the eyelids and ears begin to open, which is also when the fur begins to darken and their facial mask becomes more evident. The pups begin to wean between 2 and 4 months, becoming completely independent of maternal care at 8 months of age (Nowak 1999).

Kinkajou (*Potos flavus*): The kinkajou is the only procyonid belonging to *Potos*. Kinkajous are strictly arboreal, with nocturnal habits and a solitary lifestyle most of the time. They usually compete for space with other arboreal procyonids, such as olingos, exhibiting sympatry mainly with this species; however, they have an advantage in disputes related to food owing to their size (Julien-Laferriere 1999). With its prehensile tail and round face, this species appears and behaves more like a primate than a carnivore (Michalski and Peres 2005). The kinkajou has dense, soft hair, which mostly has a mix of yellow and brown colors, sometimes being lighter in the ventral region and darker in the dorsal region (Kays et al. 2000, Kays and Gittleman 2001, Kays 2003).

It is a species that depends on tropical forests and occurs from Mexico to Southeast Brazil. Its main habitat is Brazilian territory, mainly in the Amazon basin and northern Mato Grosso, but it is also found in areas of the Atlantic Forest. The kinkajou diet consists mainly of fruits, but they also consume seeds and insects. An important feature is their long tongues, which allow them to feed on insects and even nectar. Owing to its wide distribution and relatively high densities, this species is considered by the IUCN as a species of "Least Concern" (Helgen et al. 2016a).

During the mating period, this species changes its social behavior, ceasing to be solitary and adopting more complex methodologies, such as exhibiting a polygamous and polyandrous behavior, that is, usually two males for a female, together with their future offspring, constitute a typical social system. In a polygamous social system, dominant males mate with the females of their main group, as well as with any other female that is on the periphery of the group's territory, with the exception of females from other more distant groups. Most females are copulated by the alpha male, whereas the remaining subordinate males copulate with a smaller number of females. During the mating period, the alpha male follows the females in the estrus phase, whereas the subordinate males closely follow vocalizing and occasionally come into conflict with the dominant male. After copulation, both males and females disperse in search of food (Kays et al. 2000, Kays and Gittleman 2001, Kays 2003).

The estrus period lasts for an average of 17 days, and in this species, the possibility exists that the mating phase is not during a specific period, but extends throughout the year, mainly during periods of increased abundance of food (Kays and Gittleman 2001). The kinkajou females are the main ones to perform parental care; the males are not aggressive with the young and even occasionally interact with them. Parental care by females, from pregnancy to the end of lactation, lasts approximately 8 months; in general, up to two pups can be born, but most records account for only one individual per pregnancy. Sexual maturity is, on average, at one and a half years for males and 2 years for females (Kays et al. 2000, Kays and Gittleman 2001, Kays 2003).

Olingo (*Bassaricyon sp.*): Olingo is a procyonid of arboreal habits, and the species *Bassaricyon gabbii* was the first among olingos to be described. Although some authors consider this as the only genuine species of olingo, *Bassaricyon* was updated and currently includes a total of four species. This species is native to Central America, but there are reports of its presence in the extreme north of South America within the Amazon region (Nowak 1999).

Olingo is the largest species of *Bassaricyon*, and its anatomy is similar to that of kinkajou but smaller. In addition to the large eyes, the face is rounded, with short and round ears. The fur is thick and brown or grayish brown in color, becoming slightly darker towards the middle of the back, whereas the underparts are light cream to buff. The tail is similar in color to the body but has several faint rings of darker fur along its length. Females have a single pair of teats located on the back of the abdomen near the hind legs (Nowak 1999). According to the IUCN, it is a species of "Least Concern" (Helgen et al. 2016b).

In captivity, males are intolerant of each other; therefore, it is likely that their mating system prevents copulation with multiple partners, unlike what is observed in kinkajous. The reproduction of olingos tends to be seasonal, with gestation lasting approximately 75 days, at the end of which, on average, only one cub is born. The cub is altricial, and as in most carnivores, it is also born with its eyes closed. From 20 days onwards, the pup's eyes begin to open. Birth weight is approximately 55 g. The transition from milk to solid foods starts at 2 months. It is not known how long the young stay with their mothers, but like most carnivores that must learn to hunt their prey, the olingo young probably have some post-weaning association with their mothers. Olingos reach sexual maturity between 21 and 24 months of age (Nowak 1999).

REPRODUCTIVE ANATOMY AND PHYSIOLOGY

The procyonid reproductive anatomy is quite similar to that of domestic carnivores; in relation to males, we can observe several similarities with the domestic dog.

These animals have a penis positioned in the ventral region (Figure 6.3A), covered by a prepuce fixed to the abdomen by a preputial sheath. In the region of the preputial ostium, in pubescent males, we find extremely developed glands. In coatis, these glands can increase so much in size that it is easy to perceive them visually. The secretion released by these glands is quite consistent but has no pathological correlation (Figure 6.3B, C) (Franciolli et al. 2007). The content of these glands serves males to mark their territory, trees, and even their own bodies, such that their scent is perceived by an invading male. The type of penis that these species have is cavernous muscle, extremely vascularized, with the presence of the penile bone inside, also known as *baculum*. However, unlike dogs, a penile bulb at the base is absent; however, the cranial end of the penis is thicker than the caudal end. The testicles are maintained in a scrotum covered with hair (Queiroz et al. 2010). In coatis, it was verified that the seasons can influence the testicular physiology, mainly in relation to the volume of the testicle and the testosterone production. Soon, the occurrence of testicular recrudescence was verified, with the beginning of the reproductive period and consequently the maximum gonadal activity (Paz et al. 2012a).

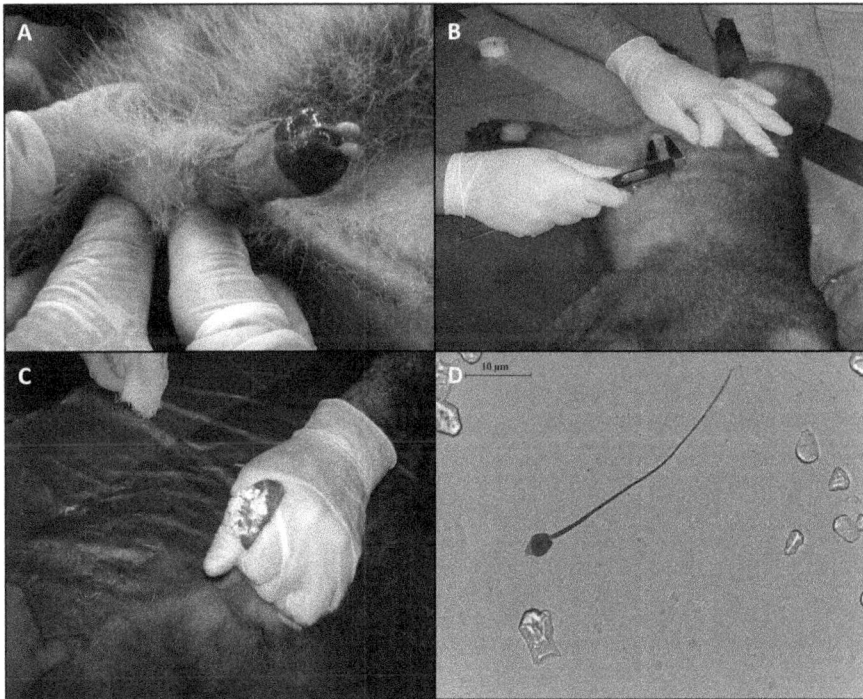

Figure 6.3 (A) Raccoon erect penis. (B) Measurement of the preputial gland in coati. (C) Secretion from the preputial gland in coati. (D) Normal coati sperm morphology. *Source*: Personal archives.

As for the female procyonids, their reproductive system and anatomical structure are quite similar to most of the carnivores already described, having two ovaries, two uterine horns, a uterine body, a cervix, and a vaginal canal. The external genitalia are very discreet and difficult to observe even when mating; however, they show a slight increase in size (Mayor et al. 2013). Regarding the ovarian structure, in coatis, in most cases, the follicles contain only a single oocyte; however, some follicles are polyovular, containing several oocytes (Mayor et al. 2013). Although most polyovular follicles contain an average of 2–3 oocytes, follicles containing up to 30 oocytes have been described. This feature of multi-oocyte follicles or polyovular follicles does not commonly occur in the ovaries of domestic dogs. In the gestational phase, as has been reported in other carnivores, we find the zonary type of placenta in procyonids (Figure 6.3B) (Mayor et al. 2013).

Semen Collection and Evaluation

Assisted reproduction in procyonids is little studied when compared to the information previously described in other families of carnivores, and much of the information on this topic focuses on descriptions of collection methods and seminal characteristics of these species. The coati is the main representative, with the largest amount of data and only some information related to raccoons is available. In contrast, the other species lack even basic information regarding reproductive anatomy.

The initial methodologies for the reproductive follow-up of procyonids started with the techniques of obtaining semen, with electroejaculation being the most common method used. It is important to emphasize that before this type of collection, the animal must be physically restrained with the use of a net and subsequently sedated; protocols have been previously described for coatis and raccoons using tiletamine/zolazepam (7 mg/kg, im.) or ketamine combined with xylazine (5 mg/kg and 1 mg/kg, respectively, im.) (Barros et al. 2009, Silva et al. 2015, 2018).

To perform collection, electroejaculation equipment with a probe for introduction into the rectum of the animal is required. The rectum must be cleaned in advance and the feces removed so that the probe, already lubricated, can be inserted with the metallic structure facing the ventral direction of the animal. It is also important to determine the protocol for electrical stimuli, which are generally divided into three series with an interval of approximately 5 min between them. Each series is also divided into three voltage cycles that are gradually increased, and in each of these electrical scales, 10 stimuli are performed. One of the series proposed is to perform the first series of 10 stimuli of 2, 3, and 4 V; the second of 10 stimuli of 3, 4, and 5 V; and the third of 10 stimuli of 5 and 6 V. Generally, the volumes obtained are very low, but the values of the kinetic parameters are quite promising. We present the values reported for coatis and raccoons on Table 6.1.

Interestingly, the procyonids' sperm cell, especially in coatis and raccoons, differs from the morphology usually described for other carnivores since it presents a peculiar characteristic, with a significant prominence in the top of the acrosomal region. By this moment, however, the function of this acrosomal protuberance remains unknown (Figure 6.3D) (Silva et al. 2014, 2015).

Table 6.1 Description of seminal parameters of coatis (*Nasua nasua*) and raccoons (*Procyon cancrivorus*)

Species	Volume (μl)	Motility (%)	Progressive motility (0–5)	Concentration (× 10⁶)/ml	Normal morphology (%)	Vitality (%)	Sperm membrane integrity (%)	References
Ring-tailed coati (*Nasua nasua*)	146	68	3.3	143	82	66	n/e	Queiroz et al. 2010
	300	79.6	3.8	403	69.6	91.4	n/e	Paz et al. 2012a
	200	44.8	2.1	131.9	63.1	66.6	n/e	Paz et al. 2012b
	60	91.3	4.5	197.5	81.2	73.1	74.3	Silva et al. 2015
	85.5	95.6	4.8	281.7	82.7	n/e	n/e	Silva et al. 2018
Crab-eating raccoon (*Procyon cancrivorus*)	550	57.5	3.5	325	3.5	n/e	58	Silva et al. 2014

n/e: not evaluated.

Semen Cryopreservation

To date, only one study on semen cryopreservation has been performed in procyonids. In 2015, a team of researchers from the Federal University of Mato Grosso (UFMT), Brazil, described the post-thawing quality of semen of coatis (*N. nasua*). Two Tris-based diluents were tested, and the results were very promising. Total motility was the only parameter that reduced after thawing, which although reduced by almost half, maintained motility above 40%. The parameters of progressive motility and integrity of the plasma membrane and acrosome remained similar to those described for fresh semen (Paz and Avila 2015). The data from this study is presented in Table 6.2.

Table 6.2   Data regarding cryopreservation of coati (*Nasua nasua*) semen (Paz et al. 2015)

Parameter	Fresh semen	Diluent 1	Diluent 2
Total motility (%)	84.28 ± 11.57^a	49.28 ± 29.99^b	44.61 ± 25.03^b
Progressive sperm motility (0–5)	3.64 ± 1.44^a	2.15 ± 1.14^a	2.07 ± 1.03^a
Plasma membrane integrity (%)	92.76 ± 3.46^a	84.69 ± 15.77^a	89.76 ± 13.97^a
Acrosome integrity (%)	94.76 ± 2.89^a	92.35 ± 4.73^a	90.58 ± 7.17^a

a,b represent difference between columns (p<0.05)

ASSISTED REPRODUCTION IN FEMALES

To date, assisted reproduction in females has received little scientific attention. To our knowledge, there are no reports of attempted ovarian stimulation, artificial insemination, or embryo technology in procyonids.

FINAL CONSIDERATIONS

Procyonids are essential animals in their respective biomes and act mainly as seed dispersers, helping maintain forests. The species that make up this family remain poorly studied, not only in terms of reproductive aspects, but also in relation to general aspects. Because they are mostly small animals and prefer to live on the tops of trees, sighting is even more difficult, so much so that a new species of procyonid was recently described.

Reproductive data are quite scarce; with regard to reproductive biotechniques, only a single study was carried out, and the others in the field of reproduction were directed at most to descriptions of anatomical aspects and seminal parameters, with the focus on only two species: coatis and raccoons. However, there is already a base of studies on the natural reproductive behavior and social dynamics of some species of procyonids, which can contribute to future studies on assisted reproduction in these species, and thus contribute to their maintenance in their natural habitat.

Procyonids are essential animals; in their respective biomes, they act mainly as seed dispersers, helping to maintain forests. The species that make up this family are still little studied in not only reproductive aspects, but also in relation to general aspects. Because they are mostly small animals and have a preference for living in the tops of trees, sighting is even more difficult, so much so that a new species of procyonid was recently described.

Reproductive data are quite scarce: with regard to reproductive biotechniques, only a single work was carried out, and the others in the field of reproduction were directed at the most to descriptions of anatomical aspects and description of seminal parameters, the focus being on only two species, that is, coatis and raccoons. However, there is already a base of studies on the natural reproductive behavior and social dynamics of some species of procyonids; such data can contribute in the future to new studies about assisted reproduction in these species and thus can contribute to the maintenance of them in their natural habitat.

REFERENCES

Alves-Costa, C.P., G.A.B. Da Fonseca and C. Christofaro. 2004. Variation in the diet of the brown-nosed coati (*Nasua nasua*) in southeastern Brazil. J. Mammal. 85: 478–482.

Alves-Costa, C.P. and P.C. Eterovick. 2007. Seed dispersal services by coatis (*Nasua nasua*, Procyonidae) and their edundancy with other frugivores in southeastern Brazil. Acta Oecologica 32: 77–92.

Barros, F.F.P.C., J.PA.F. Queiroz, A.C.M. Mota Filho, E.A.A. Santos, V.V. Paula, C.I.A. Freitas, et al. 2009. Use of two anesthetic combinations for semen collection by electroejaculation from captive coatis (*Nasua nasua*). Theriogenology 71: 1261–1266.

Beisiegel, B.M. and W. Mantovani. 2006. Habitat use, home range and foraging preferences of the coati *Nasua nasua* in a pluvial tropical Atlantic forest area. J. Zool. 269: 77–87.

Eisenberg, J.F. and K.H. Redford. 1999. Mammals of the neotropics, the central neotropics Chicago, University of Chicago v3: 609p.

Emmons, L. and K. Helgen. 2016. *Nasua nasua*. The IUCN Red List of Threatened Species 2016: e.T41684A45216227. https://dx.doi.org/10.2305/IUCN.UK.2016-1.RLTS. T41684A45216227.en. Accessed on 16 November 2022.

Franciolli, A.L.R., G.M. Costa, C.A.F. Mançanares, D.S. Martins, C.E. Ambrósio, M.A. Miglino, et al. 2007. Morfologia dos órgãos genitais masculinos de quati (*Nasua nasua* Linnaeus, 1766). Biotemas 20: 27–36.

Gompper, M.E. and D.M. Decker. 1998. *Nasua nasua*. Mammalian Species, 580: 1–9.

Helgen, K., R. Kays and J. Schipper. 2016a. Potos flavus. The IUCN Red List of Threatened Species 2016: e.T41679A45215631. https://dx.doi.org/10.2305/IUCN.UK.2016-1. RLTS.T41679A45215631.en. Accessed on 28 November 2022.

Helgen, K., R. Kays, C. Pinto, J.F. González-Maya and J. Schipper. 2016b. *Bassaricyon gabbii*. The IUCN Red List of Threatened Species 2016: e.T48637946A45196211. https:// dx.doi.org/10.2305/IUCN.UK.2016-1.RLTS.T48637946A45196211.en. Accessed on 28 November 2022.

Helgen, K., R. Kays, C. Pinto, J. Schipper and J.F. González-Maya. 2020. *Bassaricyon neblina* (amended version of 2016 assessment). The IUCN Red List of Threatened Species 2020: e.T48637280A166523067. https://dx.doi.org/10.2305/IUCN.UK.2020-1. RLTS.T48637280A166523067.en. Accessed on 28 November 2022.

Hirsch. B.T. 2007. Within-group spatial position in ring-tailed coatis (*Nasua nasua*): balancing predation, feeding success, and social competition. Ph.D. Thesis, Graduate School, Anthropological Sciences, Stony Brook University, Stony Brook, N.Y.

Hirsch, B.T. 2009. Seasonal variation in the diet of Ring-tailed Coatis (*Nasua nasua*) in Iguazu, Argentina. J. Mammal. 90: 136–143.

Hirsch, B.T. 2011. Long-term adult male sociality in ring-tailed coatis (*Nasua nasua*). Mammalia 75: 301–304.

Julien-Laferriere, D. 1999. Foraging strategies and food partitioning in the neotropical frugivorous mammals *Caluromys philander* and *Potos flavus*. J. Zool. 247: 71–80.

Kays, R., J. Gittleman and R. Wayne. 2000. Análise de microssatélites da organização social kinkajou. MolecularEcology 9: 743–751.

Kays, R. and J. Gittleman. 2001. A organização social do kinkajou *Potos flavus* (*Procyonidae*). J. Zool. 253: 491–504.

Kays, R. 2003. Poliandria social e acasalamento promíscuo em um carnívoro primata: o kinkajou (*Potos flavus*). pág. 125–137 em U Reichard, C Boesch, eds. Monogamia: estratégias de acasalamento e parcerias em pássaros, humanos e outros mamíferos. Cambridge, Nova York: Cambridge University Press.

Koepfli, K.P., M.E. Gompper, E. Eizirik, C.C. Ho, L. Linden, J.E. Maldonado, et al. 2007. Phylogeny of the Procyonidae (Mammalia: Carnivora): Molecules, morphology and the Great American Interchange. Mol. Phylogenet. Evol. 43: 1076–1095.

Mayor, P., D. Montes and C. López-Plana. 2013. Functional morphology of the female genital organs in the wild ring-tailed coati (*Nasua nasua*) in the northeastern Peruvian Amazon. Canadian J. Zool. 91: 496–504.

Michalski, F. and C.A. Peres. 2005. Anthropogenic determinants of primate and carnivore local extinctions in a fragmented forest landscape of southern Amazonia. Biol. Conserv. 124: 383–396.

Nowak, R.M. 1999. Walker's Mammals of the World (Volume 1), 6th Ed., Johns Hopkins University Press, Baltimore.

Paz, R.C.R., H.B.S. Avila, T.O. Morgado and M. Nichi. 2012a. Seasonal variation in sérum testosterone, testicular volume and semen characteristics in coatis (*Nasua nasua*). Theriogenology 77: 1275–1279.

Paz, R.C.R., T.O. Morgado, C.T.R. Viana, F.P. Arruda, D.O.B. Nascimento and L.D.A. Guimarães. 2012b. Semen collection and evaluation of captive coatis (*Nasua nasua*). Arquivos Brasileiros de Medicina Veteterinária e Zootecnia 64: 318–322.

Paz, R.C.R. and H.B.S. Avila. 2015. Coatis (*Nasua nasua*) semen cryopreservation. Braz. J. of Vet. Res. Ani. Sci. 52: 12–17.

Queiroz, J.P.A.F., F.F.P.C. Barros, G.L. Lima, T.S. Castelo, C.I.A. Freitas and A.R. Silva. 2010. Assessment of orchidometry and scrotal circumference in coatis (*Nasua nasua*). Reprod. Dom. Ani. 45: 382–386.

Reid, F., K. Helgen and J.F. González-Maya. 2016. Procyon cancrivorus. The IUCN Red List of Threatened Species 2016: e.T41685A45216426. https://dx.doi.org/10.2305/IUCN.UK.2016-1.RLTS.T41685A45216426.en. Accessed on 28 November 2022.

Silva, H.V.R., A.C. Mota-Filho, L.A. Freitas, J.N. Pinto, A.R. Silva and L.D.M. Silva. 2014. Successful semen collection in the racoon (*Procyon cancrivorus*) by electroejaculation. *In:* 47th Annual Meeting of the Society for the Study of Reproduction (SSR), 2014, Grand Rapids, MI. Proceedings of the 47th Annual Meeting of the Society for the Study.

Silva, H.V.R., F.F. Magalhães, L.R. Ribeiro, A.L.P. Souza, CI.A. Freitas, M.F. Oliveira, et al. 2015. Morphometry, Morphology and Ultrastructure of Ring-tailed Coati Sperm (*Nasua nasua* Linnaeus, 1766). Reprod. Dom. Ani. 50: 945–951.

Silva, H.V.R., P. Rodriguez-Villamil, F.F. Magalhães, T.G.P. Nunes, L.A. Freitas, L.R. Ribeiro, et al. 2018. Seminal plasma and sperm proteome of ring-tailed coatis (*Nasua nasua*, Linnaeus, 1766). Theriogenology 111: 34–42.

Chapter 7

The Puma

Gediendson Ribeiro de Araujo[1]* and Thyara de Deco-Souza[2]

[1]Laboratório Reprobiote, Biotério Central, UFMS Av. Senador Filinto Muller, 1555 Vila Ipiranga, Campo Grande – MS, Brazil, 79070–900.
Email: gediendson.araujo@ufms.br

[2]Faculdade de Medicina Veterinária e Zootecnia, Universidade Federal de Mato Grosso do Sul – UFMS Av. Sen. Filinto Müler, 2443 – Pioneiros, Campo Grande – MS, Brazil, 79070–900.
Email: thyara.araujo@ufms.br

INTRODUCTION

More than 50% of wild feline species are classified as Endangered, Vulnerable, or Near Threatened (IUCN 2022). This is particularly worrying since they are at the top of the food chain, and influences the population dynamics of other species and, consequently, the ecological balance (Redford 1997). The conservation of species depends on maintaining proper genetic variability. When a population is geographically isolated, undergoing homozygosity, several factors, such as decreased fertility, increased susceptibility to disease, and increased sperm abnormalities, combine to trigger (or accelerate) the process of extinction (O'brien and Mccullogh 1985, Wildt et al. 1987, Munson et al. 1996, Eizirik et al. 2001). Assisted reproductive technologies can be a powerful tool for conservation and deserve serious consideration, especially for *ex situ* management programs, as they can improve genetic exchange and reproductive efficiency. Studies in animal reproduction encompass a diversity of interrelated areas, including gamete biology, embryology, endocrinology, and cryobiology. To develop these technologies in felid species, the study and propagation of basic knowledge and new technologies are necessary, as species-specific variations should be considered (Swanson and Brown 2004).

*Corresponding author: gediendson.araujo@ufms.br

At least 30 subspecies of *Puma concolor* have been classified (Anderson 1983), and six of them occur in Brazil: *P. concolor anthonyi* (extreme south of Venezuela up to Pico da Neblina); *P. concolor concolor* (extreme north of the Amazon); *P. concolor borbonsis* (Amazon Basin); *P. concolor acrocodia* (open fields and forests of the Pantanal in Mato Grosso); *P. concolor greeni* (part of the Cerrado, Caatinga, and remnants of the Atlantic Forest of the Northeast); *P. concolor capricornensis* (Atlantic Forest of the Southeast region extending to the eastern limit of the Pantanal, to the extreme south of Brazil) (Fonseca et al. 1994). The puma has a wide distribution, covering several habitats from Canada to Chile, including Brazil, where it can be found in all biomes (Amazon, Cerrado, Caatinga, Pantanal, Atlantic Forest, and Campos Sulinos), both in primary and secondary areas (Redford and Eisenberg 1992, Emmons and Feer 1997, Oliveira and Cassaro 1999). This species is the second largest feline in Brazil, being smaller only than the jaguar (*Panthera onca*). Males weigh between 55 and 65 kg while females weigh between 35 and 45 kg (Logan and Sweanor 2001).

Despite its wide distribution, the puma is no longer on the National List of Endangered Species of Brazilian Fauna (Brazil 2022) but cited in Appendix II of the CITES—Convention on International Trade in Endangered Species of Wild Fauna and Flora (CITES 2022). The International Union for the Conservation of Nature (IUCN) currently cites a "decreasing" trend in the population of the puma in the wild and the main threats to the puma are the reduction of the prey population, predatory hunting, and modification and fragmentation of their habitat (IUCN 2022).

Pumas are polygamous solitary carnivores, the dominant males breed with females that interrupt their territory (Murphy 1998). Males and females have different strategies to achieve maximum reproductive success. The female's reproductive success depends on the number of viable offspring until adulthood; for this purpose, their territory must be large enough to guarantee prey to keep themselves and their offspring. Also, adult females can group to ensure the safety and survival of the young (Shaw 1987, Logan and Sweanor 2001). On the male side, reproductive success depends on the ability to fertilize as many females as possible, so the largest, strongest, and most aggressive males can compete with others and maintain long periods of dominance over the territory (Weckerly 1998).

ASSISTED REPRODUCTION IN MALES

To develop reproductive biotechnologies in the male, it is important to know the testicular morphophysiology, including the spermatogenic process. In pumas, males weighing 45 kg had a testicular mass of 7.5 g, with a gonadosomatic index of 0.03% (Leite et al 2006). Also, the adult puma has approximately 18 meters of seminiferous tubule per gram of testis (about 78% of the testicular parenchyma), and the total length of the seminiferous epithelium cycle is 9.96 days, with 44.8 days being required for sperm surge from spermatogonia. Also, according to Leite et al. (2006), puma is among the species with high spermatogenic efficiency, producing approximately 26 million spermatozoa per gram of testis per day. This information is very relevant to evaluate the efficiency of sperm collection protocols.

SPERM COLLECTION AND CONSERVATION

Semen samples from wild felines can be obtained through electroejaculation (Bonney et al 1981, Miller et al. 1990, Deco-Souza et al. 2013), collected directly from the epididymis and/or vas deferens (Cucho et al. 2016, Bento et al. 2019) or through urethral catheterization (Araujo et al. 2020a). Electroejaculation (EE) is the most used technique in wild animals, as it can be performed under anesthesia. This technique involves stimulation of the innervation of reproductive organs through low electrical currents (Bonney et al. 1981). A transducer with size and diameter compatible with the animal is introduced into the rectum, with longitudinal electrodes positioned ventrally (Platz et al. 1978). After each stimulus, the animal extends its pelvic limbs. To collect semen, the penis is exposed by applying gentle pressure to its base, and the ejaculate is collected in a preheated vial (Howard et al. 1986).

Several electroejaculation protocols have been described for wild cats (Bonney et al. 1981, Wildt et al. 1983, Howard et al. 1986). These protocols and some variations have already been used to collect semen from pumas. Bornney et al. (1981) used a sequence of 0-5-2-0 V, with a 4-second stimulation period followed by a 4-second rest period. With this protocol, several semen fractions with a total volume of 1–2 ml and a sperm concentration between 2–4 × 10^7/ml (average of 3.2 × 10^7/ml) were collected. Deco-Souza et al. (2013) used a different protocol, where they first emptied and washed the bladder to avoid contamination with urine, and then applied a maximum of 4 × 10 sets of 16 V-stimuli, with 1-minute intervals between the set. Each stimulus takes around 1 second to go from 0 V to 16 V, remaining for 2 to 3 s, followed by an abrupt return to 0 V, with the interval between stimuli being 2 to 3 s. With this protocol, they obtained ejaculates with 0.4 to 0.5 mL, a mean concentration of 205 × 10^6 sperm/mL, sperm vigor of 3.5, and sperm motility of 75%. Within this method, the authors did not report urine contamination; however, this is a frequent issue in electroejaculation procedures.

Since 2008 research has focused on pharmacological semen collection and good-quality, highly concentrated semen has been collected from felines (Zambelli et al. 2008, Araujo et al. 2018, 2020a, Jorge Neto 2019, Jorge Neto et al. 2019, Jorge-Neto et al. 2020). In pumas, medetomidine (0.08 to 0.1 mg/kg) was successfully used for semen collection, obtaining an EE-like amount of sperm (Araujo et al. 2020a). The pharmacological method is an important advance for semen collection in free-living animals since the electrical stimulation of the EE can stimulate the animal to recover from anesthesia, risking the team. Furthermore, this method collects much more concentrated samples, reducing the negative effects of seminal plasma on sperm freezing.

Another important technique, especially in animals that die or undergo orchiectomy procedures, is the collection directly from the epididymis and/or the vas deferens. Epididymal spermatozoa were collected from a wild puma right after being hit, and 100 µL of semen sample was punctioned directly from the epididymis (Carelli et al. 2017). Despite being feasible, this method probably does not collect all sperm in the epididymis and deferent duct. Epididymal spermatozoa were also collected after slicing the caudal epididymis in a Petric dish for morphologic analysis (Cucho et al. 2016), and by squeezing the caudal epididymis and deferent duct—collecting around 58×10^6 non-motile sperm (Bento et al. 2019). The rarity of studies reporting the collection of epididymal semen in cougars is worrying since it is the main method for recovering the genetics of males after they have died. Partnerships with companies that monitor highways are crucial to increase access to these animals, thus enabling the conservation of the genetics that has been lost.

Regardless of the collection method, pumas usually show high rates of morphological changes in spermatozoa, up to 91% (Barone et al. 1994, Pukazhenthi et al. 2001, Bento et al. 2019, Araujo et al. 2020b). In free-living animals, an important factor that negatively interferes with sperm quality is inbreeding. Among felines, this effect has been described in cheetah (*Acinonyx jubatus*, Wildt et al. 1983), lion (*Panthera leo*, Wildt et al. 1987), and Iberian lynx (*Lynx pardinus*, Ruiz-López et al. 2012). In puma, the high proportion of abnormal reproductive and sperm characteristics has been associated with a low level of genetic variation due to inbreeding in a geographically isolated population (Florida Panther) (Roelke et al. 1993). Corroborating this hypothesis, Penfold et al. (2022) evaluated the reproductive parameters of 65 males from this population after genetic introgression by introducing new individuals. In this study, the authors observed that mixed males, with greater genetic diversity, had higher mean testicular volume, structurally normal spermatozoa, and sperm concentration. The authors also observed that the benefits of introducing new genetics into the population were maintained even 25 years after the introduction of the new individuals.

Semen samples can be preserved by refrigeration, cryopreservation, vitrification, or freeze-drying. However, the most routinely used methods are refrigeration and cryopreservation. Cooled semen is an alternative when semen will be used shortly after collection. In cats, it was demonstrated that spermatozoa refrigerated for up to 14 days—diluted in TRIS-egg yolk medium—fertilized oocytes, with the formation of blastocysts; however, the cleavage rate of these spermatozoa was compromised (Harris et al. 2002). There are no published data on puma semen refrigeration, but this method cannot be neglected as it causes less stress to the sperm and can achieve better conception rates (Santos and Silva 2022).

Cryopreservation allows cells to be stored for longer periods, but also has a greater potential to damage their structure. Thus, adapting the composition of the medium to the species-specific needs is crucial for obtaining better thawing results. Egg yolk protects the sperm membrane from cooling damage; thus, it is the base component of many semen cryopreservation media. However, egg yolk has been replaced because it does not have a homogeneous composition, is unstable at room temperature, and is limited for exportation. In pumas, only egg-yolk-based extenders have been used, all at 20% V/V (Deco-souza et al. 2013, Araujo et al. 2020b). Another crucial component for cryopreservation is cryoprotectants, whose function is to reduce the damage caused by the ice crystals' formation during freezing. Until now, only glycerol has been used for puma semen cryopreservation with post-thaw total motility varying from 36.3 to 40% (Deco-souza et al. 2013, Araujo et al. 2020b). Although glycerol is the most used cryoprotectant in felines, studies should investigate other alternatives, such as dimethyl sulfoxide, dimethylacetamide, and dimethylformamide, since toxicity can vary between species, even within the same family.

THE FEMALE

To successfully apply reproductive biotechnologies in females, a solid knowledge of the reproductive system and reproductive physiology is necessary. Understanding ovarian physiology, for example, is a key point to develop protocols for stimulating its activity for artificial insemination (AI), laparoscopy follicular aspiration (LOPU), *in vitro* fertilization (IVF), and embryo transfer (ET). Puma's ovulation is induced in response to reflex LH release by genital somatosensory stimuli during mating (Thongphakdee et al. 2018). The estrous cycle lasts approximately 23 days, with estrus usually lasting 8 days (although 11 days of estrus have been reported) (Chapman and Feldhamer 1982, Brown 2018). In free-ranging pumas (*P. concolor coryi*), Benson et al (2012) reported a pseudo-estrus behavior, suggesting that females, even when nursing pups, mate with males to maintain friendly relations and prevent infanticide. Females give birth at any time of the year, with no seasonality interference (Chapman and Feldhamer 1982). However, a higher birth rate can be observed in the wild from June to August, possibly related to greater food availability (Jansen and Jenks 2012). Reproductive seasonality depends on several environmental factors, such as daylight length, rainfall, and humidity. Thus, seasonality is expected to vary among different puma populations given their wide geographic distribution.

According to Anderson (1983) and Logan et al. (1996), gestation lasts approximately 92 days (ranging from 84 to 98). The number of young per litter is difficult to determine in free-living populations, as females hide their young in nests (Riley 1998), but 3.0 to 3.4 pups per litter have been reported (Logan et al. 1996, Spereadbury et al. 1996). The average birth interval ranges from 18.3 to 24.3 months (Ross and Jalkotzy 1992, Lindzey et al. 1994, Spreadbury et al. 1996).

HORMONAL STIMULATION

In mammals, monitoring reproductive hormones can be performed by invasive methods, such as blood samples, or non-invasive methods, such as stool, urine, saliva, and hair sampling (Silva et al. 2017). A major advance in the hormonal study in wild species was due to measuring metabolic steroids in urine and feces, as it is less invasive than blood collection, which usually is possible only after chemical restraint (Brown 2011). In felines, most hormonal studies used fecal steroids, since almost all gonadal steroids are excreted in the feces (Brown 2006). Through these studies, it was possible to observe that felines exhibit a variety of ovulatory patterns, from almost exclusively induced to varied combinations of induced and spontaneous ovulation (Brown 2006). In pumas, spontaneous increases in progestogens following estrogen peak are absent or rare (Brown 2011). Despite this, the puma has a higher concentration of progesterone (3.5 ng/ml) compared to most felines, especially the domestic cat, but the plasma concentration of estradiol is like most felines, except the ocelot (Genaro et al. 2007).

Studies used different protocols to stimulate ovarian activity in pumas (Bonney et al. 1981, Moore et al. 1981, Miller et al. 1990, Barone et al. 1994, Baldassare et al. 2015, Carelli et al. 2017, Baldassarre et al. 2017, Jorge-Neto 2019, Cazati 2022). Equine Chorionic Gonadotropin (eCG) and Human Chorionic Gonadotropin (hCG) are routinely used mainly because they have a long half-life, with a good ovarian response, and require only one intramuscular application. On the other hand, eCG stimulates the production of anti-gonadotropin antibodies which may decrease ovarian response. Thus, a minimum interval of 6 months between applications is recommended (Swanson et al. 1995). Faced with this situation, Barone et al. (1994) tested low doses of eCG (100 IU and 200 IU) and hCG (100 IU) to assess ovarian response in pumas. They observed that even using low doses, 100% of the females ovulated after 41–50 hours. In the same paper, the authors reported that both the application of 100 and 200 IU of eCG promoted ovulation in females, but the dose of 200 IU produced four times more ($P < 0.05$) corpora lutea than 100 IU (Barone et al. 1994). Ovarian responses to hormonal stimulation can be highly variable between species and even between individuals in the same species (Brown 2011); in addition, there is a marked variation in the dosages and application interval of these hormones to induce follicular development and ovulation in pumas (Table 7.1).

Table 7.1 Hormonal protocols for induction of ovarian activity in pumas

References	Animals	eCG	Interval	hCG
Bonney et al. 1981	3	1250 IU	72 h	1000 IU
Moore et al. 1981	3	1250 IU	72 h	1000 IU
Miller et al. 1990	5	1000 IU	84 h	800 IU
	3	2000 IU	84 h	800 IU
Barone et al. 1994	6	100 IU	80 h	100 IU
	8	200 IU	80 h	100 IU
Baldassare et al. 2015	3	750 IU	75 h	500 IU
Carelli et al. 2017	3	750 IU	75 h	500 IU
Baldassarre et al. 2017	3	750 IU	75 h	500 IU
Jorge-Neto 2019	4	750 IU	85 h	300 IU
	4	750 IU		0
Cazati 2022	3	750 IU	85 h	300 IU

LAPAROSCOPIC OVUM PICK-UP (LOPU)

The LOPU is the most reliable and efficient technique for collecting high-quality oocytes from live animals, and as it is a minimally invasive technique, it facilitates their prompt recovery. According to Jorge-Neto (2019) when LOPU is well performed, it does not cause fibrosis in the ovaries or adhesions in the animal's viscera. Also, the author reported a female's pregnancy in the wild that was released months after a LOPU procedure. This reinforces the argument that LOPU is a safe procedure and that it does not negatively affect the fertility of the animal.

All studies that performed LOPU in pumas used the same hormones, but with different concentrations and application intervals (Miller et al. 1990, Baldassare et al. 2015, Jorge-Neto 2019). Miller et al. (1990), using 1000 and 2000 IU of eCG and 800 IU of hCG in eight cougars, obtained an average of 20.0 ± 5.9 oocytes per female (140 total oocytes: 77 mature, 43 immature, 20 degenerated). Baldassare et al. (2015), using 750 IU of eCG and 500 IU of hCG, obtained an average of 32.6 ± 19.3 oocytes per female in three cougars (98 oocytes, 42 mature, and 56 immature). Jorge Neto (2019), in turn, using 750 IU of eCG and 300 IU of hCG, obtained the record of oocytes collected by aspiration—106 oocytes *in vivo* matured oocytes from just one female. Cazati (2022) using 750 IU of eCG and 300 IU of hCG obtained an average of 36.3 ± 19.4 oocytes. In these studies, it was possible to observe the success in collecting *in vivo* matured oocytes, with a maturation index of 55% (Miller et al. 1990), 97.6% (Baldassare et al. 2015), 100% (Jorge-Neto 2019), and 100% (Cazati 2022).

When *in vitro* fertilization or ICSI is not immediately evaluable, the oocytes can be cryopreserved for future use. Jorge-Neto (2019) rewarmed 81 vitrified oocytes, obtaining 70 viable oocytes, and nine degenerated oocytes—two oocytes were lost during manipulation.

In Vitro Embryo Production

The difficulty of producing embryos from wild cats, both in captivity and in the wild, begins with the logistics of transporting the oocytes to the laboratory. Depending on the stimulation protocol, it is possible to obtain *in vitro* or *in vivo* matured oocytes (Jorge-Neto 2019). Studies have shown that oocytes matured *in vivo* have better quality than those matured *in vitro* since the reproductive tract provides an ideal environment for the development of oocytes (Gomez et al. 2000, Ochota et al. 2016). However, the time to perform *in vitro* fertilization after collection of the *in vivo* matured oocyte varies from 1 to 6 hours (Goodrowe et al. 1988, Miller et al. 1990, Jorge-Neto 2019), making it unfeasible in most cases due to the distance between the collection site and the laboratory. An alternative is to collect immature oocytes, as maturation can occur between 17 and 30 hours after incubation (Katska-Ksiazkiewicz et al. 2003). Most *in vitro* maturation protocols incubate the oocytes for 24 h, since after that the oocyte quality decreases, compromising the *in vitro* fertilization (Miller et al. 1990, Jin et al. 2010). However, this 24 h incubation period is enough to transport the oocyte to the laboratory, especially when using portable incubators.

One of the first studies with *in vitro* embryo production in pumas was published in 1990, when Miller et al. incubated 71 *in vivo* matured and 37 *in vitro* matured oocytes, obtaining a total of 40 fertilized oocytes, 50% of which were matured *in vivo* and 33.3% matured *in vitro*. After 24 hours of co-incubation, 10 oocytes cleaved to the 2-cell stage. In 2017, Carelli et al. performed *in vitro* fertilization on 31 vitrified puma oocytes, using epididymal semen from a male who had just been hit by a car and died. After 18 hours of IVF, 11 degenerated oocytes and 20 presumptive zygotes were observed. However, only one embryo developed to the 2-cell stage.

ARTIFICIAL INSEMINATION

The first successful artificial insemination in a wild cat occurred in 1981 in a female puma, after ovulation induction and intrauterine insemination by laparoscopy (Moore et al. 1981). The authors induced ovulation in three cougars using 1250 IU of PMSG and 1000 IU of hCG with an interval of 72 hours. Between 30–40 hours, the females were inseminated with a sperm concentration of 20–40 $\times$ 10^6 motile sperm/mL, and during the procedure, two to six ovulated follicles per female were observed, with one female giving birth (Moore et al. 1981, cited by Miller et al. 1990). Another successful puma birth through AI occurred in 1994 when Barone et al. used 200 IU eCG and 100 IU hCG to induce ovulation and inseminated the female 41 h after hCG.

The rarity of studies reporting puma births and the low fertility rate shows that it is still necessary to improve superovulation protocols, identify the best time for ovulation, improve the quality of semen collected and cryopreserved and develop new embryo production methodologies.

CONTROL OF THE REPRODUCTION

One of the major problems to develop biotechnologies in wild cats is the small number of animals kept in captivity that can reproduce. Another problem is the expressive number of vasectomized males. Many institutions choose to sterilize the male as it is a safe and less invasive method than female sterilization. From a genetic point of view, vasectomy is not very viable for animals threatened with extinction, due to the difficult and not guaranteed reversion. As a result, the genetics of this specimen is practically lost and its use in conservation programs is prevented (Cazati 2022).

In this sense, an alternative would be to perform tubal ligation in females by laparoscopy, which is a minimally invasive technique. This technique does not alter ovarian activity, allowing the aspiration of oocytes in multiple opportunities, which can be used for *in vitro* embryo production. The first study that reports female sterilization by tubal ligation using video laparoscopy in neotropical felids was performed by Cazati (2022) in pumas. According to this author, the animals had no problems after the procedure, and it was possible to aspirate oocytes after *in vivo* maturation, showing that the technique allows the maintenance of sterilized females as a living genetic bank.

BIOBANKING

Every year thousands of specimens of wild animals die directly or indirectly from human actions, such as roadkill or forest fires. In 2020, fires in the Pantanal caused the loss of 40,000 km^2, and the death of approximately 17 million vertebrates (Garcia et al. 2021, Tomas et al. 2021). Disasters like this show the urgent need for biotechnologies that can safeguard samples of genetic material from endangered species. In this sense, creating a Germplasm Bank is essential for the conservation of genetic biodiversity.

Among the available genetic sources, somatic cells have the advantage of being able to be collected from animals regardless of sex, age, and reproductive stage, and can be obtained from live animals or post-mortem (Praxedes et al. 2018, Silva et al. 2021, Santos et al. 2021). In pumas, Lira et al. (2021) demonstrated for the first time that cryopreserved tissues are viable and maintain their integrity, showing minimal changes after heating. In another study, Lira et al. (2022) reported the development of cell lines derived from cryopreserved puma fibroblasts after the 10th passage. This is the first step of a long path using the somatic cell in ART.

FINAL CONSIDERATIONS

There are considerable advances in assisted reproduction technologies in pumas, with effective protocols for semen collection and a good response to hormonal stimulation of ovarian activity, including births after artificial insemination. However, advances are needed in semen cryopreservation and refrigeration protocols, and *in vitro* embryo production. Efforts for genetic recovery from animals that died are also necessary. New technologies such as fibroblast culture, preantral follicle maturation, and testicular culture have not been effectively explored in this species.

REFERENCES

Anderson, A.E. 1983. A Critical Review of Literature on Puma (*Felis concolor*) (Special Report). Denver: Colorado Division of Wildlife, Research Section 54(8): 1–92.

Araujo, G.R. de, T.A.R. Paula, T. Deco-Souza, R.G. Morato, L.C.F. Bergo, L.C. Silva, et al. 2018. Comparison of semen samples collected from wild and captive jaguars (*Panthera onca*) by urethral catheterization after pharmacological induction. Anim. Reprod. Sci. 195: 1–7.

Araujo, G.R. de, T. Deco-Souza, L.C.F. Bergo, L.C. Silva, R.G. Morato, P.N. Jorge-Neto, et al. 2020a. Field friendly method for wild feline semen cryopreservation. J. Threat. Taxa 12: 15557–15564.

Araujo, G.R., T.A.R. Paula, T. Deco-Souza, R.G. Morato, L.C.F. Bergo, L.C. Silva, et al. 2020b. Colheita farmacológica de sêmen de onças-pardas (*Puma concolor*: Mammalia: Carnivora: Felidae). Arq. Bras. Med. Vet. Zoo. 72: 437–442.

Baldassarre, H., J.B. Carelli, L.A. Requena, M.G. Rodrigues, S. Ferreira, J. Salomão, et al. 2015. Efficient recovery of oocytes from "onça parda" (*Puma Concolor*) by laparoscopic ovum pick-up of gonadotropin-stimulated females. Anim. Reprod. 14(3): 717.

Baldassarre, H., L.A. Requena, J. Salomão, M. Rodrigues, S.A.P. Ferreira, A.S. Traldi, et al. 2017. Laparoscopic ovum pick-up is a safe procedure for the collection of oocytes for preservation efforts in Pumas (*Puma concolor*). Anim. Reprod. 14: 780.

Barone, M.A., D.E. Wildt, A.P. Byers, M.E. Roelke, C.M. Glass and J.G. Howard. 1994. Gonadotropin dose and timing of anesthesia for laparoscopic artificial insemination in the puma (Felis concolor). J. Reprod. Fertil. 101: 103–108.

Benson, J.F., M.A. Lotz, E.D. Land and D.P. Onorato. 2012. Evolutionary and practical implications of pseudo-estrus behavior in florida panthers (*Puma Concolor Coryi*). Southeast. Nat. 11(1): 149–154.

Bento, H.J., R.L.A. Vieira, G.A. Iglesias, A.H. Kuczmarski, S.M.N.D. Dias and RC.R. Paz. 2019. Coleta e avaliação de espermatozoides epididimários obtidos pela técnica de squeezing em Puma concolor de vida livre. Nat. Online 17: 82–88.

Bonney, R.C., H.D.M. Moore and D.M. Jones. 1981. Plasma concentrations of oestradiol-17 and progesterone, and laparoscopic observations of the ovary in the puma (Felis concolor) during oestrus, pseudopregnancy and pregnancy. J. Reprod. Fertil. 63(2): 523–531.

Brazil. 2022. Ministério do Meio Ambiente. PORTARIA MMA Nº 148, DE 7 DE JUNHO DE 2022: National List of Endangered Species.

Brown, J.L. 2006. Comparative endocrinology of domestic and nondomestic felids. Theriogenology 66: 25–36.

Brown, J.L. 2011. Female reproductive cycles of wild female felids. Anim. Reprod. Sci. 124(3–4): 155–162.

Brown, J.L. 2018. Comparative ovarian function and reproductive monitoring of endangered mammals. Theriogenology 109: 2–13.

Carelli, J.B., P.N. Jorge Neto, L.A. Requena, M.G. Rodrigues, J.A. Salomão Júnior, S.A.P. Ferreira, et al. 2017. *In vitro* fertilization of puma (Puma concolor) from vitrified oocytes and semen collected from epididymis of dead donor: Case report. Rev. Bras. Reprod. Anim. 41: 364.

Cazati, L. 2022. Descrição da técnica de ligadura tubária por laparoscopia em onças-pardas (*Puma concolor*). Master in Veterinary Science, Universidade Federal de Mato Grosso do Sul, Mato Grosso do Sul, BR.

Chapman, J.A. and G.A. Feldhamer. 1982. Wild mammals of North America. The Johns Hopkins University Press, Baltimore.

CITES — Convention on International Trade in Endangered Species of Wild Fauna and Flora. CITES Species Databases. Accessed on 29-08-2022.

Cucho, H., V. Alarcón, C. Ordóñez, E. Ampuero, A. Meza and C. Soler. 2016. Puma (Puma concolor) epididymal sperm morphometry assessed by the ISAS ® v1 CASA-Morph system. Asian J. Androl. 18: 879–881.

Deco-souza, T., T.A.R. de Paula, D.S. Costa, E.P. Costa, J.B.G. de Barros and G.R. de Araujo. 2013. Comparação entre duas concentrações de glicerol para a criopreservação de sêmen de suçuarana (Puma concolor). Pesq. Vet. Bras. 33: 512–516.

Eizirik, E., J.H. Kim, M.M. Raymond, P.G. Grawshaw Jr, S.J. O'brien and W.E. Johnson. 2001. Phylogeography, Population History and Conservation Genetics of Jaguars (Panthera onca, Mammalia, Felidae). Mol. Ecol. 10(1): 65–79.

Emmons, L.H. and F. Feer. 1997. Neotropical Rainforest Mammals: A Field Gide (2nd ed.). The University of Chicago Press, Chicago.

Fonseca, G.A.B., A.B. Rylands, C.M.R. Costa, R.B. Machado and Y.L.R. Leite (eds). 1994. Livro Vermelho dos Mamíferos Brasileiros Ameaçadas de Extinção. Fundação Biodiversitas, Belo Horizonte. 459 p.

Garcia, L.C., J.K. Szabo, F. de Oliveira Roque, A. de Matos Martins Pereira, C. Nunes da Cunha, G.A. Damasceno-Júnior, et al. 2021. Record-breaking wildfires in the world's largest continuous tropical wetland: Integrative fire management is urgently needed for both biodiversity and humans. J. Environ. Manage. 293: 112870.

Genaro, G., W. Moraes, J.C.R. Silva, C.H Adania and C.R. Franci. 2007. Plasma hormones in neotropical and domestic cats undergoing routine manipulations. Res. Vet. Sci. 82: 263–270.

Gomez, M.C., C.E. Pope, R. Harris, A. Davis, S. Mikota and B.L. Dresser. 2000. Births of kittens produced by intracytoplasmic sperm injection of domestic cat oocytes matured *in vitro*. Reprod. Fertil. Dev. 12(7–8): 423–433.

Goodrowe, K.L., R.J. Wall, S.J. Obrien, P.M. Schmidtm and D.E. Wildt. 1988. Developmental competence of domestic cat follicular oocytes after fertilization *in-vitro*. Biol. Reprod. 39(2): 355–372.

Harris, R., M. Gomez, S. Leibo and C. Pope. 2002. *In vitro* development of domestic cat embryos after *in vitro* fertilization of oocytes with spermatozoa stored for various intervals at 4°C (Abstract). Theriogenology 57: 365.

Howard, J.G., M. Bush and D.E. Wildt. 1986. Semen collection, analysis and cryopreservation in nondomestic mammals. pp. 1047–1053. *In:* D. Morrow (ed.). Current Therapy in Theriogenology I. W.B. Saunders Co., Philadelphia.

IUCN. 2022. The IUCN Red List of Threatened Species. Version 2022–1. https://www.iucnredlist.org. Accessed on 29-08-2022.

Jansen, B.D. and J.A. Jenks. 2012. Birth timing for mountain lions (*Puma concolor*); testing the prey availability hypothesis. PlosOne 7(9): e44625.

Jin, Y.X., X.S. Cui, X.F. Yu, Y.J. Han, I.K. Kong and N.H. Kim. 2010. Alterations of Spindle and Microfilament Assembly in Aged Cat Oocytes. Reprod. Domest. Anim. 45(5): 865–871.

Jorge-Neto, P.N. 2019. Biotecnologias reprodutivas aplicadas à produção de embriões *in vitro* de onça-parda (*Puma concolor*) e onças-pintadas (*Panthera onca*). Master in Animal Reproduction, University of São Paulo, São Paulo, Brazil.

Jorge-Neto, P.N., G.R. Araujo, T. Deco-souza, R.F. Bittencourt, A.C. Csermak Jr, C.S. Pizzutto, et al. 2019. Pharmacological semen collection of Brazilian wild felids. Rev. Bras. Reprod. Anim. 43: 704.

Jorge-Neto, P.N., G.R. Araújo, M.C.C. Silva, J.A. Salomão-Jr, A.C. Csermak-Jr, C.S. Pizutto, et al. 2020. Description of the CASA system configuration setup for jaguar (Panthera onca). Anim. Reprod. Sci. 220: 40–41.

Katska-Ksiazkiewicz, L., B. Rynska, G. Kania, Z. Smorag, B. Gajda, M. Pienkowski et al. 2003. Timing of nuclear maturation of nonstored and stored domestic cat oocytes. Theriogenology 59(7): 1567–1574.

Leite, F.L.G, T.A.R. De Paula, S.L.P. Da Matta, C.C. Fonseca, M.T.D. Das Neves, J.B.G. De Barros, et al. 2006. Cycle and duration of the seminiferous epithelium in puma (*Puma concolor*). Anim. Reprod. Sci. 91: 307–316.

Lindzey, F.G., B.B. Ackerman, D. Barnhurst, T.P. Hemker and S.P. Laing. 1994. Cougar population dynamics in southern Utah. Wildl. Soc. Bull. 58(4): 619–623.

Lira, G.P.O., A.A. Borges, M.B. Nascimento, L.V.C. Aquino, L.F.M.P. Moura, H.V.R. Silva, et al. 2021. Effects of somatic tissue cryopreservation on puma (*Puma concolor* Linnaeus, 1771) tissue integrity and cell preservation after *in vitro* culture. Cryobiology 101: 52–60.

Lira, G.P.O., A.A. Borges, M.B. Nascimento, L.V.C. Aquino, L.F.M.P. Moura, H.V.R. Silva, et al. 2022. Morphological, ultrastructural, and immunocytochemical characterization and assessment of puma (*Puma concolor* Linnaeus, 1771) cell lines after extended culture and cryopreservation. Biopreserv. Biobank.

Logan, K.A., L.L. Sweanor, T.K. Ruth and M.G. Hornocker. 1996. Cougars of the San Andres Mountains. Final Report, Federal Aid in Wildlife Restoration Project W-128-R, New Mexico Game and Fish, Santa Fe.

Logan, K.A. and L.L. Sweanor. 2001. Desert Puma: Evolutionary Ecology and Conservation of an Enduring Carnivore. Island Press, Washington, D.C.

Miller, A.M., M.E. Roelke, K.L. Goodrowe, J.G. Howard and D.E. Wildt. 1990. Oocyte recovery, maturation and fertilization *in vitro* in the puma (*Felis concolor*). J. Reprod. Fertil. 88: 249–258.

Moore, H.D.M., R.C. Bonney and D.M. Jones. 1981. Successful induced ovulation and artificial insemination in the puma (*Felis concolor*). Vet. Rec. 108: 282–283.

Munson, L., J.L. Brown, M. Bush, C. Packer, D. Janssen, S.M. Reiziss, et al. 1996. Genetic Diversity Affects Testicular Morphology in Free-Ranging Lions (*Panthera leo*) of Serengeti Plains and Ngorongoro Crater. J. Reprod. Fertil. 108: 11–15.

Murphy, K.M. 1998. The Ecology of the Cougar (Puma concolor) in the Northern Yellowstone Ecosystem: Interactions with Prey, Bears, and Humans. Ph.D. dissertation, University of Idaho, Moscow.

O'brien, M.K. and D.R. Mccullogh. 1985. Survival of black-tailed deer following relocation in California. J. Wildl. Manag. 49(1): 115–119.

Ochota, M., A. Pasieka and W. Nizanski. 2016. Superoxide dismutase and taurine supplementation improve *in vitro* blastocyst yield from poor-quality feline oocytes. Theriogenology 85(5): 922–927.

Oliveira, T.G. and K. Cassaro. 1999. Guia de Identificação de Felinos Brasileiros (2nd Ed.). Sociedade de Zoológicos do Brasil, São Paulo, Brasil.

Penfold, L.M., M. Criffield, M.W. Cunningham, D. Jansen, M. Lotz, C. Shea, et al. 2022. Long-term evaluation of male Florida panther (*Puma concolor coryi*) reproductive parameters following genetic introgression. J. Mammal. 103(4): 835–844.

Platz, C.C., D.E. Wildt and S.W.J. Seager. 1978. Pregnancy in the domestic cat after artificial insemination with previously frozen spermatozoa. J. Reprod. Fertil. 52: 279–282.

Praxedes, É.A., A.A. Borges, M.V.O. Santos and A.F. Pereira. 2018. Use of somatic cell banks in the conservation of wild felids. Zoo Bio. 37: 258–263.

Pukazhenthi, B., D.E. Wildt and J.G. Howard 2001. The phenomenon and significance of teratospermia in felids. J. Reprod. Fertil. 57: 423–433.

Redford, K.H. and J.F. Eisenberg. 1992. Mammals of the Neotropics. Vol. 2: The southern cone. University Chigago Press, Chicago.

Redford, K.H. 1997. A Floresta vazia. In Valladares Padua, C.; Bodmer, R.E.; Cullen Jr., L., Manejo e conservação de vida silvestre no Brasil (Cap. 1: 1–22). Brasília Cnpq/Belém, PA: Sociedade Mamiraua.

Riley, S.J. 1998. Integration of Environmental, Biological, and Human Dimensions for Management of Mountain Lions (*Puma concolor*) in Montana. Dissertation—Partial Fulfillment of the Requirements for the Degree of Doctor of Philosophy, Cornell University, Cornell, USA.

Roelke, M.E., J.S. Martenson and S.J. O'Brien. 1993. The consequences of demographic reduction and genetic depletion in the endangered Florida panther. Curr. Biol. 3: 340–349.

Ross, P.I. and M.G. Jalkotzy. 1992. Characteristcs of a hunted population of cougars in south-western Alberta. J. Wildl. Manag. 56(3): 417–426.

Ruiz-López, M.J., N. Gañan, J.A. Godoy, A. Del Olmo, J. Garde, G. Espeso, et al. 2012. Heterozygosity-fitness correlations and inbreeding depression in two critically endangered mammals. Conserv. Biol. 26: 1121–1129.

Santos, M.D.C.B., L.V.C. Aquino, M.B. Nascimento, M.B. Silva, L.L.V. Rodrigues, É.A. Praxedes, et al. 2021. Evaluation of different skin regions derived from a post-mortem jaguar, *Panthera onca* (Linnaeus, 1758), after vitrification for the development of cryobanks from captive animals. Zoo Biol. 40: 280–287.

Santos, R.P. dos and A.R. Silva. 2022. Sperm cooling as an assisted reproduction tool for wildlife: an underrated technology. Biopreserv. Biobank. 2022;10.1089/bio.2022.0025.

Shaw, H.G. 1987. Mountain Lion: Field Guide (Special Report, 9, 3rd Printing). Arizona: Ariziona Game & Fish.

Silva, A.R., N. Moreira, A.F. Pereira, G.C.X. Peixoto, K.M. Maia, L.B. Campos, et al. 2017. Estrus cycle monitoring in wild mammals: challenges and perspectives. pp. 22–45. *In:* R.P. Carreira (ed.). Theriogenology [Internet]. London: Intech Open; Available from: https://www.intechopen.com/chapters/55696

Silva, M.B., É.A. Praxedes, A.A. Borges, L.R. Oliveira, M.B. Nascimento, H.V. Silva, et al. 2021. Evaluation of the damage caused by *in vitro* culture and cryopreservation to dermal fibroblasts derived from jaguars: An approach to conservation through biobanks. Zoo Biol. 40: 288–296.

Spreadbury, B.R., K. Musil, J. Musil, C. Kaisner and J. Koviak. 1996. Cougar population characteristics in southeastern British Columbia. J. Wildl. Manag. 60(4): 962–969.

Swanson, W.F., D.W. Horohov and R.A. Godke. 1995. Production of exogenous gonadotrophin-neutralizing immunoglobulins in cats after repeated eCG hCG treatment and relevance for assisted reproduction in felids. J. Reprod. Infertil. (1): 35–41.

Swanson, W.F. amd J.L. Brown. 2004. International training programs in reproductive sciences for conservation of Latin American felids. Anim. Reprod. Sci. 82–83: 21–34.

Thongphakdee, A., W. Tipkantha, C. Punkong and K. Chatdarong. 2018. Monitoring and controlling ovarian activity in wild felids. Theriogenology 109: 14–21.

Tomas, W.M., C.N. Berlinck, R.M. Chiaravalloti, G.P. Faggioni, C. Strüssmann, R. Libonati, et al. 2021. Distance sampling surveys reveal 17 million vertebrates directly killed by the 2020 wildfires in the Pantanal, Brazil. Sci. Rep. 11: 23547.

Weckerly, F.W. 1998. Sexual-size dimorphism: influence of mass and matting systems in the most dimorphic mammals. J. Mammal. 79: 33–52.

Wildt, D.E., M. Bush, J.G. Howard, S.J. O'brien, D. Meltzer, A. Van Dyk, et al. 1983. Unique seminal quality in the south african cheetah and a comparative evaluation in the domestic cat. Biol. Reprod. 29: 1019–1025.

Wildt, D.E., M. Bush, K.L. Goodrowe, C. Packer, A.E. Pusey, J.L. Brown, et al. 1987. Reproductive and genetic consequences of founding isolated lion populations. Nature 329(6137): 328–331.

Zambelli, D., F. Prati, M. Cunto, E. Iacono and B. Merlo. 2008. Quality and *in vitro* fertilizing ability of cryopreserved cat spermatozoa obtained by urethral catheterization after medetomidine administration. Theriogenology 69: 485–490.

Chapter 8

The Small Felids

Dieferson da Costa Estrela

Federal Institute of the Triângulo Mineiro – Campus Uberlândia, Uberlândia,
Minas Gerais, Brazil. Postal address: Fazenda Sobradinho s/n⁰,
Zona Rural, Postal Code: 1020, ZIP Code: 38400-970.
Email: diefersonestrela@gmail.com

INTRODUCTION

South America is the continent hosting the largest number of species belonging to family Felidae—felids can be found in all its countries. Small felids (subfamily Felinae) account for most of such iversity in the continent (Kitchener et al. 2017). The physical size of small felids living in the wilds ranges from 1.5–3 kg in adult guiñas (*Leopardus guigna*) to 45–65 kg in adult pumas (*Puma concolor*), which are the smallest and the largest South American Felinae species, respectively (Nowell and Jackson 1996, Currier 1983). Although current phylogenetic classifications categorize pumas as small felids, the present book has followed the body mass criterion (Brasil 2018), which led to its classification as large felid in Chapter 5. Thus, the current chapter presents South American representatives of subfamily Felinae, except for puma.

The coat color of small South American felids ranges from monochromatic patterns with slight variations at the ends of the body, as seen in adult jaguarundis [*Herpailurus yagouaroundi* (Geoffroy Saint-Hilaire 1803)], to complex polychromatic patterns including spots, rosettes and stripes, as seen in ocelots (*Leopardus pardalis*) and margays (*Leopardus wiedii*). Felids' beauty is overall acknowledged by human population, due to spiritual and mythological beliefs associated with them; however, it also leads to the slaughter of a larger number of animals in order to sell their skin, teeth and claws, among others (Roe 2002, Mendonça et al. 2012, Herrmann et al. 2013).

Small South American felids are widely distributed in different biomes and at different temperature ranges on the continent; Andean cats can be found at altitudes ranging from sea level to 5,000 meters (Vilalba et al. 2016). Some felid species, such as guiñas, can be found at narrow ranges estimated at approximately 300,000 km^2, whereas others, such as jaguarondis, can be observed across almost the entire continent (Caso et al. 2015, Napolitano et al. 2015). Most of these species share wide sympatry regions with other small felids, and it makes it hard to explain phylogenetic relationships taking place in this group.

Small South American felids are currently acknowledged as representatives of puma (jaguarundi) and ocelot (all species belonging to genus *Leopardus*) lineages. Recent ocelot lineage radiation presents species that emerged in less than 1 Ma (Johnson et al. 2006), as well as well-documented cases of hybridization zones between Geoffroy's cat (*Leopardus geoffroyi)* and Southern tigrina (*Leopardus gutullus*), and between Northern tigrina (*Leopardus tigrinus*) and pampas cat (*Leopardus colocola*), as reported by Trigo et al. (2008, 2013, 2014).

Species belonging to puma and ocelot lineages are currently threatened due to human actions, such as habitat loss and fragmentation, hunting, wild animals' trafficking, as well as trading of skins, bones and claws, among others (Mendonça et al. 2012, Zanin et al. 2015, Peters et al. 2016). According to extinction-risk assessments conducted by experts from the International Union for the Conservation of Nature (IUCN), Geoffroy's cat, jaguarundi and ocelot species are classified as of 'least concern'; margay and pampas cat are classified as 'near threatened'; guiña, Northern tigrina and Southern tigrina are categorized as 'vulnerable'; and Andean cat is classified as "endangered" species (Caso et al. 2015, Napolitano et al. 2015, Oliveira et al. 2015, Paviolo et al. 2015, Pereira et al. 2015, Lucherini et al. 2016, Oliveira et al. 2016, Payan and Oliveira 2016, Villalba et al. 2016).

The predatory nature of these animals is the main factor generating conflicts with owners of both farm animals and pets, which are eventually preyed upon by felids (Palmeira and Barrella 2007, Mendonça et al. 2012, Peters et al. 2016). Peters et al. (2016) have evidenced slaughter events involving individuals belonging to all felid species observed in the Atlantic Forest and Pampa biomes, in Brazil. The main motivations for hunting these animals comprised retaliation for killing domestic animals, prevention of new predation events or fear of likely damage to the physical integrity of individuals who come across larger animals (Peters et al. 2016).

In light of these threats to the conservation of small South American felids, the reproductive management of animals *in situ* and *ex situ*, based on Assisted Reproduction Techniques (ART), is an important line of study to help conserving these species. The main techniques used in the reproductive monitoring and assisted reproduction of male and female animals will be presented in the next sections, with emphasis on the most used techniques and on lack of studies about some species. However, constant revisions applied to this group's taxonomy have direct impacts on studies focused on investigating it. Therefore, it is necessary to address part of these revisions to help better understand the extent to which published studies conducted with some of these small felids comprise one or two species, or even a hybrid sample.

SMALL FELIDS' TAXONOMY

Small South American felids form a group with complex phylogenetic relationships and a history of frequent taxonomic changes. Taxonomic uncertainty about the most suitable genus to be attributed to jaguarundi has led to different conclusions, based on morphological, molecular and paleontological data, in recent years (Kitchener et al. 2017). Thus, the current chapter follows the consensus among felid experts addressed by Kitchener et al. (2017) and it addresses the investigated species as *H. yagouaroundi*, despite uncertainties that may change its genus again in the future.

The ocelot lineage comprises the largest number of South American species and of recent taxonomic changes. Nowadays, all species belonging to this lineage are classified as genus *Leopardus*, and their total number ranges from seven to thirteen, according to the herein used taxonomic references. Species known as tigrina was divided into Northern [*L. tigrinus* (Schreber 1775)] and Southern tigrina [*L. guttulus* (Hensel 1872)]; this classification was well-accepted by the scientific community (Trigo et al. 2013, Trindade et al. 2021). However, a recent study has pointed out that Northern tigrina populations likely comprise two species, rather than one, a fact that may lead to the acknowledgement of a new species in the future (Trindade et al. 2021).

Similar condition is observed for pampas cat species *L. colocola* (Molina 1782), since analysis focused on pooling morphological, molecular and ecological niche models has indicated that the currently acknowledged taxon represents a complex of five species, namely: *L. braccatus*, *L. colocola*, *L. garleppi*, *L. munoai* and *L. pajeros* (Nascimento et al. 2021). If these classifications, or part of them, are established, studies conducted with pampas cat and tigrina may represent complexes comprising species other than *L. colocola* and *L. tigrinus*, or even mixed populations comprising species belonging to each complex. Therefore, data are presented in the current chapter, based on the way the aforementioned authors have named the species, i.e. *L. colocola* or *L. tigrinus*, although such data may represent complexes of species rather than a single species.

Other *Leopardus* species comprise Andean cat [*L. jacobita* (Cornalia 1865)], ocelot [*L. pardalis* (Linnaeus 1758)], margay [*L. wiedii* (Schinz 1821)], Geoffroy's cat [*L. geoffroyi* (d'Orbigny and Gervais 1844)] and guiña [*L. guigna* (Molina 1782)]. These species also present well-known subspecies that are considered stable, so far.

MONITORING AND CONTROL OF THE REPRODUCTION IN MALES

Male individuals belonging to small South American felid species appear to reach puberty at the age of approximately one year, as recently observed for jaguarundi and Southern tigrina based on monitoring fecal androgens—both species reached puberty between the age of 10 and 12 months (Souza et al. 2021). Studies based on semen collection and longitudinal hormonal monitoring have indicated that sperm production in small South American felids, such as jaguarundi, ocelot, Northern tigrina and margay, takes place throughout the year, despite production peaks observed in summer (Moraes et al. 2002, Estrela 2022).

Spermatogenesis process in jaguarundi lasts approximately 37.8 days, whereas that of ocelot lasts 56.3 days (Silva et al. 2010, Silva 2014). Daily sperm production estimated for jaguarundi reached 51.52 million per gram of testis, whereas the one estimated for ocelot reached 18.3 million/gram of testis (Silva et al. 2010, Silva 2014). Jaguarundi presents long reproductive longevity, with emphasis on the record of an 18-year-old male who presented normal seminal parameters, high qualitative levels and fathered a cub at this age (Estrela 2022).

A set of studies has monitored serum testosterone, as well as cortisol and fecal androgen levels, in Geoffroy's cat, jaguarundi, margay, Northern tigrina, ocelot and pampas cat (Nogueira and Silva 1997, Morais et al. 2002, Swanson et al. 2003, Erdmann 2014). Swanson et al. (2003) have punctually assessed mean testosterone and cortisol serum concentrations (±SEM) in ocelot (2.00±0.27 ng/mL; 35.2±4.8 ng/mL, respectively), margay (2.07±0.33; 54.4±6.9), Northern tigrine (0.76±0.14; 91.1±11.3), Geoffroy's cat (0.47±0.10; 109.0±10.8), pampas cat (0.18±0.07; 125.3±16.3) and jaguarundi (0.96±0.19; 223.3±18.2), during semen collection procedures. Swanson et al. (2003) recorded the largest sample (in number) in small South American felid individuals and species, but they apparently resulted from a single collection conducted in each individual.

However, few of the herein mentioned studies carried out longitudinal monitoring, and it enabled tracing the baseline for a particular hormone, as well as identifying changes, based on it. Hormone concentrations are influenced by individuals' age, climate season, containment, handling and human contact frequency, among others (Norkaew et al. 2019, Fazio et al. 2020). Thus, longitudinal hormonal sampling is the ideal type, based on collections performed over a period-of-time capable of representing the investigated phenomenon, as the one performed in the study conducted by Morais et al. (2002) with margay, Northern tigrine and ocelot.

Morais et al. (2002) recorded similar mean (±SEM) serum testosterone concentrations among ocelot, margay and Northern tigrina (3.7±0.6, 4.8±0.7 and 3.4±0.6 nmol/L, respectively) across seasons; they also observed that fecal androgens in Northern tigrina and margay were not affected by seasons, as well as that ocelot presented higher concentrations of them in summer.

Studies focused on investigating reproduction control in males were not available in the literature, most likely because there are few studies about reproductive monitoring and because the investigated species produce semen throughout the year, a fact that does not require interventions to obtain sperm. Cases of reproductive activity suppression in males were not found; the technique mainly adopted for this purpose lies on separating males from females by keeping them in different enclosures.

SPERM COLLECTION AND CONSERVATION

Semen collection is the ART mostly applied to small South American felids; it is mainly performed based on the electroejaculation method. Howard's protocol (1993) is the electroejaculation method mostly applied to this group, based on the application of 80 electrical stimuli split in three series (30, 30 and 20 stimuli) ranging from 2 to 5 V. Several studies have successfully obtained semen from almost all felid species; however, studies conducted with guiña and Andean cat were not found in the literature. Some examples of studies that have successfully collected semen in Geoffroy's cat, jaguarundi, margay, Northern and Southern tigrina, ocelot and pampas cat species comprise Swanson et al. (2003), Erdmann (2014), Leite (2009), Iglesias et al. (2020), Estrela (2022) and Tebet et al. (2022).

In addition to electroejaculation, the urethral catheterization collection technique, which is also known as pharmacological ejaculation, was recently applied to jaguarundi, margay (Madrigal-Valverde et al. 2019, Estrela 2022) and Southern tigrina (Iglesias et al. 2020); results have validated this technique as viable alternative to the aforementioned method. However, comparisons between urethral catheterization and electroejaculation have shown heterogeneous results, since results recorded for urethral catheterization applied to jaguarundi were less satisfactory than the ones recorded when this very same technique was applied to Southern tigrina (Iglesias et al. 2020, Estrela 2022). Since the literature only reported two studies focused on comparing semen collection techniques, it is necessary to conduct further studies to help better understand the effectiveness of urethral catheterization in comparison to electroejaculation.

Collected semen is often assessed through qualitative parameters, such as sperm vigor, total motility, progressive motility, motility index, vitality (intact sperm), based on using eosin/nigrosin staining; acrosomal integrity, based on using rose bengal/fast greem staining; and sperm morphology using eosin/nigrosin staining or rose bengal. Although these parameters are classic in the andrology field, their almost exclusive use shows lack of complementary analyses. Among them, one finds computerized semen analysis system (CASA), which was only used in the study conducted by Madrigal-Valverde et al. (2019) with jaguarundi and margay, as well as sperm DNA integrity assessment, which was only conducted in study carried out by Barros (2007) and Estrela (2022) with Northern tigrina and jaguarundi.

Using fluorescent probes to assess membrane integrity, mitochondrial activity, DNA fragmentation, reactive oxygen species, among other parameters, is an interesting analytical follow-up; however, only the study conducted by Araujo (2012) assessed some of these parameters in ocelot sperm. These probes can be used in manual analyses conducted in fluorescence microscope or in automated analyses based on flow cytometry.

Expensive equipment and laboratory facilities are required at semen collection site, or very close to it, in order to apply some of these techniques to fresh semen;

however, it is not the case of most studies conducted with small South American felids. Some alternatives have emerged in light of these limitations; among them, one finds portable systems for CASA and the comet assay protocol standardized by Estrela (2022) to be used in jaguarundi. This protocol enables assessing DNA damage by preparing the sample a few hours after semen collection. In addition, the previous preparation of solutions to be used in the test enables performing it in the field due to its simplified infrastructure and easy transport.

Furthermore, it is also possible to prepare slides with fluorescent probes in the field to carry out microscopic analysis, although it is manually performed and comprises fewer parameters than flow cytometry. These and other analytical advances should be applied to seminal analysis to be performed in small South American felids in the future to help better understand the seminal features of different species, as well as to contribute to genetic resource storage and to species conservation.

Overall, after semen collection, most studies carry out its cryopreservation, based on using manual cooling and freezing protocols. The extenders used so far present remarkably similar formulation; they mainly comprise Tris (hydroxymethyl-aminomethane), some carbohydrates (fructose, glucose or lactose), sodium citrate, egg yolk and cryoprotectant glycerol (Swanson et al. 2003, Iglesias et al. 2020, Estrela 2022, Tebet et al. 2022). In addition to proprietary formulations, several commercial extenders used for other species were assessed, namely: AndroMed® CSS single step (Minitub, Tiefenbach, Germany), Bioxcell® (IMV Technologies, L'Aigle, France), BotuCRIO®, BotuDOG®, BotuBOV® (Botupharma, Botucatu, Brazil), and Test Yolk Buffer® (Irvine Scientific, Santa Ana, USA), which presented results similar to the ones observed by Erdmann (2014), Iglesias et al. (2020), Estrela (2022), and Tebet et al. (2022).

Sperm motility, vigor and vitality losses, as well as increased number of sperm pathologies, were recorded after cryopreservation and thawing (Baudi et al. 2008, Erdmann 2014, Estrela 2022, Tebet et al. 2022). Motility losses in the analyzed sample ranged from approximately 20% to 100% (Baudi et al. 2008, Erdmann 2014, Estrela 2022, Tebet et al. 2022). Overall, most of the evaluated extenders and species have shown motile sperms after thawing. The number of sperms was sufficient to perform artificial insemination (AI), *in vitro* fertilization (IVF) or intracytoplasmic sperm injection (ICSI), and it has evidenced the importance of cryopreserving semen, as well as its potential to be used.

In some cases, semen was collected and used for AI or IVF (Swanson et al. 1995, 2002, Rodriguez et al. 2021a); however, the way the semen was packed and handled was briefly described. Estrela (2022) used the commercial BotuSemen® medium (Botupharma, Botucatu, Brazil) indicated for semen refrigeration and transport purpose and he did not observe increased sperm DNA damage after five-hour refrigeration. The aforementioned study did not perform AI or IVF, and other parameters, such as post-refrigeration vitality and motility, were not assessed; however, this alternative can be investigated in AI and IVF applied to small felids.

TESTICULAR TISSUE CRYOPRESERVATION AND CULTURE

Studies focused on testicular tissue culture and cryopreservation in small South American felids remain scarce, so far. Only the study conducted by Silva et al. (2012) investigated the likelihood of transplanting spermatogonial stem cells (SSC) from ocelot's testis to a recipient domestic cat. Based on this technique, the aforementioned authors were able to induce SSC multiplication in the cat's testis; ocelot's spermatozoa production in the investigated cat was observed 13 weeks after transplantation.

This technique enabled maintaining donor's germplasm in the recipient individual, as well as producing wild felids' sperm in the analyzed domestic cat. In the future, this technique may be associated with SSC cryopreservation to enable freezing genetic resources collected from endangered animals and, subsequently, induce sperm production in domestic recipients.

ARTIFICIAL INSEMINATION

Few studies available in the literature performed artificial insemination in small South American felids; however, some cases did not describe the adopted procedure in details. Javorouski (2003) induced ovulation in female Northern tigrina, margay and ocelot individuals based on using eCG and hCG; intrauterine AI was performed in cases presenting recent ovulation points. However, the aforementioned study did not describe the AI details or its results. According to Swanson et al. (1995, 1996), a female ocelot subjected to the highest gonadotropin dose (500 IU eCG/225 IU hCG) adopted in their study was inseminated with 7.5×10^6 thawed motile sperm (47% presenting intact acrosomes), got pregnant and delivered a healthy male calf after 78-day pregnancy.

Small seminal volume and relatively low motile sperm concentration are limitations to perform AI procedures in small felids. This factor may lead to preferential use of IVF, which requires fewer motile sperm (50,000/IVF) than AI (10×10^6/AI), despite being a more labor-intensive procedure (Swanson et al. 2007). For example, the study conducted by Stoops et al. (2007) with ocelots recorded mean concentration of 124.9×10^6 spermatozoa per fresh ejaculate, with 74.9% motility. This value enabled performing 6 uterine AIs/ejaculate (10×10^6/AI) or 1,309 IVFs/ejaculate (5×10^4/IVF); motility loss by 30%, which represented the extensor showing the best result in the aforementioned study, was taken into consideration in both cases.

The semen deposition site in female individuals is another important criterion, since 10×10^6 and 2×10^6 of motile sperm/AI are used when AI is performed in the uterus or oviduct, respectively; this method increases the potential use of each ejaculate in comparison to the vaginal deposition, which requires 40×10^6 of motile sperm/AI (Swanson et al. 2007).

Monitoring and Control of the Reproduction in Females

Female reproductive monitoring aims at determining the estrous cycle stage, which was herein investigated based on the combination of vaginal cytology to exogenous hormonal stimulation (eCG/hCG and FSH/LH) and on the evaluation of ovarian structures through video laparoscopy applied to ocelot, margay and Northern tigrina (Moreira et al. 2001, Paz et al. 2010). Vaginal cytology carried out by Moreira et al. (2001) has evidenced prevalence of basal or intermediate cells (>70%) in anestrus (presence of corpus luteum (CL) or lack of follicles), whereas cells with pyknotic nuclei or enucleated cells were more prevalent (>80%) during the follicular phase (lack of CL and presence of follicles).

Moreira et al. (2001) have evidenced that fecal steroid concentrations corresponded to laparoscopic observations and vaginal cytology results reported in all three species, with emphasis on increased fecal estrogen levels during the follicular phase; they were 50% higher than baseline concentrations, but remained at baseline in anestrus. Fecal progestin levels increased to levels higher than the baseline in margays showing CL (Moreira et al. 2001). Vaginal cytology in ocelot was effective in detecting the estrous cycle phase; thus, it is a valuable tool used for reproductive monitoring and management purposes (Paz et al. 2010, Freire 2017).

In addition to identifying the estrous cycle phases and their duration, it is essential to understand whether the investigated species show spontaneous or induced ovulation. Information about this parameter is available for few small South American felid species, such as margay, which presents spontaneous ovulation, whereas ocelot and Northern tigrina present induced ovulation (Moreira et al. 2001). Species presenting spontaneous ovulation are more challenging at the time to perform stimulation with exogenous gonadotropins. However, there are records of cases showing both spontaneous and induced ovulation in the same species; this finding indicates that, although there is a prevalent pattern, both ovulation forms should be taken into consideration in the programmed reproductive management of each species (Moreira et al. 2001, Brown 2006).

Ovulation control was initially assessed in ocelot by Swanson et al. (1995, 1996), who investigated ovarian response to combined administration of exogenous eCG and hCG at dose gradient ranging from 100 to 500 IU of eCG and from 75 to 225 IU of hCG, through AI, after stimulation. Increased antral follicles and corpus luteum production was observed as gonadotropin doses increased; the highest dose of it (eCG + hCG) provided the highest ovarian response rate (Swanson et al. 1995, 1996). However, the number of ocelots responding to ovarian stimulation did not increase as gonadotropin doses increased (Swanson et al. 1995, 1996).

Performing ovarian stimulation with exogenous hormones, mainly in repetitive reproductive management at short-time intervals between stimuli, can lead to risk of hyperestrogenism, as well as of reduced, or even interrupted, ovarian cycling. This factor has encouraged studies focused on investigating these risks in ocelot and Northern tigrina. However, Paz et al. (2005, 2006) and Swanson et al. (1995, 1996) did not observe reduced ovarian activity in ocelot and Northern tigrina subjected to eCG/hCG and/or pFSH/pLH stimulation sequences. On the other hand, Paz et al. (2005, 2006) observed increased anti-gonadotropin antibodies in some of the analyzed females.

Ocelots and Northern tigrinas subjected to eCG/hCG and/or pFSH/pLH stimulation recorded from 6.8 to 8.8 and from 2.3 to 2.5 follicles or corpora lutea per stimulation, on average, respectively (Swanson et al. 1996, Paz et al. 2005, 2006). Paz et al. (2005, 2006) have shown that alternating different gonadotropins (eCG/hCG or pFSH/pLH) can enable remarkably intensive ocelot and Northern tigrina reproduction management, which still needs to be confirmed in other small felids.

Another line of research, focused on preventing follicular hyperstimulation and hyperestrogenism from happening, has investigated the likelihood of interrupting ovarian activity before stimulation with exogenous gonadotropins. Accordingly, Zimmermann (2012) used etonogestrel (Implanon®, Organon, São Paulo, Brazil) in ocelot, whereas Micheletti et al. (2015) used altrenogest (Regumate®, Intervet, São Paulo, Brazil) in Southern tigrina; both drugs led to reduced ovarian activity and fecal estrogens' concentration in most females. Thus, these drugs are an interesting alternative to help reducing risks of hyperstimulation and hyperestrogenism in these species.

Conservation of Female Gametes

Follicles were mainly collected through aspiration in laparoscopic procedures, after stimulation with eCG/hCG, in small felids. Paz et al. (2009) performed ocelot and Northern tigrina aspiration and assessed oocytes' maturation stage (metaphase II) based on morphology and cytogenetics. In total, 12% (4/33) and 36% (4/11) of oocytes deriving from the 33 ocelot and 11 Northern tigrina individuals presented condensed chromosomes, respectively (Paz et al. 2009). However, no metaphase II oocytes were observed; this finding led to the conclusion that aspiration resulted in immature oocytes, which require cultivation in specific media to reach the Metaphase II stage (Paz et al. 2009).

Swanson et al. (2002) obtained mature oocytes ($\geq$2 mm) through laparoscopic aspiration in ocelot and Northern tigrina, after ovulation induction with eCG (400 IU/ocelot; 200 IU/Northern tigrina) and hCG (200 IU/ocelot; 150 IU/Northern tigrina). The aforementioned authors recovered, on average, ($\pm$SEM) 9.9 $\pm$ 1.7 and 10.3 $\pm$ 2.1 ovarian follicles, as well as 6.8 $\pm$ 1.6 and 8.7 $\pm$ 4.0 high-quality oocytes from ocelot and Northern tigrina, respectively (Swanson et al. 2002). These oocytes were incubated with 5×10^5 motile sperm/ml; they recorded fertilization rate of 57.1% (60/105) in ocelot and 62.8% (49/78) in Northern tigrina, which resulted in the cryopreservation of 64 ocelot and 52 Northern tigrina embryos (Swanson et al. 2002).

EMBRYO TECHNOLOGY

Embryo production based on IVF was assessed and embryos were identified in some small South American felids. As reported in the previous session, Swanson et al. (2002) obtained 64 ocelot embryos, which were cryopreserved. Years later, 24 of these embryos were transferred to 8 ocelot females (3/1) by Blank et al. (2022) and this process resulted in three offspring born from three different births (12.5% of embryos and 37.5% of females). These findings support the great potential for long-term frozen TE production in ocelot conservation processes; however, some hormonal changes observed throughout pregnancy may be associated with low conception and parturition rates and dystocia cases in felids (Blank et al. 2022).

Studies conducted by Swanson et al. (2002) and Blank et al. (2022) were the only homologous IVF and ET cases found in the literature. Other studies performed heterologous IVF with oocytes collected from domestic cats and semen deriving from Geoffroy's cat and ocelot (Stoops et al. 2007, Sestelo et al. 2019, Rodrigues et al. 2021a, b). Stoops et al. (2007) performed assays with heterologous IVF, based on using fresh and cryopreserved ocelot semen. They obtained 73% (38/52) and 64% (32/50) fertilized oocytes with nuclear maturation, respectively, based on the extender with the best result.

Sestelo et al. (2019) and Rodrigues et al. (2021a, b) obtained Geoffroy's cat hybrid embryos from cat oocytes and cryopreserved semen cultured, either alone or aggregated with homologous cat embryos. Sestelo et al. (2019) performed heterologous IVF and obtained embryos from cryopreserved semen, which resulted in cleavage rate of 49% (117/238), as well as in development to morula of 40% (96/238) and to blastocyst stage of 26% (61/238). Rodriguez et al. (2021a, b) performed heterologous IVF, as well as aggregation of Geoffroy's cat hybrid embryos with cat homologs; they recorded increased number of blastocyst cells and nuclei expressing OCT4+ gene (related to multipotency), in comparison to diploid cat embryos (2n) and to non-aggregated tetraploids (4n).

OTHER ASSISTED REPRODUCTION TECHNIQUES

In addition to conventional ARTs, Carreiro et al. (2017) have used the ovarian slicing technique for *post-mortem* oocyte retrieval in a run-over Northern tigrina specimen. The aforementioned authors have sliced the animal's ovaries right after its death and collected 88 oocytes (19 grade-I, 42 grade-II, and 27 grade-III oocytes) and demonstrated the viability of the technique.

Sperm fertilization capacity test conducted with heterologous oocytes deriving from domestic cats, as well as with chicken egg yolk perivitelline membrane, were performed and validated based on using ocelot and Norther tigrina semen (Stoops et al. 2007, Baudi et al. 2008, Garay 2012, Araujo 2012, Araujo et al. 2015). These tests were added to traditional seminal analysis techniques to enable inferring the fertilization capacity of fresh and thawed semen; however, it is necessary to validate these techniques in other species.

SOMATIC RESOURCE BANKS AND CLONING

Somatic cell cryopreservation is a research field that has recently started to be explored in small felids. Arantes (2018) has assessed the cryopreservation of fibroblasts collected in skin biopsy samples deriving from Northern tigrina and pampas cat, based on using 2.5%, 5% and 10% of cryoprotectant DMSO; all three DMSO concentrations were equally efficient and thawed fibroblasts have shown proliferation in culture. Echeverry et al. (2019a) induced multipotency in pampas cat dermal fibroblasts based on using growth factors and reprogramming molecules; this finding can enable further studies to advance in germ cell reprogramming or nuclear transfer processes.

Echeverry et al. (2019b) investigated the growth properties of mesenchymal stem cells deriving from guiñas' adipose tissue; the observed features have evidenced that these cells can be cultivated for different biotechnological uses, such as cloning and nuclear transfer. Gallegos (2017) and Veraguas et al. (2017) were able to manipulate the cell cycle of guiñas' fibroblasts, as well as to induce G0/G1 state, and it enabled producing embryos through interspecific somatic cell nuclear transfer (ISCNT) in studies conducted by Veraguas et al. (2018, 2020).

The aforementioned authors observed that the culture of queen embryos aggregated with guiña (from ISCNT) has increased morula rate, but it did not increase blastocyst rate (Veraguas et al. 2018, 2020). It is necessary to conduct this type of study with other species, as well as improving protocols proposed for guiña and pampas cat, to enable the next steps towards obtaining embryos based on cloning and ISCNT.

FINAL CONSIDERATIONS

The herein conducted review has evidenced lack of knowledge about guiña and Andean cat species, as well as lack of studies focused on investigating the application of the main ARTs in either male or female specimens. Moreover, no studies focused on investigating ARTs application in female jaguarundi, pampas cat and Geoffroy's cat were found in the literature. The lack of studies about assisted reproduction in more than 50% of small South American felids counteracts the high risk of extinction faced by these species. Similarly, if *L. colocola* and *L. tigrina* populations were promoted to species, some of them would be highly threatened and lack studies based on ARTs.

Although most herein analyzed studies were conducted with ocelot, Northern tigrina and margay species, few of them conducted experiments based on using AI, IVF and ET, and recorded low success levels. This finding has evidenced the need of conducting further research to enable improving reproductive control and increasing the success of these and other ARTs. The set of available ARTs has great potential to be applied in the conservation of small South American felids; although some valuable advances have already been achieved, it is necessary to conduct further studies focused on assessing ARTs in different species, as well as getting more funding to help improving ARTs' success levels.

REFERENCES

Arantes, L.G. 2018. Efeitos da criopreservação na viabilidade de fibroblastos de felinos silvestres. Dissertation, Universidade de Brasília, Brasília, Brazil.

Araujo, G.R. 2012. Using fluorescent probes and sperm binding to heterologus oocytes and to periviteline membrane, using chiken (*Gallus gallus*) eggs, for evaluate fresh and frozen-thawed ocelot (*Leopardus pardalis*) sperm. Dissertation Universidade Federal de Viçosa, Viçosa, Brazil.

Araujo, G.R., T.A. Paula, T.D. Deco-Souza, R.M. Garay, L.C.F. Bergo, A.C. Csermak-Júnior, et al. 2015. Ocelot and oncilla spermatozoa can bind hen egg perivitelline membranes. Anim. Reprod. Sci. 163: 56–62.

Barros, P.M.H. 2007. Estresse oxidativo e integridade do DNA em sêmen resfriado de gato-do-mato-pequeno (*Leopardus tigrinus*, Schreber, 1775). Ph.D. Thesis Universidade de São Paulo, São Paulo, Brazil.

Baudi, D.L., K. Jewgenow, B.S. Pukazhenthi, K.M. Spercoski, A.S. Santos, A.L. Reghelin, et al. 2008. Influence of cooling rate on the ability of frozen-thawed sperm to bind to heterologous zona pellucida, as assessed by competitive in vitro binding assays in the ocelot (*Leopardus pardalis*) and tigrina (*Leopardus tigrinus*). Theriogenology 69(2): 204–211.

Blank, M.H., C.H. Adania, W.S. Swanson, D.S.R. Angrimani, M. Nichi, A.B.V. Guimarães, et al. 2022. Comparative fecal steroid profile during pregnancy, parturition, and lactation between natural fertilization and embryo transfer in ocelots (*Leopardus pardalis*). Theriogenology 182: 26–34.

Brasil, Ministério do Meio Ambiente/Instituto Chico Mendes de Conservação da Biodiversidade. 2018. Portaria nº 612, de 22 de junho de 2018. Plano de Ação Nacional para a Conservação dos Grandes Felinos—PAN Grandes Felinos. Diário Oficial da União, edição 121, seção 1, p. 45, 26 de jun. de 2018.

Brown, J.L. 2006. Comparative endocrinology of domestic and nondomestic felids. Theriogenology 66(1): 25–36.

Carreiro, A.N., A.Y.F. De La Salles, B.M.R. Falcão, J.G. Souza, D.V.F. Araújo, N.L. Souza, et al. 2017. Obtenção de oócitos *post mortem* em *Leopardus tigrinus* Schreber, 1775– Relato de Caso. Rev. Bras. Reprod. Anim. 3(41): 688–690.

Caso, A., T. de Oliveira and S.V Carvajal. 2015. *Herpailurus yagouaroundi*. The IUCN Red List of Threatened Species 2015: e.T9948A50653167. https://dx.doi.org/10.2305/ IUCN.UK.2015-2.RLTS.T9948A50653167.en. Accessed on 22 September 2022.

Currier, M.J.P. 1983. Felis concolor. Mamm. Species 200: 1–7.

Echeverry, D., D. Rojas, C. Aguilera, L. Rodriguez-Alvarez and F. Castro. 2019a. Effect of growth factors and reprogramming molecules on induction to multipotency of dermal fibroblasts from colocolo (*Leopardus colocolo*). Reprod. Fertil. Dev. 32: 232–232.

Echeverry, D.M., D.M. Rojas, C.J. Aguilera, D.M. Veraguas, J.G. Cabezas, L. Rodríguez-Álvarez, et al. 2019b. Differentiation and multipotential characteristics of mesenchymal stem cells derived from adipose tissue of an endangered wild cat (*Leopardus guigna*). Austral J. Vet. Sci. 51(1): 17–26.

Erdmann, R.H. 2014. Protocolos de criopreservação de sêmen em felídeos do gênero *leopardus* e quantificação de metabólitos fecais de andrógenos e glicocorticoides. 2014. 142 f. Ph.D. Thesis Universidade Federal do Paraná, Curitiba, Brazil.

Estrela, D.C. 2022. Parâmetros andrológicos e criopreservação de sêmen em onça-parda (*Puma concolor*—Linnaeus, 1771) e jaguarundi (*Puma yagouaroundi*—É. Geoffroy Saint-Hilaire, 1803). Ph.D. Thesis Universidade Federal do Paraná, Curitiba, Brazil.

Fazio, J.M., E.W. Freeman, E. Bauer, L. Rockwood, J.L. Brown, K. Hope, et al. 2020. Longitudinal fecal hormone monitoring of adrenocortical function in zoo housed fishing cats (*Prionailurus viverrinus*) during institutional transfers and breeding introductions. PLoS One 15(3): e0230239.

Freire, L.M.P. 2017. Monitoramento da atividade reprodutiva de fêmeas de jaguatirica (*Leopardus pardalis*) mantidas em cativeiro. Dissertation Universidade Estadual do Ceará, Brazil.

Gallegos, P.F. 2017. Inducción de la fase G0/G1 del ciclo celular en cultivo de fibroblastos de piel de gatos domésticos (*Felis silvestris catus*) y güiñas (*Leopardus guigna*). Monography—Universidad de Concepción, Chilán, Ghile.

Garay, R.M. 2012. Use of fluorescent probes to assess seminal ejaculate of oncilla (*Leopardus tigrinus*) and binding assay with perivitelline layer of chicken´s egg (*Gallus gallus*) as a tool for prediction of sperm fertility. Dissertation Universidade Federal de Viçosa, Viçosa, Brazil.

Herrmann, T.M., E. Schüttler and P. Benavides. 2013. Values, animal symbolism, and human-animal relationships associated to two threatened felids in Mapuche and Chilean local narratives. J. Ethnobiol. Ethnomed. 9: 41.

Howard, J.G. 1993. "Semen collection and analysis in nondomestic carnivores." pp. 390–399. *In*: M.E. Fowler and R.E. Miller (eds). Zoo and Wildlife Animal Medicine: Current Therapy III. Philadelphia.

Iglesias, G.A. 2019. Comparação entre coletas de sêmen em gato-do-mato-pequeno (*Leopardus guttulus*) pelos métodos de cateterismo uretral e eletroejaculação. Dissertation Universidade Federal de Mato Grosso, Cuiabá, Brazil.

Iglesias, G.A., H.J. Bento, A.H. Kuczmarski, T.L.C. Costa, J. Ribeiro, S. Pimentel, et al. 2020. Coleta de sêmen em *Leopardus guttulus* pelo método do cateterismo uretral. Arq. Bras. Med. Vet. Zootec. 72(03).

Javorouski, M.L. 2003. Comparação da resposta adrenocortical de fêmeas de felídeos submetidas a anestesia, laparoscopia e manipulação genital. Dissertation Universidade Federal do Paraná, Curitiba, Brazil.

Johnson, W.E., E. Eizirik, J. Pecon-Slattery, W.J. Murphy, A. Antunes, E. Teeling, et al. 2006. The late Miocene radiation of modern Felidae: a genetic assessment. Science 311(5757): 73–77.

Kitchener, A.C., C.H. Breitenmoser-Würsten, E. Eizirik, A. Gentry, L. Werdelin, A. Wilting, et al. 2017. A revised taxonomy of the Felidae. The final report of the Cat Classification Task Force of the IUCN/SSC Cat Specialist Group. Cat News 11, pp. 80.

Leite, D.K.V.H. 2009. Avaliações das características histológicas, citológicas, clínicas e seminais de felinos domésticos (*Felis catus*, Linnaeus, 1758) e selvagens (*Leopardus tigrinus*, Schreber, 1775), *Leopardus geoffroyi*, d'Orbign & Gervais, 1843 e *Puma yagouaroundi*, E. Geoffroyi, 1803). Felidae Carnivora. Ph.D. Thesis Universidade Federal Fluminense, Niterói, Brazil.

Lucherini, M., E. Eizirik, T. de Oliveira, J. Pereira and R.S.R. Williams. 2016. *Leopardus colocolo*. The IUCN Red List of Threatened Species 2016: e.T15309A97204446.

Madrigal-Valverde, M., R.F. Bittencourt, A. de Lisboa Ribeiro Filho, M.P. Lents, M.C. de Azevedo and A. Valverde-Abarca. 2019. Biometría testicular y características seminales en felinos neotropicales (Carnivora: Felidae) sometidos a cateterismo uretral. Rev. Biol. Trop. 67(4): 975–988.

Mendonça, L.E.T., C.M. Souto, L.L. Andrelino, W.M.S. Souto, W.L.S. Vieira and R.R.N. Alves. 2012. Conflicts between people and wild animals in semiarid areas of Paraíba and their implications for conservation. SCB 11(2): 185–199.

Micheletti, T., J.L. Brown, S.L. Walker, Z.S. Cubas, P.V. Furtado, S.B. Putman, et al. 2015. The use of altrenogest to avoid hyperestrogenism after eCG-hCG ovulation induction in southern tigrina (*Leopardus guttulus*). Theriogenology 84(4): 575–582.

Morais, R.N., R.G. Mucciolo, M.L. Gomes, O. Lacerda, W. Moraes, N. Moreira, et al. 2002. Brown. Seasonal analysis of semen characteristics, serum testosterone and fecal androgens in the ocelot *(Leopardus pardalis)*, margay (*L. wiedii*) and tigrina (*L. tigrinus*). Theriogenology 57(8): 2027–2041.

Moreira, N., E.L. Monteiro-Filho, W. Moraes, W.F. Swanson, L.H. Graham, O.L. Pasquali, et al. 2001. Reproductive steroid hormones and ovarian activity in felids of the *Leopardus* genus. Zoo Biol. 20(2): 103–116.

Napolitano, C., N. Gálvez, M. Bennett, G. Acosta-Jamett and J. Sanderson. 2015. *Leopardus guigna*. The IUCN Red List of Threatened Species 2015: e.T15311A50657245.

Nascimento, F.O., J. Cheng and A. Feijó. 2020. Taxonomic review of the pampas cat *Leopardus colocola* complex (Carnivora: Felidae): an integrative approach. Zool. J. Lin. Soc. 191(2): 575–611.

Nogueira, G.P. and J.C.R Silva. 1997. Plasma cortisol levels in captive wild felines after chemical restraint. Braz. J. Med. Biol. Res. 30: 1359–1361.

Norkaew, T., J.L. Brown, P. Bansiddhi, C. Somgird, C. Thitaram, V. Punyapornwithaya, et al. 2019. Influence of season, tourist activities and camp management on body condition, testicular and adrenal steroids, lipid profiles, and metabolic status in captive Asian elephant bulls in Thailand. PLoSOne 14(3): e0210537.

Nowell, K. and P. Jackson. 1996. Wild cats: status survey and conservation action plan. Gland: IUCN.

Oliveira, T., A. Paviolo, J. Schipper, R. Bianchi, E. Payan and S.V. Carvajal. 2015. *Leopardus wiedii*. The IUCN Red List of Threatened Species 2015: e.T11511A50654216.

Oliveira, T., T. Trigo, M. Tortato, A. Paviolo, R. Bianchi and M.R.P. Leite-Pitman. 2016. *Leopardus guttulus*. The IUCN Red List of Threatened Species 2016: e.T54010476A54010576.

Palmeira, F.B.L. and W. Barrella. 2007. Conflitos causados pela predação de rebanhos domésticos por grandes felinos em comunidades quilombolas na Mata Atlântica. Biota Neotrop. 7: 119–128.

Paviolo, A., P. Crawshaw, A. Caso, T. de Oliveira, C.A. Lopez-Gonzalez, M. Kelly, et al. 2015. *Leopardus pardalis* (errata version published in 2016). The IUCN Red List of Threatened Species 2015: e.T11509A97212355.

Payan, E. and T. de Oliveira. 2016. *Leopardus tigrinus*. The IUCN Red List of Threatened Species 2016: e.T54012637A50653881.

Paz, R.C., W.F. Swanson, E.A. Dias, C.H. Adania, V.H. Barnabe and R.C. Barnabe. 2005. Ovarian and immunological responses to alternating exogenous gonadotropin regimens in the ocelot (*Leopardus pardalis*) and tigrina (*Leopardus tigrinus*). Zoo Biol. 24(3): 247–260.

Paz, R.C., E.A. Dias, C.H. Adania, V.H. Barnabe and R.C. Barnabe. 2006. Ovarian response to repeated administration of alternating exogenous gonadotropin regimens in the ocelot (*Leopardus pardalis*) and tigrinus (*Leopardus tigrinus*). Theriogenology 66(6–7): 1787–1789.

Paz, R.C., C.H. Adania, V.H. Barnabe and R.C. Barnabe. 2009. Cytogenetic analyses of ocelot (*Leopardus pardalis*) and tigrinus (*Leopardus tigrinus*) oocytes collected after ovarian stimulation. Braz. J. Vet. Res. Ani. Sci. 46(4): 309–316.

Paz, R.C., C.H. Adania, V.H. Barnabe and R.C. Barnabe. 2010. Detecção de estro em jaguatirica (*Leopardus pardalis*) utilizando citologia vaginal. Arq. Bras. Med. Vet. Zootec. 62(6): 2010.

Pereira, J., M. Lucherini and T. Trigo. 2015. *Leopardus geoffroyi*. The IUCN Red List of Threatened Species 2015: e.T15310A50657011.

Peters, F.B., F.D. Mazim, M.O. Favarini, J.B. Soares, T.G. Oliveira, C. Castanõ-Uribe et al. 2016. Caça preventiva ou retaliativa de felinos por humanos no extremo sul do Brasil. II. Conflictos entre felinos y Humanos em América Latina. Castaño-Uribe, Serie Editorial Fauna Silvestre Neotropical. Instituto de Investigación de Recursos Biológicos Alexander von Humboldt (IAvH), Bogotá, DC, Colombia: 311–325.

Rodriguez, M.D., A. Sestelo, C. Buemo, L.D. Ratner, R. Fernandez-Martin and D.F. Salamone. 2021a. Development and Oct4/Cdx2 gene expression of *Puma concolor*, *Leopardus geoffroyi*, and *Panthera onca* hybrid embryos produced using domestic cat oocytes. Reprod. Fertil. Dev. 33: 117–117.

Rodriguez, M.D., A. Gambini, L.D. Ratner, A.J. Sestelo, O. Briski, C. Gutnisky, et al. 2021b. Aggregation of *Leopardus geoffroyi* hybrid embryos with domestic cat tetraploid blastomeres. Reproduction 161(5): 539–548.

Roe, D. 2002. Making a killing or making a living: wildlife trade, trade controls, and rural livelihoods. IIED.

Sestelo, A.J., M.D. Rodriguez, N. Gañan, D.F. Salamone, L. Barañao and E.R.S. Roldan. 2019. Functionality evaluation of two extenders for *Leopardus geoffroyi* sperm cryopreservation by interspecific IVF with domestic cat oocytes. Reprod. Fertil. Dev. 31: 178–179.

Silva, R.C., G.M. Costa, L.M. Andrade and L.R. França. 2010. Testis stereology, seminiferous epithelium cycle length, and daily sperm production in the ocelot (*Leopardus pardalis*). Theriogenology. 73(2): 157–167.

Silva, R.C., G.M. Costa, S.M. Lacerda, S.R. Batlouni, J.M. Soares, G.F. Avelar, et al. 2012. Germ cell transplantation in felids: a potential approach to preserving endangered species. J. Androl. 33(2): 264–276.

Silva, V.H.D. 2014. Avaliação do processo espermatogênico de Gatos-mouriscos (*Puma yagouarundi*, Lacépède, 1809) adultos. Dissertation Universidade Federal de Viçosa, Viçosa, Brazil.

Souza, A., E. Boffo, F. Azzolini, N. Moreira, M.J. Oliveira, R.N. Ribeiro, et al. 2021. Monitoramento endócrino para identificação da puberdade em gato-mourisco (*Herpailurus yagouaroundi*) e gato-do-mato-pequeno (*Leopardus guttulus*). PUBVET 15(2): 1–8.

Stoops, M.A., J.B. Bond, H.L. Bateman, M.K. Campbell, G.P. Levens, T.R. Bowsher et al. 2007. Comparison of different sperm cryopreservation procedures on post-thaw quality and heterologous in vitro fertilisation success in the ocelot (*Leopardus pardalis*). Reprod. Fertil. Dev. 19(5): 685–694.

Swanson, W.F., J.G. Howard, T.L. Roth, K. Kaemmerer, T. Alvarado, M. Burton, et al. 1995. Laparoscopic intrauterine insemination in the ocelot (*Felis pardalis*) after ovarian stimulation with exogenous gonadotropins. Theriogenology 43(1): 331.

Swanson, W.F., J.G. Howard, T.L. Roth, J.L. Brown, T. Alvarado, M. Burton, et al. 1996. Responsiveness of ovaries to exogenous gonadotrophins and laparoscopic artificial insemination with frozen-thawed spermatozoa in ocelots (*Felis pardalis*). J. Reprod. Fertil. 106(1): 87–94.

Swanson, W.F., R.C.R. Paz, R.N. Morais, M.L.F. Gomes, W. Moraes and C.H. Adania. 2002. Influence of species and diet on efficiency of *in vitro* fertilization in two endangered brazilian felids—the ocelot (*Leopardus pardalis*) and tigrina (*Leopardus tigrinus*). Theriogenology 57: 593.

Swanson, W.F., W.E. Johnson, R.C. Cambre, S.B. Citino, K.B. Quigley, D.M. Brousset, et al. 2003. Reproductive status of endemic felid species in Latin American zoos and implications for *ex situ* conservation. Zoo. Biol. 22(5): 421–441.

Swanson, W.F., G.M. Magarey and J.R. Herrick. 2007. Sperm cryopreservation in endangered felids: developing linkage of in situ-ex situ populations. Soc. Reprod. Fertil. Suppl. 65: 417–432.

Tebet, J.M., F. Ferreira de Souza, M.I. Mello Martins, V.H. Chirinéa, J. Candido de Carvalho, F.O. Papa, et al. 2022. Assessment of thawed sperm quality from feline species: Ocelot (*Leopardus pardalis*) and oncilla (*Leopardus gutullus*). Theriogenology 177: 56–62.

Trigo, T.C., T.R. Freitas, G. Kunzler, L. Cardoso, J.C. Silva, W.E. Johnson, et al. 2008. Inter-species hybridization among Neotropical cats of the genus *Leopardus*, and evidence for an introgressive hybrid zone between *L. geoffroyi* and *L. tigrinus* in southern Brazil. Mol. Ecol. 17(19): 4317–4333.

Trigo, T.C., A. Schneider, T.G. de Oliveira, L.M. Lehugeur, L. Silveira, T.R. Freitas, et al. 2013. Molecular data reveal complex hybridization and a cryptic species of neotropical wild cat. Curr. Biol. 23(24): 2528–2533.

Trigo, T.C., F.P. Tirelli, T.R.O. de Freitas and E. Eizirik. 2014. Comparative Assessment of Genetic and Morphological Variation at an Extensive Hybrid Zone between Two Wild Cats in Southern Brazil. PLoSOne 9(9): e108469.

Trindade, F.J., M.R. Rodrigues, H.V. Figueiró, G. Li, W.J. Murphy and E. Eizirik. 2021. Genome-wide SNPs clarify a complex radiation and support recognition of an additional cat species. Mol. Biol. Evol. 38(11): 4987–4991.

Veraguas, D., P.F. Gallegos, F.O. Castro and L. Rodriguez-Alvarez. 2017. Cell cycle synchronization and analysis of apoptosis-related gene in skin fibroblasts from domestic cat (*Felis silvestris catus*) and kodkod (*Leopardus guigna*). Reprod. Dom. Ani. 52(5): 881–889.

Veraguas, D., C. Aguilera, D. Echeverry, D. Saez-Ruiz, F.O. Castro and L. Rodriguez-Alvarez. 2018. Effect of the zona-free aggregation on the developmental competence of kodkod (*Leopardus guigna*) embryos generated by interspecies somatic cell nuclear transfer. Reprod. Fertil. Dev. 31: 134–135.

Veraguas, D., C. Aguilera, D. Echeverry, D. Saez-Ruiz, F.O. Castro and L. Rodriguez-Alvarez. 2020. Embryo aggregation allows the production of kodkod (*Leopardus guigna*) blastocysts after interspecific SCNT. Theriogenology. 158: 148–157.

Villalba, L., M. Lucherini, S. Walker, N. Lagos, D. Cossios, M. Bennett, et al. 2016. *Leopardus jacobita*. The IUCN Red List of Threatened Species 2016: e.T15452A50657407.

Zanin, M., F. Palomares and D. Brito. 2015. What we (don't) know about the effects of habitat loss and fragmentation on felids. Oryx 49(1): 96–106.

Zimmermann, M.L. 2012. Concentrações fecais de estrógenos e progestágenos em jaguatiricas (*Leopardus pardalis*) submetidas a protocolo hormonal para inseminação artificial. Dissertation Universidade Federal do Paraná, Curitiba, Brazil.

Part III

Primates

Part Co-Editor: Sheyla F.S. Domingues
Federal University of Pará
Castanhal, Pará, Brazil
shfarha@ufpa.br or shfarha@gmail.com

Chapter 9

The Aotidae

Frederico Ozanan Barros Monteiro[1]*, Gessiane Pereira da Silva[1],
Thyago Habner de Souza Pereira[1], Rafaela S.C. Takeshita[2],
Julio Cesar Ruiz[3] and Pedro Mayor[1,4]

[1]Postgraduate Program in Animal Health and Production in Amazonia (PPGSPAA),
Federal Rural University of the Amazon (UFRA), Brazil.
Email: frederico.monteiro@ufra.edu.br, gessyane05@hotmail.com,
thyagohabner1@gmail.com

[2]Department of Anthropology, Kent State University, Kent, Ohio, USA.
Email: rtakeshi@kent.edu

[3]Department of Comparative Medicine,
The University of Texas MD Anderson Cancer Center, USA.
Email: ruizjulioc@yahoo.com.ar

[4]Departament de Sanitat i d'Anatomia Animals,
Universitat Autònoma de Barcelona (UAB), Barcelona, Spain.
Email: mayorpedro@hotmail.com

INTRODUCTION

The genus *Aotus* belongs to the Order Primates, family Aotidae, which is the only family within anthropoids characterized by monkeys with nocturnal habit and a vision adapted for low light, including disproportionately large eyes. According to Martins-Junior et al. (2022), they still have a poorly understood taxonomy and geographic range. The number of species varies from one to nine, and there are at least 18 known karyotypes, varying from 2n = 46 to 2n = 58. Traditionally, Hershkovitz (1983), Ford (1994), and Groves (2001) divided the *Aotus* genus into grey and red necked groups. However, a recent study found evidence for nine different taxa, and rejected the existence of the grey and red necked groups (Martins-Junior et al. 2022). These authors suggest a taxonomic reclassification of the genus, which is divided in Northern and Southern groups with respect to

* Corresponding author: frederico.monteiro@ufra.edu.br

the Amazon River. The Northern group includes eight species (*A. lemurinus*, *A. griseimembra*, *A. vociferans*, *A. trivirgatus*, *A. brumbacki*, *A. zonalis*, *A. miconax* and *A. nancymae*) and the Southern group includes four species (*A. nigriceps*, *A. boliviensis*, *A. azarae* and *A. infulatus*) (Figure 9.1). The diversification of owl monkeys into this taxonomic classification is thought to be a result of the Andean uplift and subsequent changes in the Amazon landscape, as well as reproductive isolation due to geographic barriers such as the Amazon-Solimoes and Tapajos Rivers, though most of the Amazonian rivers are small and seemingly do not prevent dispersal (Figure 9.1A). There is further evidence that the Southern group was a results from recent diversification event, and its home range spans through Tapajos, Xingu, Tocantins, and Guapore Rivers to the Cerrado vegetation (Martins-Junior et al. 2022).

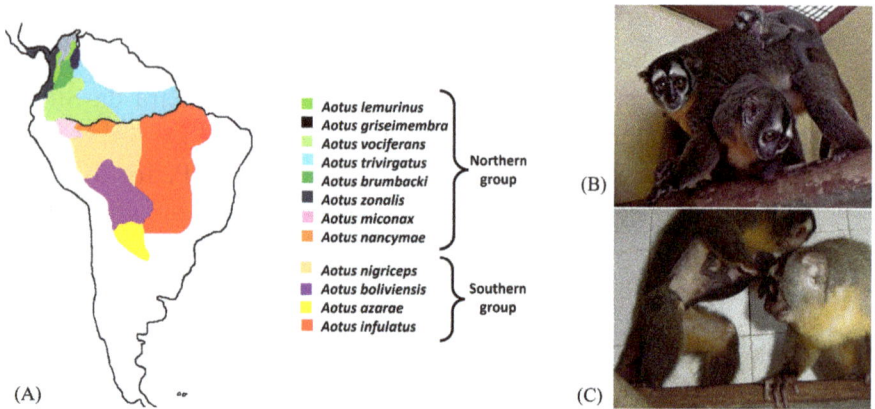

Figure 9.1 (A) Taxonomic distribution of the owl monkey in Northern and Southern groups with respecto to the Amazon-Solimoes River (Martins-Junior et al. 2022). Monogamous family of (B) *Aotus nancymae* (Northern group), from the Instituto Veterinario de Investigaciones Tropicales y de Altura (IVITA, Iquitos, Peru) and (C) *Aotus infulatus* (Southern group) from the National Primate Center (CENP, Ananindeua Pará, Brazil).

These species are known as owl monkeys (Figure 9.1B and 9.1C), night monkeys or douroucoulis (Hershkovitz 1983, Rylands et al. 2012). Owl monkeys can be found in primary and secondary forests, tropical rainforests, and cloud forests, occurring in Panama, Colombia, Venezuela, Ecuador, Peru, Brazil, Bolivia, Paraguay and Argentina. These animals are primarily frugivorous, but can supplement their diet with leaves and flowers, nectar, insects and fungi (Ayres and Deutsch 1982, Fernández-Duque 2012).

Among these species, *Aotus azarae* has cathemeral habits (i.e. day and night activities). During the day, they are hidden in trunks or in the treetops and they start their activities at around 18:00 until 5:00–6:00 the following morning (Ayres and Deutsch 1982, Fernández-Duque 2007, Wright 2011). The temperature interferes with their activity pattern, being lower in colder nights (below 15°C), except for *A. azarae*, which tolerates a wider temperature range. The activity of these primates is related to the availability of lunar brightness, being greater on nights with a full moon (Fernández-Duque 2007, Porto et al. 2015).

Physical characteristics include lacking hair around the eyes and mouth, head-body length of 24 to 47 cm with a large non-prehensile tail in relation to its body size (ranging from 22 a 42 cm), and a bag under the chin, containing the laryngeal air sacs which, when inflated, gives resonance to the sound produced (Ankel-Simons 2007, Porto et al. 2015). These primates do not show visible sexual dimorphism in relation to their size or general conformation, weighing around one kilogram (kg) for both sexes (Ford, 1994). In Brazil, *Aotus infulatus* raised in captivity shows no significant differences between the body mass of males and females 0.980 ± 0.06 and 0.946 ± 0.09 kg (mean $\pm$ standard deviation), respectively (Castro et al. 2003). However, in Argentina, wild females of *Aotus azarae* has a larger weight of 1.246 ± 0.114 and 1.254 ± 0.118 kg in females and males, respectively (Fernández-Duque 2004).

Males and females reach sexual maturity between 12–48 months (Dixson 1994, Fernández-Duque 2007, 2012), or when they weigh between 0.700 and 0.800 kg (Hertig et al. 1976). In captivity, the offspring may remain in the natal group even after reaching sexual maturity, with generally no agonistic behavior from the parents. However, in some groups, serious fights can occur as infants approach sexual maturity, between 18 and 24 months of age. In the wild, individuals who reach sexual maturity migrate from their family groups before the onset of aggression between members (Dixson 1994).

Among females, age at first birth averages around 3, while males sire their first offspring between 5–6.25 years (Juárez et al. 2003, Fernández-Duque 2004, Huck et al. 2011). According to the species and the methodology utilized, the females have a gestation period of 117 to 159 days and an interbirth interval of 370 days (Fernández-Duque and Huntington 2002, Fernández-Duque 2007, Coutinho et al. 2013). Owl monkeys are altricial, and give birth to usually one offspring per pregnancy (Fernández-Duque et al. 2012), which weighs 90–105 g at birth (Auricchio 1995). The average lifespan in captivity for this genus is 20 years (Ross 1991, Aquino and Encarnación 1994), but *A. nancymae* and *A. vociferans* are sexually active up to 12 years of age. These animals have a monogamous mating system (Fernández-Duque et al. 2012), but they may form a small social group composed of six individuals, including the male, female, and the last two offspring. In this group, males have an important involvement in parental care (Porto et al. 2015).

Owl monkeys have been widely used in research on infectious diseases, but studies to assess their reproductive biology are scarce, which limits the use of biotechnologies to increase the reproductive performance of this genus in captivity. Nevertheless, some primate and biomedical facilities have demonstrated effort in maintaining and breeding owl monkeys in captivity, including the National Primate Center (CENP, Ananindeua, Brazil), the Keeling Center for Comparative Medicine and Research at The University of Texas MD Anderson Cancer Center (Bastrop, Texas, USA) and the Instituto Veterinario de Investigaciones Tropicales y de Altura (IVITA, Iquitos, Peru).

This chapter reviews the reproductive biology of the male and female owl monkey, and describes assisted reproduction techniques developed for these species, as well as their limitations.

Reproductive Aspects of the Male Owl Monkey

The testicle has a thick tunica albuginea, poorly vascularized, and with moderately loose collagen (Takeshita et al. 2014). The seminiferous tubules present varied amount of Sertoli cells, spermatogonia, spermatids, and spermatocytes, according to the tubular portion. In all seven animals evaluated, a small amount or no sperm was observed in the seminiferous tubules. Leydig cells have a well-defined rounded morphology, with pale cytoplasm. The epididymis is made up of pseudostratified epithelium with stereocilia, in which large myoid cells and moderate loose connective tissue are observed in the periphery. Several epididymal tubules has cellular debris in the lumen.

According to Martinez and Garcia (2020), the primate sperm diversity reflects the evolutionary divergences of the different primate species and the impact of a combination of variables exerting selection pressures on sperm form, function, and competition. Thus, mating systems, life cycle or ecological variables are some of the important factors driving sperm diversity and explaining variation in terms of sperm morphology, parameters or male sexual characters. In this context, considering the accessories glands, in monogamous species like in *Aotus griseimembra*, in which females mate with one or a few males per breeding season, vesicles tend to be significantly small compared with other species that have multimale/multifemale mating systems, such as observed in *Ateles*, *Brachyteles*, *Cebus*, *Lagothrix*, and *Saimiri* (Dixson 1998).

The study of spermatic production and morphology in Platyrrhine primates, considering evolutive aspects in sperm morphology, shows that in *Aotus* (also *Callithrix* and *Callimico* as a basal genus in the Cebidae family), the spermatozoon has a narrower head and reduced total length, which is proposed as the oldest form for Cebidae. Furthermore, the genus *Callimico* and *Aotus* have the narrowest sperm heads among the Cebidae, in which *Aotus lemurinus* presents the highest elongation and ellipticity values. These aspects are not expected for species that are monogamous and have low sperm competition. On the other hand, they exhibit the shorter spermatozoa when compared to other evaluated species such as *Saimiri collinsi* (Steinberg et al. 2019). However, more evaluations are needed to deepen on the role of sperm morphometry and selection pressure on sperm in the co-evolution of male and female genital morphology.

Abnormalities in sperm morphology in *Aotus lemurinus* have been reported (Nakazato et al., 2015), more frequently on the head than on the neck and tail, and not related to age. Instead, 41.8% of abnormal spermatozoa cases were found in young animals (7 years old) and 36.5% in adults (Nakazato et al. 2015). However, the animals used had not mated with females for a long time, which may be related to the poor conditions of their cauda epididymis and high rates of abnormal spermatozoa. In relation to the endocrine profile of circulating testosterone concentrations in owl monkeys, Dixson and Gardner (1981) described a marked circadian rhythm, with highest levels during the day (24.8 ± 5.3 ng/mL) when the animals are inactive or asleep, and lowest levels during the night (4.7 ± 1.2 ng/mL) when the animals are most active.

Reproductive Biotechniques Applied to the Male Owl Monkey

Despite scarce studied assisted reproductive technology in the male owl monkey, techniques for sperm collection used in closely related species can also be applied in this group. Among platyrrhines, penile vibrostimulation and electrostimulation stand out as the most efficient methods for semen collection (Martinez and Garcia 2020). In addition, epididymal sperm extraction is an alternative technique for obtaining gametes in wild species, and has been described in *Aotus lemurinus* and *Saimiri boliviensis* (Nakazato et al. 2015). However, the choice of technique depends on its practicality and viability in the target group and the current condition of the animals (wildlife x captivity or alive x dead), as it influences the seminal volume obtained and the success of semen collection (Leão et al. 2020).

One cryopreservation technique in *Aotus lemurinus* has been tested using a commercial kit (FERTIUP, Kyudo Co., LTD.) designed for mouse and its efficacy was tested based on sperm motility (Nakazato et al. 2015). Despite the low motility of fresh sperm, the authors demonstrate the efficacy of semen freezing in this species, not being affected by a freeze-thaw cycle. The poor sperm motility and low sperm amount in owl monkeys is typical of monogamous species, which tend to produce fewer sperm than those living in a polygamous system due to the lack of male-male competition. In contrast, Watanabe et al. (2015) studied intracytoplasmic sperm injection (ICSI) into oocytes matured *in vitro*, and observed that sperm motility and viability were very poor. Thus, methods of spermatic cryopreservation of owl monkey semen and sperm viability should be studied more closely.

Reproductive Morphophysiology of the Female Owl Monkey

Regarding the morphology of the female genital organs, the ovaries of owl monkeys are ovoid, creamy-white, and with a smooth surface (Figure 9.2A). Ovarian length varies from 0.47 to 0.89 cm and the width varies between 0.30 and 1.00 cm according to the activity of the corpus luteum (CL), with no differences in the volume between the left and right ovaries (Hertig et al. 1976, Mayor et al. 2015b).

Inside the ovaries, it is possible to observe ovarian follicles (Figure 9.2B) and CL (Figure 9.2C) can be observed at different stages of development. Ovarian follicles can be identified in non-pregnant and pregnant females, with a diameter from 0.25 to 3.45 mm, varying due to different waves of follicular growth. Preantral follicles have smaller diameters and show progressive growth until reaching the minimum preovulatory size of 3.5 cm (Mayor et al. 2015b). However, antral follicles smaller than 2 mm hav e also been observed in females in the luteal and gestational phase. After ovulation, the follicle is rapidly transformed into the CL, responsible for the production of progesterone, increasing the volume of the ovaries to occupy almost the entire stroma, with a diameter ranging from 0.70 mm to 7.90 mm in luteal females, and 2.30 mm to 9.10 mm in pregnant females. Non-pregnant females may have atretic CL and interstitial tissue, while pregnant female of owl monkey does not have accessory CL (Mayor et al. 2015b), unlike Poeppig's woolly monkey (Mayor et al. 2013).

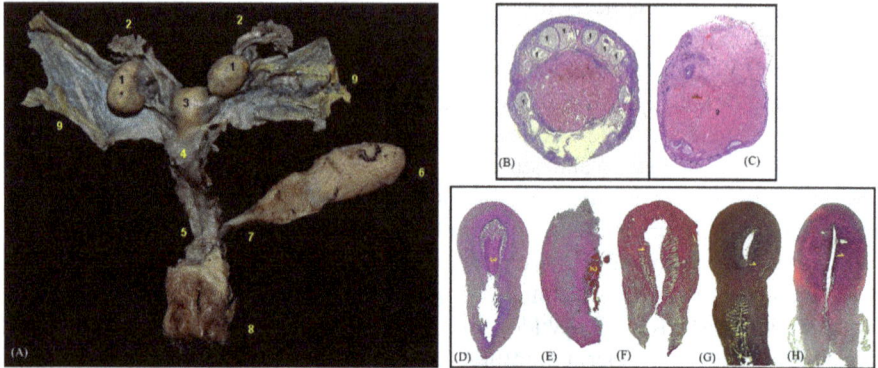

Figure 9.2 (A) Dissection of urogenital organs (dorsal view) in the female owl monkey. 1. Ovaries; 2. Uterine tubes and infundibulum; 3. Uterine body; 4. Cervix; 5. Vagina; 6. Urinary bladder; 7. Urethra; 8. Vulva; 9. Broad ligament of the uterus. (B) Histological sections of the ovary in (B) follicular phase and (C) Luteal phase of the owl monkey. 1. Ovarian follicles; 2. Corpus luteum. (D–H) Uterus histological sections in owl monkey: (D) Pregnant female; (E) Menstruation phase; (F) Advanced follicular-proliferative phase; (G) Luteal-secretory phase; (H) Early follicular-proliferative phase. Progressive growth of the endometrium and endometrial glands is observed in cyclic females, until reaching the menstruation phase in which the functional endometrium sheds. The endometrium of the pregnant female is invaded by fetal membranes, facilitating maternal-fetal contact through placental discs. 1. Functional endometrium; 2. Blood clots; 3. Placental disc. Adapted from Mayor and Plana (2021).

The uterine tube of *Aotus* spp. is very simple and the mesosalpinx does not form the ovarian bursa. The average length of the uterine tubes is 1.60 ± 0.51 cm, with no significant changes during the reproductive phases of the female. The mucosa has columnar ciliated epithelium and is highly pleated. The musculature is formed by two layers, the inner circular and the outer with cells arranged longitudinally (Mayor et al. 2015a).

In non-pregnant females, the uterus is elongated and has a single compartment with a globular bottom and no division into uterine horns, classified as simplex. In these females, the uterus is located entirely in the pelvic cavity. However, in pregnant uterus grows projecting into the abdominal cavity in the fetal phase of pregnancy (Hertig et al. 1976, Mayor et al. 2015a). Concerning cyclic uterine changes (Figure 9.2), females in the follicular phase have a lower density and secretory activity of the endometrial glands (Figure 9.2F). In the luteal phase (Figure 9.2G), under influence of progesterone, the thickness of the endometrium increases, presenting greater vascularization, and proliferation and secretory activity of endometrial glands (Mayor et al. 2015a), responsible for preparing the environment for embryonic development (Mayor et al. 2012).

In the absence of pregnancy, the menstrual phase begins, and the endometrium undergoes degradation with vascular rupture and high remodeling activity. Initially, the blood vessels in the lamina propria of the endometrium are dilated. Then, desquamated endometrial cells shed, and erythrocytes are released into the uterine lumen. However, much of the tissue and blood resorption occurs in the subepithelial region and gross evidence of hemorrhagic loss is mildly observed (Mayor et al. 2019). Thus, the presence of collagen, fibrin, and hemosiderin beneath the endometrial epithelium indicates endometrial epithelial tissue regeneration and remodeling after the menstrual period (Figure 9.2E). As the physiological changes during this period are usually not observed macroscopically, this process is denominated micromenstruation (Mayor et al. 2019). During pregnancy, the endometrial glands shows increased secretive activity, enabling the implantation and nutrition of the embryo in the endometrium (Figure 9.2D). The placenta of the female owl monkey is classified as hemochorial, deciduous, and bi-discoidal, as they have two discs as the maternal-fetal attachment region (Mayor et al. 2015a).

The cervix is small, with thick, firm musculature, and projects into the vaginal canal. The average length of the cervix ranges from 0.18 to 0.65 cm and is shorter in nulliparous females compared to multiparous females (Mayor et al. 2015a). The mucosa has a simple columnar epihtelium with several primary folds and mucus-secreting cells, whose activity is most evident in the early stage of pregnancy (Hertig et al. 1976, Mayor et al. 2015a).

The vagina is the female copulatory organ and extends from the cervix (external orifice of the uterus) to the place where the urethra empties (external urethral orifice). It is a tubular organ with thin walls, long longitudinal folds, and stratified squamous epithelium, which varies according to the reproductive stage. In females in the follicular phase, the vaginal epithelium is thicker and more cornified, which develops as the dominant follicle grows. Thus, this variation allows the use of colpocytology to evaluate the sexual phase in female of the owl monkeys, contributing to the management of the species and the development of assisted reproduction techniques (Mayor et al. 2015a).

Gynecological and Obstetrical Ultrasound in the Female Owl Monkey

Ultrasonography has become a useful tool for the gynecological and obstetrical examination of non-human primates maintained in animal facilities. From the examination of the female reproductive system, it is possible to identify diseases that may interfere with female fertility, diagnose pregnancy and evaluate the development of the offspring throughout the intrauterine period (Seier et al. 2000, Monteiro et al. 2009).

To perform ultrasonography in owl monkeys, it is important to maintain the females' calm through the use of chemical restraint to reduce stress and prevent urination, which would prevent the formation of an acoustic window that is used to conduct sound waves and to locate the uterus (Monteiro et al. 2006). Alternatively, the use of positive reinforcement by offering fruits during the exams has been demonstrated efficient in reducing stress without the use of anesthesia (Monteiro et al. 2011b). The type of transducer and the wave frequency used in the exam is important for the proper visualization of the region to be examined. Owl monkeys have a small abdominal area, and therefore, the use of micro convex transducers is recommended, as the linear type requires a large area of contact with the body surface and forms image artifacts when used in these species (Monteiro et al. 2006, 2009). In owl monkeys, the frequency of 7.5 MHz is suitable for taking measurements and for evaluating the shape, position, and echogenic texture of the uterus (Monteiro et al. 2006, Coutinho et al. 2011).

In the ultrasound examination, the females are placed in the supine position and the abdominal region is shaved to assist in the contact of the transducer with the skin (Coutinho et al. 2013). Generally, the ovaries are located cranial-laterally to the uterus, close to the internal iliac arteries and veins (Coutinho et al. 2013). However, the ovarian position varies according to the bladder and intestinal tract volume, with a high degree of mobility of these organs being observed during the examination, and careful screening is necessary to avoid repeated evaluations of the same ovary (Monteiro et al. 2009). The identification of the follicle is possible due to its anechoic appearance in contrast to the ovarian tissue. In the absence of follicles or the impossibility of their identification, the ovarian parenchyma presents a homogeneous and hypoechoic texture to the adjacent tissue. Concerning the CL, the luteal tissue is isoechoic to the ovarian parenchyma and, therefore, is difficult to identify on ultrasound (Monteiro et al. 2009).

The uterus of owl monkeys can be identified in the center of the pelvic cavity, ventral and longitudinally to the urinary bladder. The uterine body and fundus can be seen on ultrasound; however, the cervical ostium and the vaginal canal are difficult to visualize (Monteiro et al. 2006). In non-pregnant females, it is possible to observe the uterine body with an elongated shape and with homogeneous, isoechoic, or slightly hypoechoic echogenicity compared to the adjacent tissues. On sagittal and transverse planes, a hyperechoic line can be identified in the center of the uterus, corresponding to the inner surfaces of the endometrium (Monteiro et al. 2009).

Early and developmental ultrasonographic signs of pregnancy were described in *Aotus azarae* by ultrasonography (Monteiro et al. 2011a). The early signs of pregnancy were defined as endometrial thickness (ET) and uterine volume (UV) increases. The ET evaluation attempted to detect an echogenicity increase in the uterine fundus region at the thickest point between the two basal layers on the anterior and posterior uterine walls (Figure 9.3A). According to these authors, the ET and UV increase were considered early signs of pregnancy, but were not sufficient to establish a precise gestational diagnosis because these variables may undergo pathological and physiological alterations that affect the endometrial echogenicity and uterine volume when the female is not necessarily pregnant. The larger density of the endometrial glands and the endometrial thickness in *A. nancymae* females (Figure 9.3B) in females in the follicular phase was related to the preparation of embryo implantation (Mayor et al. 2019).

The early diagnosis of pregnancy can be performed between 28 and 38 days post-mating, with the observation of the gestational and the yolk sac in the uterine fundus region (Monteiro et al. 2011a). From the 7th week of pregnancy onwards, the diameter of the gestational sac and the craniocaudal length (CCC) are measurable. Cranial measurements of biparietal and occipitofrontal diameters, along with skull area and circumference, can be measured from the eighth week. Abdominal area and circumference are obtained from the ninth week onwards and the femur length is measured from the tenth week onwards (Monteiro et al. 2011a).

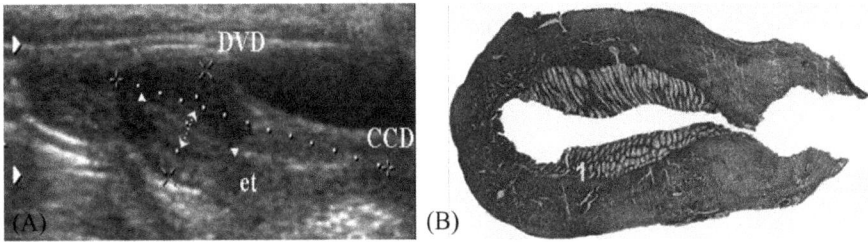

Figure 9.3 (A) Ultrasound image showing the hyperechogenic increase of the endometrial thickness (et – arrows) in *Aotus infulatus*, which are considered early gestational signs. (B) Photomicrography of the uterine body section in the follicular phase with a mid-developed endometrial growth (1) in *Aotus nancymae* female. In both situations, the endometrium was in proliferative phase waiting the embryonic implantation if fecundation occurred.

In the case of embryo mortality, the gestational sac disappears and the uterine volume is reduced, resulting in the presence of intrauterine echoes. In more advanced pregnancies, signs of fetal death are the absence of heartbeat, loss of fetal characteristics on examination, and placenta with reduced echogenicity and areas of spider-web-like placental separation (Schuler et al. 2007).

Reproductive Biotechniques Applied to the Female Owl Monkey

Despite the greater information on the reproductive morphophysiology in the owl female monkeys with respect to males, the application of reproductive techniques biotechnologies in these species is still quite limited. The ultrasound-guided technique for obtaining oocytes was described in owl monkey and is considered a minimally invasive protocol in platyrrhines. It consists of abdominal evaluation by ultrasound, with females maintained in the supine position to identify ovarian follicles larger than 2 mm. After viewing the follicles, the ultrasound gel is removed and an antiseptic with a chlorhexidine-based solution followed by 70% alcohol is applied to the abdominal region. Then, a 1 mL syringe containing 0.1 mL of saline solution is attached to the intravenous needle for follicular puncture with light suction (Rech et al. 2021).

In vitro maturation (IVM) of oocytes has been applied in *Aotus lemurinus*. Human tubal fluid (HTF) medium was used for the species with an incubation time of 22 hours. To perform the IVM, the oocytes were incubated in 100 μl droplets of HTF medium containing 10% fetal bovine serum in 35 mm diameter culture plates kept in an incubator at 37°C (5% CO_2). This technique was successful in owl monkey oocytes, as it resulted in 83.3% and 50% of matured oocytes in excellent and good condition, respectively (Matsumoto et al. 2015).

In addition, intracytoplasmic injection of spermatozoa into owl monkey oocytes has been used to attempt *in vitro* fertilization. Despite the success in achieving 10-cell cleaved embryos, there was no further advancement of embryonic development and maturation to the blastocyst stage. Alternative suggestions to improve this technique include further research on superovulation techniques, selection of better-quality sperm and the culture medium (Watanabe et al. 2015).

FINAL CONSIDERATIONS

In summary, assisted reproduction techniques in the owl monkeys are under-developed and need further testing, particularly in males, which are characterized by poor sperm motility and low sperm count typical of monogamous species. The female reproductive biology has been well described through anatomical and histological studies and provide the basis for assisted reproduction. The advance in ultrasound techniques has been useful in monitoring female gestation, and are promising as a tool to obtain oocytes, however, *in vitro* maturation and fertilization need refinement.

ACKNOWLEDGMENTS

We thank the staff from the National Primate Center (CENP), Federal Rural University of the Amazon (UFRA), National Council of Technological and Scientific Development (CNPq Nos. 305821/2017-2, 400881/2019-5 and 316750/2021-2), and the Coordination for the Improvement of Higher Education Personnel (CAPES-Procad Amazônia No. 21/2018).

REFERENCES

Ankel-Simons, F. 2007. Survey of living primates. pp. 47–160. *In*: F. Ankel-Simons (ed.). Primate Anatomy. Academic Press, Massachusetts, USA.

Aquino, R. and F. Encarnación. 1994. Owl monkey populations in Latin America: field work and conservation. pp. 59–95. *In*: J.F. Baer, R.E. Weller and I. Kakoma (eds). Aotus: The Owl Monkey. Academic Press, Massachusetts, EUA.

Auricchio, P. 1995. Primatas do Brasil. Terra Brasilis, São Paulo.

Ayres, J.M. and L.A. Deutsch. 1982. Os macacos da região amazônica. Rev. Geogr. Univ. pp. 71–82.

Castro, P.H.G., R.R. Valle, H.S. Ferreira, F.O.B. Monteiro, C.R. Valle, R.A. Carvalho, et al. 2003. Biometria e colocação de microchips em macacos da noite (*Aotus azarai infulatus*) no Centro Nacional de Primatas. *In:* Congresso da Associação Brasileira de Veterinários de Animais Selvagens, 7., 2003, São Paulo, SP. Anais. São Paulo: Abravas, 2003. 1 CD-Rom.

Coutinho, L.N., F.O.B. Monteiro, R.S.C. Takeshita, F.L. Miranda Lins e Lins, G.A. da Silva, C. Faturi, et al. 2011. Effect of age and number of parturitions on uterine and ovarian variables in owl monkeys. J. Med. Primatol. 40: 310–316.

Coutinho, L.N., M.B.S. de Brito, F.O.B. Monteiro, R.S. de Andrade, M.E.B. da Conceição, M.R. Feliciano, et al. 2013. Analysis of follicular events in owl monkeys (*Aotus azarai infulatus*) using B-mode and Doppler ultrasound. Theriogenology 80: 99–103.

Dixson, A.F. and J.S. Gardner. 1981. Diurnal variations in plasma testosterone in a male nocturnal primate, the owl monkey (*Aotus trivirgatus*). J. Reprod. Fert. 62: 83–86.

Dixson, A.F., 1994. Reproductive biology of the owl monkey. pp. 113–132. *In:* J.F. Baer, R.E. Weller and I. Kakaoma (eds). Aotus: The Owl Monkey. Academic Press, Massachusetts, USA.

Dixson, A.F. 1998. Sexual selection and the evolution of the seminal vesicles in primates. Folia Primatol. 68: 300–306.

Fernández-Duque, E. and C. Huntington. 2002. Disappearances of individuals from social groups have implications for understanding natal dispersal in monogamous owl monkeys (*Aotus azarai*). Am. J. Primatol. 57: 219–225.

Fernández-Duque, E. 2004. High levels of intrasexual competition in sexually monomorphic owl monkeys (*Aotus azarai*). Folia Primatol. 75: 260.

Fernández-Duque, E. 2007. Aotinae: Social monogamy in the only nocturnal haplorhine. pp. 139–154. *In:* C.J. Campbell, A. Fuentes, K.C. Mackinnon, M. Panger and S.K. Bearder (eds). Primates in Perspective. Oxford University Press, Oxford.

Fernández-Duque, E. 2012. Owl Monkeys *Aotus* spp. in the wild and in captivity. Int. Zoo Yearb. 46: 80–94.

Ford, S.M. 1994. Taxonomy and distribution of the owl monkey. pp. 1–53. *In:* J.F. Baer, R.E. Weller and I. Kakaoma (eds). Aotus: The Owl Monkey. Academic Press, San Diego, California, EUA.

Groves, C. 2001. Primate Taxonomy. Smithsonian Series in Comparative Evolutionary Biology. Smithsonian, Washington, 1.

Hershkovitz, P. 1983. Two new species of night monkeys, genus *Aotus* (Cebidae, Platyrrhini): a preliminary report on *Aotus* taxonomy. Am. J. Primatol. 4: 209–243.

Hertig, A.T., B.R. Barton and J.J. MacKey. 1976. The female genital tract of the owl monkey (*Aotus trivirgatus*) with special reference to the ovary. Lab. Anim. Sci. 26: 1041–1067.

Huck, M.A., M. Rotundo and E. Fernández-Duque. 2011. Growth, development, and age categories in male and female wild monogamous owl monkeys (*Aotus azarai*) of Argentina. Int. J. Primatol. 32: 1133–1152.

Juárez, C., M. Rotundo and E. Fernández-Duque. 2003. Behavioral sex differences in the socially monogamous night monkeys of the Argentinean Chaco. Rev. Etol. 5: 174.

Leão, D.L., W.V. Sampaio, H.L. Queiroz and S.F.S. Domingues. 2020. Biotecnologias da reprodução sob a perspectiva dos machos de primatas neotropicais: contribuições para a conservação de espécies ameaçadas de extinção. Ciênc Anim. 30: 10–22.

Martins-Junior, A.M.G., I. Sampaio, A. Silva, J. Boubli, T. Hrbek, I. Farias, et al. 2022. Out of the shadows: Multilocus systematics and biogeography of night monkeys suggest a

Central Amazonian origin and a very recent widespread southeastward expansion in South America. Mol. Phylogenet. Evol. 170: 107426.

Martinez G. and C. Garcia. 2020. Sexual selection and sperm diversity in primates. Mol. Cell. Endocrinol. 518: 110974.

Matsumoto, T., K. Isobe, K.T. Kusakabe, T. Kuraishi, S. Hattori, C. Nakazato, et al. 2015. Morphological characterization and *in vitro* maturation of follicular oocytes from the owl monkey (*Aotus lemurinus*). J. Mamm. Ova. Res. 32: 103–108.

Mayor, P., M. Bowler and C. López-Plana. 2012. Anatomicohistological characteristics of the tubular genital organs of the female woolly monkey (*Lagothrix poeppigii*). Am. J. Primatol. 74: 1006–1016.

Mayor, P., M. Bowler and C. López-Plana, C. 2013. Ovarian functionality in Poeppig's woolly monkey (*Lagothrix poeppigii*). Anim. Reprod. Sci. 136: 310–316.

Mayor, P., R.S.C. Takeshita, L.N. Coutinho, N. Sánchez, H. Gálvez, C. Ique, et al. 2015a. Functional morphology of the tubular genital organs in the female owl monkey (*Aotus* spp.). J. Med. Primatol. 44: 158–167.

Mayor, P., R.S. Takeshita, L.N. Coutinho, N. Sánchez, H. Gálvez, C. Ique, et al. 2015b. Ovarian function in captive owl monkeys (*Aotus nancymaae* and *A. vociferans*). J. Med. Primatol. 44:187–193.

Mayor, P., W. Pereira, V. Nacher, M. Navarro, F.O.B Monteiro, H.R. El Bizri, et al. 2019. Menstrual cycle in four new word primates: Poeppig's woolly (*Lagothrix poeppigii*), red uacari (*Cacajao calvus*), large-headed capuchin (*Sapajus macrocephalus*) and nocturnal monkey (*Aotus nancymaae*). Theriogenology 123: 11–21.

Mayor, P. and C. López-Plana. 2021. Atlas de Anatomia de Espécies Silvestres Amazônicas: Volume III—Mamíferos. Edufra, Belém.

Monteiro, F.O.B., M.B. de Koivisto, W.R.R Vicente, R. de Amorim Carvalho, C.W. Whiteman, P.H.G. Castro, et al. 2006. Uterine evaluation and gestation diagnosis in owl monkey (*Aotus azarai infulatus*) using the B mode ultrasound. J. Med. Primatol. 35: 123–130.

Monteiro, F.O.B., L.N. Coutinho, E.S.S. Pompeu, P.H.G. de Castro, C.E. Maia, W.L.A. Pereira, et al. 2009. Ovarian and Uterine Ultrasonography in *Aotus azarai infulatus*. Int. J. Primatol. 30: 327–336.

Monteiro, F.O.B., L.N. Coutinho, G.A. Silva, P.H.G. Castro, C.E. Maia, K.S.M. Silva, et al. 2011a. Ultrasound evaluation of pregnancy in owl monkey (*Aotus azarai infulatus*). Anim. Reprod. 8: 40–46.

Monteiro, F.O.B. L.N. Coutinho, R.S.C. Takeshita, G.A. da Silva, K.S.M. da Silva, C.W. Whiteman, et al. 2011b. A protocol for gynecological and obstetric examination of owl monkeys using ultrasound. Rev. Ciênc Agron. 54: 5–11.

Nakazato, C., M. Yoshizawa, K. Isobe, K.T. Kusakabe, T. Kuraishi, S. Hattori, et al. 2015. Morphological characterization of spermatozoa of the night monkey. J. Mamm. Ova. Res. 32: 37–40.

Porto, G.S., M.O. Bordignon, N.R. Reis, A.L. Peracchi and G.L.M. Rosa. 2015. Gênero *Aotus*. pp. 164–176. *In*: N.R. Reis, A.L. Peracchi, C.B. Batista and G.L.M. Rosa (eds). Primatas do Brasil Guia de Campo. Technical Books Editora, Rio de Janeiro.

Rech, F., M.P. Souto, J.W.M. Oliveira, S.K.S.M. da Silva, P.V. Furtado, A.A. Imbeloni, et al. 2021. Ultrasonography-guided oocyte recovery in owl monkeys (*Aotus azarai infulatus*). J. Med. Primatol. 50: 134–137.

Ross, C. 1991. Life history pattern of new world monkeys. Int. J. Primatol. 12: 481–502.

Rylands, A.B., R.A. Mittermeier and J.S. Silva-Júnior. 2012. Neotropical primates: taxonomy and recently described species and subspecies. Inter. Zoo Yearb. 46: 11–21.

Schuler, A.M., V.L. Parks, C.R. Abee and J.G. Scammell. 2007. Ultrasonographic monitoring of a spontaneous abortion in an owl monkey (*Aotus nancymaae*). J. Am. Assoc. Lab Anim. Sci. 46: 74–76.

Seier, J.V., G. Horst, M. Kock and K. Chwalisz. 2000. The detection and monitoring of early pregnancy in the vervet monkey (*Cercopithecus aethiops*) with the use of ultrasound and correlation with reproductive steroid hormones. J. Med. Primatol. 29: 70–75.

Steinberg, E.R., A.J. Sestelo, M.B. Ceballos, V. Wagner, A.M. Palermo and M.D. Mudry. 2019. Sperm morphology in neotropical primates. Animals 9: 839.

Takeshita, R.S.C., F.O.B. Monteiro, F.L.M.L. Lins, R.S. Andrade, G.A. Silva, A.M.C. Cardoso, et al. 2014. Aspectos histológicos dos ovários e testículos de macacos-da-noite, *Aotus azarai infulatus* (Kuhl, 1820). pp. 303–314. *In*: F.C. Passos and J.M.D. Miranda (eds). A Primatologia no Brasil, Editora UFPE, Curitiba, Brasil.

Watanabe, H., T. Matsumoto, M. Nishi, K.T. Kusakabe, T. Kuraishi, S. Hattori, et al. 2015. Intracytoplasmic sperm injection into oocytes matured *in vitro* and early embryonic development in the owl monkey (*Aotus lemurinus*). Reprod. Med. Biol. 15: 183–186.

Wright, P.C. 2011. The neotropical primate adaptation to nocturnality: feeding in the night (*Aotus nigriceps* and *A. azarae*). pp. 369–382. *In*: M. Norconk, A.L. Rosenberger, P.A. Garber (eds). Adaptive Radiations of Neotropical Primates. Plenum Press, New York, EUA.

Chapter 10

The Atelids

Paloma Rocha Arakaki

São Paulo State Wildlife Coordinating Office
Secretariat for the Environment, Infrastructure and Logistics
Avenida Miguel Stéfano, 4241, São Paulo, SP, Brasil, 04301-905
Email: paloma.arakaki@gmail.com.

INTRODUCTION

The Atelidae Family (Gray 1825) is composed of two Subfamilies: Alouattinae and Atelinae. Subfamily Alouattinae has only one genus—*Alouatta,* and Subfamily Atelinae has four—*Ateles*, *Lagothrix*, *Oreonax* and *Brachyteles* (Rylands and Mittermeier 2008). Among Platyrrhines or Neotropical Primates, the Atelids present the heaviest body weights, from over 5 kg to more than 10 kg in the largest species—*Lagothrix* and *Brachyteles* (Ferrari 2008), *Brachyteles* being the largest primate in the Americas.

SUBFAMILY ALOUATTINAE

Howler monkeys (genus *Alouatta*) are the most widely distributed group of Neotropical primates, occurring from 21°N to 30°S in Central and South America (Kowalewski et al. 2015, Doyle et al. 2020). They inhabit diverse forest types, like closed-canopy tropical rain forests, flooded forests, highly seasonal deciduous and semi deciduous woodlands and gallery forests (Crockett 1998). Most species live in cohesive groups with less than 15 animals, usually with just one adult male per group and rarely more than three (Fernandez-Duque et al. 2012).

Howlers, at the genus level, are primarily folivorous. Nonetheless, they are also considered folivore–frugivore primates, whose diets may be affected by rainfall patterns, group size, and forest size that a particular species or population faces, and thus their dietary habits may range from higher folivory to higher frugivory (Dias and Rangel-Negrin 2015). Studies about their anatomy have been published, which focused on dental and cranial anatomy, the hyoid apparatus, and the prehensile tail (Kowalewski et al. 2015). According to the Primate Specialist Group–International Union for Conservation of Nature/Species Survival Commission (IUCN/SSC), there are currently 23 recognized taxa in the genus *Alouatta* (IUCN Primate Specialist Group 2021).

SUBFAMILY ATELINAE

Spider monkeys (genus *Ateles*), wooly monkeys (genera *Lagothrix* and *Oreonax*) and muriquis (genus *Brachyteles*) live in generally large groups that typically contain multiple reproductive—age animals of both sexes (Fernandez-Duque et al. 2012). Unlike howler monkeys that live in cohesive groups, spider monkeys typically live in "fission-fusion" societies, in which social groups split on a daily basis into smaller subgroups that change size and composition frequently (Symington 1990, Wallace 2008). Fission–fusion social dynamics has also been reported in Southern muriquis (*Brachyteles arachnoides*), but not in wooly monkeys, although some within-group association patterns seem to be dynamic, like two or three traveling subgroups that generally did not move independently of each other (Fiore et al. 2009, Coles et al. 2012).

Atelines are more frugivorous than howler monkeys, and among Atelines, *Brachyteles* are the most folivorous. Still in comparison to the howlers, Atelines present larger home ranges and, as already mentioned, a more fluid group association pattern, these different strategies are a characteristic mark of the divergent evolution of these genera (Strier 1992). There are 15 recognized taxa in the genus *Ateles*, seven taxa in the genus *Lagothrix*, one species in the genus *Oreonax*, and two species in the genus *Brachyteles* (Rylands and Mittermeier 2008, IUCN Primate Specialist Group 2021).

REPRODUCTION AND CONSERVATION

Nonhuman primates are our closest living biological relatives; they provide comprehension about human evolution, biology, behavior, and the threat of emerging diseases, besides occupying a central place in tropical biodiversity (Estrada et al. 2017). Unsustainable human activities are now the main cause leading to primate populations' decline and driving primate species to extinction (Estrada et al. 2017, Estrada and Garber 2022). More than 65% of all primate species are listed as Vulnerable, Endangered or Critically Endangered, and 93% of species have declining populations (Estrada and Garber 2022). Regarding South American Primates (173 species), 21.4% are listed as Vulnerable, 12.7% as Endangered and 9.2% as Critically Endangered (IUCN 2022). There is an urgent need to develop Assisted reproductive technologies (ARTs) for endangered wild species (Herrick 2019), including Nonhuman Primates, considering that conservation breeding can be optimized with their application, causing a considerable impact on conservation efforts.

SEMEN COLLECTION

Several methods are described for semen collection in nonhuman primates, but for Atelids, only one has been described so far, as we will see below.

Rectal Probe Electrostimulation (RPE)

Rectal probe electrostimulation (RPE) is the most used method of semen collection for wild animals, and this is also true for primates. RPE is a technique that presents advantages—it is an effective technique, and no training is required. However, the need for anesthesia is a drawback to the application of this method, since it presents significant health risks for the animals, limiting the sampling frequency (Martinez and Garcia 2020). Furthermore, the application of the technique requires trained technicians, with the proper placement of the rectal probe essential to obtain ejaculation.

In Atelids, RPE was used to collect semen in the following species: *Alouatta caraya*, *A. guariba*, *Ateles geoffroyi*, *A. marginatus*, *A. paniscus* and *Brachyteles arachnoides* (Moreland et al. 2001, Hernández-López et al. 2002a, b 2007, 2008a, b, Cerda-Molina et al. 2009, Flores-Herrera et al. 2012, Silva et al. 2013, Valle et al. 2013, 2004, Carvalho et al. 2014, Swanson et al. 2016, Arakaki, unpublished data, Arakaki et al. 2019c).

Finally, although RPE is the most widely applied method for semen collection, especially in medium and large primate species, a safer alternative method is being applied for small species (see the Chapter 11 "The Callitrichids").

SEMEN ANALYSIS AND CHARACTERISTICS

In general, simple methods are used to perform semen analysis of South American primates, many of which are also applied for domestic and other wild animals. Standard semen evaluation and sperm function tests, using techniques like non fluorescent staining, can assess sample characteristics in an accurate and affordable manner, due to their simplicity, cost, and portability (Arakaki et al. 2019a). Although all reported seminal collections and analyses of Atelids have been performed with animals under human care, it is important to note that the methods used to perform the semen analyses could be used in locations with little infrastructure, including research with *in situ* populations.

Data from six Atelidae species' sperm parameters are summarized in Table 10.1. The difference in values found between species and within species is noteworthy. Also noteworthy is the presence of coagulum in seminal samples. Most primate species have seminal samples that coagulate during or after ejaculation, with viscosity ranging from gelatinous semen without the formation of a distinct clot to the formation of a copulatory plug that is compact and maintains its shape (Dixson and Anderson 2002, Valle et al. 2004). This characteristic hampers the handling, analysis, and application of seminal samples from these animals.

Among Neotropical primates in which seminal characteristics have been studied, only *Alouatta caraya* and *A. guariba* present coagulum-free ejaculates, i.e., howler monkeys do not present semen coagulum (Arakaki, unpublished data, Grabner 2016, Moreland et al. 2001, Valle et al. 2004). At the other extreme, the ejaculate of *Brachyteles arachnoides* forms an extremely solid, spongy-looking coagulum (Arakaki et al. 2019c) (Figure 10.1). The method of semen collection can also influence the ejaculates' characteristics (see the Chapter 11 "The Callitrichids").

Different methods for attempts to dissolve the seminal coagulum of neotropical primates have been tested, including in Atelids. In *Ateles geoffroyi*, the enzyme trypsin was tested in different concentrations (0.1%, 1.0% and 5.0%), which led to coagulum dissolution at different periods (0.1% (8.45 h), 1.0% (32 min), 5.0% (21 min), control (16.6 h)), but the higher concentrations led to fragmentation of sperm head and midpiece (Flores-Herrera et al. 2012), causing loss of seminal quality. Currently, there is no effective protocol for dissolving the seminal coagulum in nonhuman primates, which is an obstacle in obtaining the maximum amount of sperm cells contained in the ejaculates of these species.

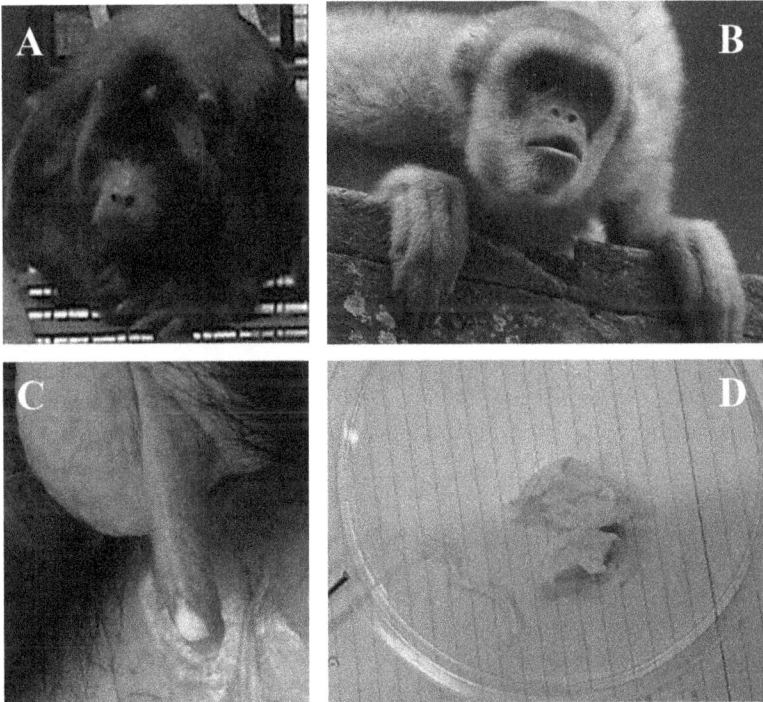

Figure 10.1 Appearance of ejaculates in two South American primates. On the left, (A) A male *Alouatta caraya* subject and (C) semen sample from *A. caraya* in the tip of the penis without coagulum formation. On the right, (B) A male *Brachyteles arachnoides* subject and (D) a copulatory plug on a Petri dish right after ejaculation, from *B. arachnoides*. Photos A: Leonardo Henriques, B: Paulo Gil/FPZSP, C and D: Paloma Arakaki/FPZSP.

Table 10.1 Semen analysis of Atelidae primates

Common name Scientific name	# of animals [# of samples]	Seminal coagulation§	Volume (µl)		pH		Motility (%)		Sperm conc. ($\times 10^6$/mL)		Plasma membrane integrity (%)		References
			Mean	Range	Mean	Range	Mean	Range	Mean	Range	Mean	Range	
Black howler monkey (*Alouatta caraya*)	6 [18]	CFE	81.68	–	8.9	–	73.0	–	97.96	7.0–583	–	–	Moreland et al. 2001
	9 [58]	CFE	90	10–200	8.1	6.5–9	75.7	30–95	649.5	7.5–5400.00	68.3	21–90	Valle et al. 2004
	6 [26]	–	86.80	20–218	7.45	6.6–7.8	60	5–90	726	52.29–2489.36	47.38	19–82	Carvalho et al. 2014
	8 [28]	–	146.31	28–490	7.98	7.3–8.7	72.50	1–90	646.68	32.80–2436.00	62.62	24–94	Carvalho 2012
	2 [2]	CFE	22/30	–	8.1/7.8	–	75/70	–	729.6/3197.7	–	59/70	–	Grabner 2016
Southern brown howler monkey (*Alouatta guariba*)	2 [2]	CFE	20/27	–	8.1/7.2	–	0/60	–	65/1727.3	–	45/81	–	Grabner 2016
	1 [1]	CFE	25	–	8.15	–	30	–	513.75	–	21	–	Arakaki, unpublished data[a]
Black-handed spider monkey (*Ateles geoffroyi*)	3 [20]	4	3300	1000–8000	–	–	64.4	10–94	62.5	18–225	78.5	56–93	Hernández–López et al. 2002a
	3 [44]	4	–	–	–	–	35.5/65.3	–	48.6/61.7	–	60.4/79.3	–	Hernández–López et al. 2002b[b]

(Contd.)

Table 10.1 Semen analysis of Atelidae primates (*Contd.*)

Common name Scientific name	# of animals [# of samples]	Seminal coagulation§	Volume (µl) Mean	Range	pH Mean	Range	Motility (%) Mean	Range	Sperm conc. (×10⁶/mL) Mean	Range	Plasma membrane integrity (%) Mean	Range	References
	3 [169]	4	–	–	8.0	–	52.02	–	–	–	–	–	Hernández–López et al. 2008a
	3 [76]	4	–	–	–	–	53.7/20.2[†]	–	137.9/82.56	–	–	–	Hernández–López et al. 2008b[c]
	3 [129]	–	–	–	–	–	≈23/≈11[*]	–	157.87/145.57	–	–	–	Cerda–Molina et al. 2009[c]
	7 [18]	–	210/181.5/ 297.1/315.7	–	–	–	10.1/15.3/ 40.1/30.4	–	57.8/69.5/ 33.3/33.3	–	29.1/46.5/ 70.8/72.5	–	Flores–Herrera et al. 2012[d]
White–cheeked spider monkey (*Ateles marginatus*)	1 [1]	4	1940	–	8	–	60[a]	–	3.02	–	ND	–	Silva et al. 2013
Black spider monkey (*Ateles paniscus*)	2 [2]	4	1940	–	8	–	20[a]/30[a]	–	3.02	–	ND	–	Silva et al. 2013
Southern muriqui (*Brachyteles arachnoides*)	5 [5]	4	368.33	303–402	7.7	7.4–8.0	80	70–95	144[a]	–	64.5	47–82	Arakaki et al. 2019c

CFE – Coagulum–free ejaculates

§ Ratings of seminal coagulation: 1 – no coagulation; 2 – gelatinous semen with no distinct coagulum; 3 – semen coagulum; 4 – copulatory plug formation, according to Dixon and Anderson 2002

† Semen retrieved from uterus after AI

* Values calculated from a graph

a Absolute values

b Values from different coagulum dissolution treatments. First – stir, second – Trypsin 0.25%

c Values from different seasons of the year. First – dry season, second – rainy season

d Values from different coagulum dissolution treatments. First – control (without Trypsin), second – Trypsin 0.1%, third – Trypsin 1.0%, fourth – Trypsin 5.0%

Semen Cryopreservation

Cryopreservation of semen from South American primates is usually performed using the slow freezing method, with egg-yolk based extenders being widely used, and glycerol, the cryoprotectant of choice. As for all other nonhuman primates, there is no standard protocol for semen cryopreservation, and the variety of protocols and their respective outcomes makes it difficult to reach a consensus regarding an ideal method designed to routinely cryopreserve sperm from these animals (Arakaki et al. 2019b). Among the Atelids, studies have investigated sperm freezing in *Alouatta caraya*, *Ateles paniscus* and *A. marginatus* (Carvalho 2012, Silva et al. 2013). It is important to note that the number of South American primate species studied for semen cryopreservation is still very low. Besides the species mentioned in this chapter, a few studies have investigated sperm freezing in species like marmosets, lion tamarins, capuchin monkeys, and squirrel monkeys (see Chapters 9, 10 and 11).

Testicular Morphometry/Testes Size

Testes produce spermatozoa, hormones, and secretions that compose seminal plasma. Relative testes size (testes mass divided by body mass) may reflect selection pressures related to these functions (Møller 1988). In mating systems where the female has multiple partners, such as in multimale-multifemale systems, the competition between males for reproductive success does not end after copulation. When the female copulates with several partners during the same fertile cycle, competition may continue in the reproductive tract of the female (Harcourt 1995), that is, sperm from two or more males compete to fertilize the oocytes of the same female, or a given set of ova. This phenomenon is known as sperm competition (Parker 1970, 1998).

Larger testes size seems to be one of the evolutionary strategies evolved by males of species where sperm competition occurs (Dixson and Anderson 2004), because they usually contain a greater volume of seminiferous tissue and, therefore, should be associated with the production of more spermatozoa and ejaculates with higher sperm concentrations (Dixson and Anderson 2004, Møller 1988). Another striking trait under positive selection via sperm competition is the copulatory plug/semen coagulum formation (Dixson 2018, Dixson and Anderson 2002)—see the section "Semen analysis and characteristics".

Ideally, testicular morphometry should be performed in every andrological examination, preceding semen collections (Figure 10.2). Total testicular mass of Atelidae primates is summarized in Table 10.2. In South American primates, the reports of testicular volume aimed at describing the testes morpho-physiology, their correlation with mating systems, the presence or absence of reproductive seasonality, and the reproductive strategies of males (Møller 1988, Harcourt et al. 1995, Garber et al. 1996, Guimarães et al. 2003, Araújo and Sousa 2008, Arakaki et al. 2019c, Cardoso et al. 2021).

Table 10.2 Mean values for total testicular mass of Atelidae primates

Common name Scientific name	Total weight (g)	References
Howler monkey (*Alouatta caraya*)	13.65	This chapter
Brown howler monkey (*Alouatta guariba*)	8.32	This chapter
Mantled howler monkey (*Alouatta palliata*)	24.79	(Hrdlicka 1925) (Harcourt et al. 1981)
Black-handed spider monkey (*Ateles geoffroyi*)	13.4	(Schultz 1938) (Harcourt et al. 1981)
Southern muriqui (*Brachyteles arachnoides*)	78.35	(Arakaki et al. 2019c)
Common woolly monkey (*Lagothrix lagotricha*)	11.2	(Schultz 1938) (Harcourt et al. 1981)
Mean Atelidae Family	24.95	

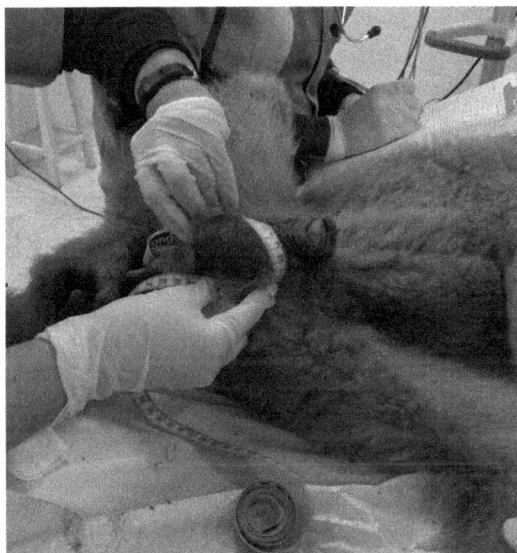

Figure 10.2 *Brachyteles arachnoides* under testicular morphometry during andrological examination, preceding semen collection, at São Paulo Zoo Foundation. Photo: Paulo Gil/FPZSP.

Testicular Tissue Cryopreservation and Culture

Gonadal tissues (ovarian and testicular tissues) are important components of a biobank. Cryopreservation of testicular tissue can be performed from mature or prepubertal males, immediately postmortem or after orchiectomy (Silva et al. 2020). Regarding South American primates, most reports concern Callitrichids (see the Chapter 11 "The Callitrichids"), largely due to studies in a human-relevant preclinical model in cases of boys who receive chemotherapy or radiation therapy to eradicate cancer and face the possibility of being infertile when they reach adulthood due to the gonadotoxicity of the regimens used (Jahnukainen et al. 2012).

MONITORING AND CONTROL OF THE REPRODUCTION

Non-invasive Reproductive Endocrine Monitoring

The application of assisted reproductive technologies to control female reproduction involves ovarian cycle synchronization, between oocyte donors and embryo recipients, for example, and ovarian stimulation, to increase the population of oocytes prior to retrieval (Kropp et al. 2017).

Ovarian cycle control has not been applied yet in Atelid primate species. However, researchers studying *in situ* and *ex situ* populations using non-invasive techniques for reproductive endocrine monitoring have provided information on female reproductive biology, essential for the application of ARTs. Steroid hormones produced in the gonads and adrenals circulate in blood, are filtered by liver and kidneys and are excreted as hormones metabolites in urine and feces (Palme 2005, Taylor 1971, Ziegler et al. 2009). The use of non-invasive techniques allows the measurement of hormones indirectly, via feces and urine, the two most used matrices for non-invasive reproductive endocrine monitoring. Ovarian cycle lengths for Atelidae primates are described in Table 10.3.

Table 10.3 Ovarian cycle lengths for Atelidae primates

Common name Scientific name	Days (mean ± SD)	Range	References
Brown howler monkey (*Alouatta guariba*)	16 ± 0.52	10–22	(Silvestre et al. 2017)
Colombian red howler monkey (*Alouatta seniculus*)	29.5 ± 1.5	29–31	(Herrick et al. 2000)
Mantled howler monkey (*Alouatta palliata*)	16.1 ± 4.3	–	(Jones 1985)
Brown-headed spider monkey (*Ateles fusciceps*)	–	20–22	(Hodges et al. 1981)
Black-handed spider monkey (*Ateles geoffroyi*)	25.3 ± 3.0	22–28	(Hernández-López et al. 1998)
Southern muriqui (*Brachyteles arachnoides*)	21.0 ± 5.4	16–38	(Strier and Ziegler 1997)
Northern muriqui (*Brachyteles hypoxanthus*)	21.3 ± 5.2	–	(Strier et al. 2003)
Woolly monkey (*Lagothrix lagotricha poeppigii*)	21.9 ± 3.0	17–39	(Abondano et al. 2022)

ARTIFICIAL INSEMINATION

Artificial insemination, one of the simplest ARTs, consists of semen deposition into the female tract. It does not involve oocyte manipulation and male gametes can be fresh or frozen-thawed. In Atelids, artificial insemination was successfully applied in *Ateles geoffroyi*. Five females from thirteen to twenty-seven years old were inseminated intravaginally with fresh semen. The thinnest part of a conic sterilized plastic tube was introduced into the vagina and advanced until it reached the cervix. Semen fractions were transferred to the tube—the liquid fraction flowed down the tube wall and the coagulum was pushed down inside the tube using a cotton swab, until it reached the end of the tube.

One female became pregnant after the first round of inseminations and delivered a healthy infant. She was inseminated again, became pregnant once more, and subsequently aborted. A second female aborted, apparently due to an intramural uterine leiomyoma. Other two females stopped menstruating for a few months, then restarted menstruating—these females may have been pregnant and aborted (Hernández-López et al. 2007).

Other Assisted Reproductive Technologies

In vitro fertilization, intracytoplasmic sperm injection (ICSI), embryo transfer, and nuclear transfer are ARTs that have been successfully applied in some Old World primate species, such as *Macaca mulatta, M. fascicularis, M. fuscata,* and *Gorilla gorilla* (Bavister et al. 1984, Pope et al. 1997, Torii et al. 2000, St. John and Schatten 2004, Morichika et al. 2012, Liu et al. 2018). However, in South American primates, not all of these techniques have been described, and the described ones have been applied to a small number of species, mainly those of interest in biomedical research (see the Chapter 11 "The Callitrichids").

FINAL CONSIDERATIONS

ARTs in non-human primates, in both Old and New World primates, are especially applied in model species, like macaques and marmosets. These techniques are widely applied in laboratory rodents, livestock and humans. It is important to mention that investments on basic reproductive biology research, in its various aspects—anatomy, physiology, endocrinology, were necessary, as well as knowledge in cryobiology and molecular biology involving the target species. The complexity of the accomplishment of these techniques added to the diversity of species found in the Order Primates and particularities found in each one of them are a great challenge to those who work with Assisted Reproduction aiming at the conservation of endangered species. Even artificial insemination, which is one of the most basic ARTs, is dependent on a deep knowledge of reproductive aspects involving both males and females. We must invest in research to apply ARTs in the conservation of South American Primates. Finally, we emphasize that no ART can save a species from extinction, rather ARTs are auxiliary tools for conservation that can promote the maintenance of genetic diversity in both *in situ* and *ex situ* populations of a particular species.

ACKNOWLEDGMENTS

I thank São Paulo Zoo Foundation for providing the pictures for this chapter and Dr. Barbara R.A. Lindsey for English proofreading.

REFERENCES

Abondano, L.A., T.E. Ziegler and A. Di Fiore. 2022. Reproductive endocrinology of wild female woolly monkeys (*Lagothrix lagotricha poeppigii*) during puberty, ovarian cyclicity, and pregnancy. Am. J. Primatol. 84: 1–13. https://doi.org/10.1002/ajp.23303

Arakaki, P.R., P.A.B. Salgado, J.D.A. Losano, M.H. Blank, M. Nichi and R.J.G. Pereira 2019a. Assessment of different sperm functional tests in golden-headed lion tamarins (*Leontopithecus chrysomelas*). Am. J. Primatol. 81: 1–9. https://doi.org/10.1002ajp.23034.

Arakaki, P.R., P.A.B. Salgado, J.D.A. Losano, D.R. Gonçalves, R.R. Valle, R.J.G. Pereira, et al. 2019b. Semen cryopreservation in golden-headed lion tamarin, *Leontopithecus chrysomelas*. Am. J. Primatol. 81: e23071. https://doi.org/10.1002ajp. 23071

Arakaki, P.R., P.A.B. Salgado, R.H.F. Teixeira, F.B. Rassy, M.A.B.V. Guimarães and R.R. Valle. 2019c. Testicular volume and semen characteristics in the endangered southern muriqui (*Brachyteles arachnoides*). J. Med. Primatol. 48: 244–250. https://doi.org/10.1111/jmp.12418

Araújo, A. and M.B.C. Sousa. 2008. Testicular Volume and Reproductive Status of Wild *Callithrix jacchus*. Int. J. Primatol. 29: 1355–1364. https://doi.org/10.1007/s10764-008-9291-4

Bavister, B.D., D.E. Boatman, K. Collins, D.J. Dierschke and S.G. Eisele. 1984. Birth of rhesus monkey infant after *in vitro* fertilization and nonsurgical embryo transfer. Proc. Natl. Acad. Sci. U.S.A. 81, 2218–2222. https://doi.org/10.1073/pnas.81.7.2218

Cardoso, D.L., D.A.A. Guimarães, P. Mayor, M.A.P. Ferreira, H.L.T. Dias, R.F. Espinheiro, et al. 2021. Reproductive biology of owl (*Aotus* spp.) and capuchin (*Sapajus* spp.) monkeys. Anim. Reprod. Sci. 227: 106732. https://doi.org/10.1016/j.anireprosci.2021.106732

Carvalho, F.M. 2012. Evaluation of egg yolk-based and soy-lecithin-based extenders for the cryopreservation of semen from Alouatta caraya [thesis]. Universidade Estadual Paulista.

Carvalho, F.M., P.R. Arakaki, M. Nichi, J.A.P.C. Muniz, J.M.B. Duarte and R.R. Valle. 2014. Evaluation of sperm quality in successive regular collections from captive black-and-gold howler monkeys (*Alouatta caraya*). Anim. Reprod. 11: 11–18.

Cerda-Molina, A.L., L. Hernández-lópez, R. Chavira-Ramírez, M. Cárdenas and R. Mondragon-Ceballos. 2009. Seasonality of LH, testosterone and sperm parameters in spider monkey males (*Ateles geoffroyi*). Am. J. Primatol. 71: 427–431. https://doi.org/10.1002/ajp.20671

Coles, R.C., P.C. Lee and M. Talebi. 2012. Fission—fusion dynamics in southern Muriquis (*Brachyteles arachnoides*) in continuous Brazilian Atlantic Forest. Int. J. Primatol. 33: 93–114. https://doi.org/10.1007/s10764-011-9555-2

Crockett, C.M. 1998. Conservation biology of the genus alouatta. Int. J. Primato 19: 549–578. https://doi.org/10.1023/A:1020316607284

Dias, P.A. and A. Rangel-Negrin. 2015. Diets of howler monkeys. pp. 21–56. *In:* M.M. Kowalewski, P.A. Garber, L. Córtes-Ortiz, B. Urbani and D. Youlatos (eds). Howler Monkeys: Behavior, Ecology, and Conservation. Springer New York, NY, USA.

Dixson, A.F. and M.J. Anderson. 2002. Sexual selection, seminal coagulation and copulatory plug formation in Primates. Folia Primatol. 73: 63–69. https://doi.org/10.1159/000064784

Dixson, A.F. and M.J. Anderson. 2004. Sexual behavior, reproductive physiology and sperm competition in male mammals. Physiol. Behav. 83: 361–371. https://doi.org/10.1016/j.physbeh.2004.08.022

Dixson, A.F. 2018. Copulatory and postcopulatory sexual selection in primates. Folia Primatol. 89: 258–286. https://doi.org/10.1159/000488105

Doyle, E.D., I. Prates, I. Sampaio, C. Koiffmann, W. Araujo, Jr, S., Carnaval, et al. 2020. Molecular phylogenetic inference of the howler monkey radiation (Primates: *Alouatta*). Primates 62: 177–188. https://doi.org/10.1007/s10329-020-00854-x

Estrada, A., P.A. Garber, A.B. Rylands, C. Roos, E. Fernandez-duque, A. Di Fiore, et al. 2017. Impending extinction crisis of the world's primates: Why primates matter. Sci. Adv. 3: 1–16. https://doi.org/10.1126/sciadv.1600946

Estrada, A. and P.A. Garber. 2022. Principal drivers and conservation solutions to the impending primate extinction crisis: introduction to the special issue. Int. J. Primatol. 43: 1–14. https://doi.org/10.1007/s10764-022-00283-1

Fernandez-Duque, E., A. Di Fiore and M. Huck. 2012. The behavior, ecology, and social evolution of new world monkeys. pp. 43–64. *In:* J.C. Mitani, J. Call, P.M. Kappeler, R.A. Palombit and J.B. Silk (eds). The Evolution of Primate Societies. The University of Chicago Press, Chicago, IL, USA.

Ferrari, S.F. 2008. Predation risk and antipredator strategies. pp. 251–277. *In:* P.A. Garber, A. Estrada, J. Bicca-Marques, E.W. Heymann and K.B. Strier (eds). South American Primates—Comparative Perspectives in the Study of Behavior, Ecology and Conservation. Springer New York, NY, USA.

Fiore, A. Di, A. Link, C.A. Schmitt and S.N. Spehar. 2009. Dispersal patterns in sympatric woolly and spider monkeys: integrating molecular and observational data. Behaviour 146: 437–470. https://doi.org/10.1163/156853909X426345

Flores-Herrera, H., D.G. Acuña-Hernández, J.A. Rivera-Rebolledo, M.A. González-Jiménez, A.Z. Rodas-Martínez and W.F. Swanson. 2012. Effect of increasing trypsin concentrations on seminal coagulum dissolution and sperm parameters in spider monkeys (*Ateles geoffroyi*). Theriogenology 78: 612–619. https://doi.org/10.1016/j.theriogenology.2012.03.007

Garber, P.A., L. Moya, J.D. Pruetz and C. Ique. 1996. Social and seasonal influences on reproductive biology in male moustached tamarins (*Saguinus mystax*). Am. J. Primatol. 38: 29–46. doi.org/10.1002/(SICI)1098-2345(1996)38:1<29::AID-AJP4>3.0.CO;2-V

Grabner, A.P. 2016. Análise comparativa dos aspectos da ultraestrutura do espermatozoide de Mico-Leão-de-Cara-Dourada (*Leontopithecus chrysomelas*) e Bugio (Alouatta caraya e Alouatta guariba clamitans) 1–125.

Guimarães, M.A.B.V., C.A. De Oliveira and R.C. Barnabe. 2003. Seasonal variation in the testicular volume of capuchin monkeys (*Cebus apella*) in captivity. Folia Primatol. 74: 54–6. https://doi.org/10.1159/000068394

Harcourt, A.H., P.H. Harvey, S.G. Larson and R.V. Short. 1981. Testis weight, body weight and breeding system in primates. Nature 293: 55–56.

Harcourt, A.H. 1995. Sexual selection and sperm competition in primates: What are male genitalia good for? Evol. Anthropol. Issues, News, Rev. 4: 121–129. https://doi.org/10.1002/evan.1360040404

Harcourt, A.H., A. Purvis and L. Liles. 1995. Sperm competition: mating system, not breeding season, affects testes size of primates. Funct. Ecol. 9: 468. https://doi.org/10.2307/2390011

Hernández-López, L., L. Mayagoitia, C. Esquivel-Lacroix, S. Rojas-Maya and R. Mondragón-Ceballos. 1998. The menstrual cycle of the spider monkey (*Ateles geoffroyi*). Am. J. Primatol. 44: 183–195. https://doi.org/10.1002/(SICI)1098-2345 (1998)44:3<183::AID-AJP1>3.0.CO;2-S

Hernández-López, L., G.C. Parra, A.L. Cerda-Molina, S.C. Pérez-Bolaños, V. Díaz Sánchez and R. Mondragón-Ceballos. 2002a. Sperm quality differences between the rainy and dry seasons in captive black-handed spider monkeys (*Ateles geoffroyi*). Am. J. Primatol. 57: 35–41. https://doi.org/10.1002/ajp.1086

Hernández-López, L., G.C. Parra, A.L. Cerda-molina, S.C. Pérez-bolaños and R. Mondragón-ceballos. 2002b. Digestion by trypsin enhances assessment of sperm parameters in the black-handed spider monkey (*Ateles geoffroyi*). Lab. Primate Newsl. 41.

Hernández-López, L., A.L. Cerda-Molina, D.L. Páez-Ponce, S. Rojas-Maya and R. Mondragón-Ceballos. 2007. Artificial insemination in black-handed spider monkey (*Ateles geoffroyi*). Theriogenology 67: 399–406. https://doi.org/10.1016/j.theriogenology.2006.06.016

Hernández-López, A.L. Cerda-Molina, D.L. Páez-Ponce and R. Mondragón-Ceballos. 2008a. The seminal coagulum favours passage of fast-moving sperm into the uterus in the black-handed spider monkey. Reproduction 136: 411–421. https://doi.org/10.1530/REP-08-0135

Hernández-López, L., A.L. Cerda-Molina, L.D. Páez-Ponce and R. Mondragón-Ceballos. 2008b. Seasonal emission of seminal coagulum and in vivo sperm dynamics in the black-handed spider monkey (*Ateles geoffroyi*). Theriogenology 69: 466–472. https://doi.org/10.1016/j.theriogenology.2007.10.016

Hodges, J.K., B.A. Gulick, N.M. Czekala and B.L. Lasley. 1981. Comparison of urinary oestrogen excretion in South American primates. Reproduction 61: 83–90. https://doi.org/10.1530/jrf.0.0610083

Herrick, J.R., G. Agoramoorthy, R. Rudran and J.D. Harder. 2000. Urinary progesterone in free-ranging red howler monkeys (*Alouatta seniculus*): Preliminary observations of the estrous cycle and gestation. Am. J. Primatol. 51: 257–263. https://doi.org/10.1002/1098-2345(200008)51:4<257::AID-AJP5>3.0.CO;2-6.

Herrick, J.R. 2019. Assisted reproductive technologies for endangered species conservation: developing sophisticated protocols with limited access to animals with unique reproductive mechanisms. Biol. Reprod. 100: 1158–1170. https://doi.org/10.1093/biolre/ioz025

Hrdlicka, A. 1925. Weight of the brain and of the internal organs in American monkeys. With data on brain weight in other apes. Am. J. Phys. Anthropol. 8: 201–211. https://doi.org/10.1002/ajpa.1330080207

IUCN Primate Specialist Group. 2021. Primates of the Neotropics, March 2021. PSG/IUCN/SSC. 2021 [WWW Document]. URL http://www.primate-sg.org/primates_of_neotropics/ (accessed 7.15.22).

IUCN. 2022. The IUCN Red List of Threatened Species. Version 2022-1 [WWW Document]. URL https://www.iucnredlist.org/search/stats?query=PRIMATES&searchType=species (accessed 7.15.22).

Jahnukainen, K., J. Ehmcke, M. Nurmio and S. Schlatt. 2012. Autologous ectopic grafting of cryopreserved testicular tissue preserves the fertility of prepubescent monkeys that receive sterilizing cytotoxic therapy. Cancer Res. 72: 5174–5178. https://doi.org/10.1158/0008-5472.CAN-12-1317

Jones, C. 1985. Reproductive patterns in mantled howler monkeys: estrus, mate choice and copulation. Primates 26: 130–142.

Kowalewski, M.M., P.A. Garber, L. Cortés-Ortiz, B. Urbani and B. Youlatos. 2015. Why is it important to continue studying the anatomy, physiology, sensory ecology, and evolution of Howler Monkeys? pp. 3–17. *In*: M.M. Kowalewski, P.A. Garber, L. Cortés-Ortiz, B. Urbani and D. Youlatos (eds). Howler Monkeys—Adaptive Radiation, Systematics, and Morphology. Springer New York, NY, USA.

Kropp, J., A. Di Marzo and T. Golos. 2017. Assisted reproductive technologies in the common marmoset: an integral species for developing nonhuman primate models of human diseases. Biol. Reprod. 96: 277–287. https://doi.org/10.1095/biolreprod.116.146514

Liu, Z., Y. Cai, Y. Wang, Y. Nie, C. Zhang, Y. Xu, et al. 2018. Cloning of macaque monkeys by somatic cell nuclear transfer. Cell 172: 881–887.e7. https://doi.org/10.1016/j.cell.2018.01.020

Martinez, G. and C. Garcia. 2020. Sexual selection and sperm diversity in primates. Mol. Cell. Endocrinol. 518. https://doi.org/10.1016/j.mce.2020.110974

Møller, A.P. 1988. Ejaculate quality, testes size and sperm competition in primates. J. Hum. Evol. 17: 479–488. https://doi.org/10.1016/0047-2484(88)90037-1

Moreland, R.B., M.E. Richardson, N. Lamberski and J.A. Long. 2001. Characterizing the reproductive physiology of the male southern black howler monkey, *Alouatta caraya*. J. Androl. 22: 395–403.

Morichika, J., C. Iwatani, H. Tsuchiya, S. Nakamura, T. Sankai and R. Torii. 2012. Triplet pregnancy in a cynomolgus monkey (*Macaca fascicularis*) after double embryo transfer. Comp. Med. 62: 69–72.

Palme, R. 2005. Measuring fecal steroids: Guidelines for practical application. Ann. N. Y. Acad. Sci. 1046: 75–80. https://doi.org/10.1196/annals.1343.007

Parker, G.A. 1970. Sperm competition and its evolutionary consequences in the insects. Biol. Rev. 45: 525–567. https://doi.org/10.1111/j.1469-185X.1970.tb01176.x

Parker, G.A. 1998. Sperm competition and the evolution of ejaculates: towards a theory base. pp. 3–54. *In*: Sperm Competition and Sexual Selection. Academic Press. https://doi.org/10.1016/b978-012100543-6/50026-x

Pope, C.E., BL. Dresser, N.W. Chin, J.H. Liu, N.M. Loskutoff, E.J. Behnke, et al. 1997. Birth of a western lowland gorilla (*Gorilla gorilla*) following *in vitro* fertilization and embryo transfer. Am. J. Primatol. 41: 247–260. https://doi.org/10.1002/(SICI)1098-2345(1997)41:3<247::AID-AJP6>3.0.CO;2-X

Rylands, A.B. and R.A. Mittermeier. 2008. The diversity of the New World primates—an annotated taxonomy. pp. 23–54. *In:* P.A. Garber, A. Estrada, J.C. Bicca-Marques, E.W. Heymann and K.B. Strier (eds). South American Primates—Comparative Perspectives in the Study of Behavior, Ecology and Conservation. Springer New York, NY, USA.

Schultz, A.H. 1938. The relative weight of the testes in primates. Anat. Rec. 72: 387–394. https://doi.org/10.1002/ar.1090720310

Silva, A.M. da, A.F. Pereira, P. Comizzoli and A.R. Silva. 2020. Cryopreservation and culture of testicular tissues: an essential tool for biodiversity preservation. Biopreserv. Biobank. 18: 235–243. https://doi.org/10.1089/bio.2020.0010

Silva, K., H. RIbeiro, R.R. Valle, J, Sousa, A. Silva and E. Barbosa. 2013. Efeitos do trimetilaminoetano (TES) e ringer lactato em sêmen de macacos-aranha mantidos em cativeiro (Ateles paniscus e A. marginatus). Arq. Bras. Med. Vet. e Zootec. 65: 934–937. https://doi.org/10.1590/S0102-09352013000300044

Silvestre, T., E.S. Zanetti, J.M.B. Duarte, F.G. Barriento, Z.M.B. Hirano, J.C. Souza, et al. 2017. Ovarian cycle of southern brown howler monkey (*Alouatta guariba clamitans*) through fecal progestin measurement. Primates 58: 131–139. https://doi.org/10.1007/s10329-016-0561-z

St. John, J.C. and G. Schatten. 2004. Paternal mitochondrial DNA transmission during nonhuman primate nuclear transfer. Genetics 167: 897–905. https://doi.org/10.1534/genetics.103.025049

Strier, K.B. 1992. Atelinae adaptations: behavioral strategies and ecological constraints. Am. J. Phys. Anthropol. 88: 515–524. https://doi.org/10.1002/ajpa.1330880407

Strier, K.B. and T.E. Ziegler. 1997. Behavioral and endocrine characteristics of the reproductive cycle in wild muriqui monkeys, *Brachyteles arachnoides*. Am. J. Primatol. 42: 299–310. https://doi.org/10.1002/(SICI)1098-2345(1997)42:4<299::AID-AJP5>3.0.CO;2-S

Strier, K.B., J.W. Lynch and T.E. Ziegler. 2003. Hormonal changes during the mating and conception seasons of wild northern muriquis (*Brachyteles arachnoides hypoxanthus*). Am. J. Primatol. 61: 85–99. https://doi.org/10.1002/ajp.10109

Symington, M.M. 1990. Fission-fusion social organization in ateles and pan. Int. J. Primatol. 11: 47–61.

Taylor, W. 1971. The excretion of steroid hormone metabolites in bile and feces. Vitamins & Hormones 29: 201–285. https://doi.org/10.1016/S0083-6729(08)60050-3

Torii, R., Y. Hosoi, Y. Masuda, A. Iritani and H. Nigi. 2000. Birth of the Japanese Monkey (*Macaca fuscata*) infant following *in-vitro* fertilization and embryo transfer. Primates 41: 39–47. https://doi.org/10.1007/BF02557460

Valle, R.R., M.A.B.V. Guimarães, J.A.P.C. Muniz, R. Barnabe and W. Vale. 2004. Collection and evaluation of semen from captive howler monkeys (*Alouatta caraya*). Theriogenology 62: 131–138. https://doi.org/10.1016/j.theriogenology.2003.08.004

Valle, R.R., P.R. Arakaki, F.M. Carvalho, J.A.P.C. Muniz, C.L.V. Leal and M. García-Herreros. 2013. Identification of sperm head subpopulations with defined pleiomorphic characteristics in ejaculates of captive Goeldi's monkeys (*Callimico goeldii*). Anim. Reprod. Sci. 137. https://doi.org/10.1016/j.anireprosci.2012.12.007

Wallace, R.B. 2008. Towing the party line: territoriality, risky boundaries and male group size in spider monkey fission—fusion societies. Am. J. Primatol. 70: 271–281. https://doi.org/10.1002/ajp.20484

Ziegler, T.E., K.B. Strier and S. Van Belle. 2009. The reproductive ecology of south american primates: ecological adaptations in ovulation and conception. pp. 191–210. *In*: P.A. Garber, A. Estrada, J.C. Bicca-Marques, E.W. Heymann and K.B. Strier (eds). South American Primates, Developments in Primatology: Progress and Prospects. Springer New York, NY, https://doi.org/10.1007/978-0-387-78705-3

Chapter 11

The Callitrichids

Paloma Rocha Arakaki
São Paulo State Wildlife Coordinating Office
Secretariat for the Environment, Infrastructure and Logistics
Avenida Miguel Stéfano, 4241, São Paulo, SP, Brasil, 04301-905.
Email: paloma.arakaki@gmail.com

INTRODUCTION

The Callitrichidae Family (Gray 1821) is composed of 8 species—*Callithrix, Cebuella, Callibella, Mico, Saguinus, Leontopithecus, Callimico* and *Leontocebus* (Rylands et al. 2016, Rylands and Mittermeier 2008). Among Platyrrhines or Neotropical Primates, the Callitrichids have the smallest body weights, and the smallest New World monkeys are the pygmy marmosets (*Cebuella pygmaea*), that have an average body mass of 100 g (Fernandez-Duque et al. 2012).

Certainly, the most studied species of the Callitrichidae family is the *Callithrix jacchus*, the common marmoset, due to its importance as a model for several areas of biomedical research. Therefore, this chapter will focus mainly on this species, and will mention the other Callitrichids opportunely.

Callithrix jacchus

Common marmosets are endemic to Northeast Brazil. Like other Callitrichids, they live in cohesive social groups of 3 to 15 individuals (Ferrari and Lopes Ferrari 1989, Schiel and Souto 2017). In populations under human care, sexual maturity begin around one year of age for both males and females, the average age of first conception is 2.49 years, and the average gestation period is 143–144 days (Tardif et al. 2003).

The mating system of marmosets and tamarins is characterized as non-exclusive monogamy; extrapair copulations may occur, with both polyandry and polygyny (Digby 1999). Sexual behavior within groups is typically restricted to breeding individuals—the socially dominant male and female in the group (Digby 1999). This is maintained through the inhibition of reproductive behavior and physiology (suppression of ovulation) in subordinate females by the dominant female (Digby 1999, Fernandez-Duque et al. 2012, Schiel and Souto 2017).

Marmosets and tamarins are the only anthropoid primates that ovulate multiple oocytes during a single cycle, and litter size generally ranges from 1 to 4, twins being the most common litter size (Tardif et al. 2003). Within Callitrichids, parents, other relatives like older siblings, and other group members share the care of the offspring, and group members often compete for the opportunity to carry them. This cooperative breeding has not been reported for any other primate except humans (Fernandez-Duque et al. 2012).

Another interesting characteristic is the presence of postpartum ovulation in marmosets, which typically results in conception and successful delivery. The first ovulation generally occurs 10 to 20 days after delivery (Tardif et al. 2003).

REPRODUCTION AND CONSERVATION

As seen in the chapter "The Atelids" Chapter 10, from the 173 species, 21.4% are listed as Vulnerable, 12.7% as Endangered and 9.2% as Critically Endangered (IUCN 2022). In the Family Callitrichidae, some species are considered Endangered, like *Callithrix aurita, Leontopithecus caissara* and *L. chrysopygus* (Rezende et al. 2020, de Melo et al. 2021b, Ludwig et al. 2021), and others are considered Critically Endangered, *C. flaviceps, Saguinus bicolor* and *S. oedipus* (de Melo et al. 2021a, Gordo et al. 2021, Rodríguez et al. 2021). Little or nothing is known about the reproductive biology of these species and the vast majority of other Callitrichid species. Again, there is an urgent need to study, develop and apply Assisted Reproductive Technologies (ARTs) not only for the Primates of this Family, but also for all Primates of the Order.

SEMEN COLLECTION

Several methods are described for semen collection in nonhuman primates. Beyond doubt, Rectal Probe Electrostimulation (RPE) is the most used method for primates. However, for small-bodied primates like Callitrichids, the Penile Vibrostimulation Technique (PVS) is an alternative to RPE which offers many benefits. PVS is a technique that was originally developed for men with spinal cord injuries (Sønksen et al. 1994). Only 3 years later, the technique was already being applied in *Saimiri boliviensis,* a New World primate from the Cebidae Family (Yeoman et al. 1997), and soon after, the technique was used in *C. jacchus* (Kuederling et al. 2000).

For 25 years, PVS has been used to collect semen from South American Primates. This method provides ejaculates through normal ejaculation via activation of the ejaculatory reflex in the thoracolumbar area of the spinal cord

(Sønksen and Ohl 2002). Vibration is applied to the penis of physically restrained animals, so no sedation is required. However, animals must be conditioned to the procedure (Figure 11.1).

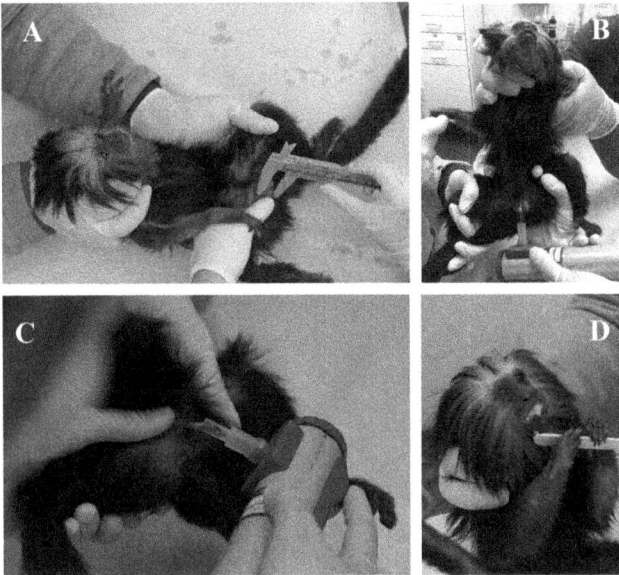

Figure 11.1   Penile vibrostimulation in *Leontopithecus chrysomelas.* (A) Physical restraint of the male to perform testicular morfometry (B) application of vibratory stimuli on the penis (C) the moment of ejaculation and seminal coagulum inside the tube and (D) reward between stimuli. Photos: Paulo Gil/FPZSP.

In Callitrichids, PVS has been applied in marmosets and tamarins (Table 11.1). Besides the benefit of not using sedation, the production of natural ejaculations, free from contamination (from urine, which may happen during RPE), provides higher quality samples, with better motility and total number of sperm, compared to RPE (Yeoman et al. 1998, Schneiders et al. 2004). The main challenge with this technique is its use in larger species, but this possibility should be tested (Arakaki et al. 2019d). Other challenges are the need for trained technicians to apply the technique and the training of animals through positive reinforcement technique, which would limit its use for animals kept under human care.

Spermatozoa can also be obtained by epididymal extractions, mainly collected opportunistically following castrations or in *post-mortem* contexts. For the conservation of endangered species, this might be the last chance to keep the genetics of an important individual, by freezing its germ cells in a biobank. In *C. jacchus*, studies on the physiology of epididymal spermatozoa were performed, like the structure and maturation, maturation of sperm motility, and profile of changes in sperm kinematics during maturation in the epididymis (Moore et al. 1984, Yeung et al. 1996). Besides that, epididymal spermatozoa were used for *in vitro* fertilization and for artificial insemination, the first reporting births following AI in this species (Wilton et al. 1993, Morrell et al. 1997).

Table 11.1 Semen analysis of Callitrichidae primates

Common name Scientific name	# of animals [# of samples]	Semen collection method	Seminal coagulation§	Volume (l)		pH		Motility (%)		Sperm conc. (× 10⁶/mL)		Plasma membrane integrity (%)		References
				Mean	Range	Mean	Range	Mean	Range	Mean	Range	Mean	Range	
Goeldi's monkey (*Callimico goeldii*)	6 [13]	RPE	2	26.9	–	7.61	–	83.33	–	143.18	–	36.38	–	(Arakaki et al. 2017)
Common marmoset (*Callithrix jacchus*)	16 [77]	RPE	2	30	8–85	7.6	7.0–8.1	48*	10–76	–	–	–	–	(CUI et al. 1991)
	9 [45]	RPE	2	40.2	8–170	7.51	7.0–8.0	47.4	–	27.3×10⁴	2.2–71×10⁴	–	–	(Cui 1996)
	11 [40]	VW	2	–	–	–	–	68.5	20–95	6.07	0.3–40.1 × 10⁶/100 L	84.1	38–94.5	(Kuederling et al. 1996)
	10 [31]	PVS	2	31.9	–	–	–	59.6	–	33.7	–	74.6	–	(Kuederling et al. 2000)
	10 [40]	VW	2	–	–	–	–	74.8	50–95	31.94	1–114	84.7	61–97	(Morrell et al. 1996)
	10 [15]	RPE	2	–	–	–	–	70.7	40–90	20	1–66	80.96	47–95	(Morrell et al. 1996)
	13 [99]	PVS	2	≈38	–	–	–	≈58	–	≈625	–	≈72	–	(Schneiders et al. 2004)
	10 [46]	RPE	2	≈21	–	–	–	≈42	–	≈125	–	≈60	–	(Schneiders et al. 2004)
	14 [41]	PVS	2	26.6	5–55	7.6	7.4–8.0	82.7	20–95	1062.59	36.18–3557.3	84.7	21–98	(Valle et al. 2014)
	8 [18]	PVS	2	15.6	–	7.5	–	51.11	–	1238.58	–	48.56	–	(Arakaki et al. 2019a)

(Contd.)

Table 11.1 Semen analysis of Callitrichidae primates (*Contd.*)

Common name Scientific name	# of animals [# of samples]	Semen collection method	Seminal coagulation§	Volume (l) Mean	Volume (l) Range	pH Mean	pH Range	Motility (%) Mean	Motility (%) Range	Sperm conc. (× 10^6/mL) Mean	Sperm conc. (× 10^6/mL) Range	Plasma membrane integrity (%) Mean	Plasma membrane integrity (%) Range	References
Black–tufted–ear marmoset (*Callithrix penicillata*)	10 [6]	PVS	2	16.8	–	7.5	–	56.67	–	1473.85	–	62.83	–	(Arakaki et al. 2019a)
Golden–headed lion tamarin (*Leontopithecus chrysomelas*)	7 [39]	PVS		–	–		–	91.90	85–95	–	–	93	85–99	(Arakaki et al. 2019c)
	7 [54]	PVS	2	76.7	–	7.5	–	88.52	–	284.31	–	92.76	–	(Arakaki et al. 2020)
White–footed tamarin (*Saguinus leucopus*)	15 [10]	PVS	2	24	–	7.5	–	97.1*	–	87.617 × 10^4	–	93.7	–	(Poches et al. 2013)
Moustached tamarin (*Saguinus mystax*)	17 [–]	RPE	2	–	–	–	–	28.4	–	179.8	–	–	–	(Harrison and Wolf 1985)
	6 [–]	RPE	2	–	–	–	–	45.8	–	265.1	–	–	–	(Harrison and Wolf 1985)
	18 [–]	RPE	2	–	–	–	–	56.0	–	183.2	–	–	–	(Harrison and Wolf 1985)

§ Ratings of seminal coagulation: 1 – no coagulation; 2 – gelatinous semen with no distinct coagulum; 3 – semen coagulum; 4 – copulatory plug formation, according to Dixon and Anderson 2002.

RPE – Rectal Probe Electrostimulation, VW – Vaginal washing, PVS – Penile vibratory stimulation

* Values of progressive motility

Obtaining ejaculated sperm after natural mating by vaginal washing (VW) in *C. jacchus* was described by one group of researchers in Germany (Kuederling et al. 1996, Morrell et al. 1996). Copulations at specific times were achieved by separating the males from their females for a certain period and subsequently putting them back together. After observed mating, vaginal washing was performed on non-sedated females. When comparing RPE with VW, the proportions of motility (74.8 vs. 70.7%), plasma membrane integrity (84.7 vs. 81%)—see Table 11.1, and morphologically normal sperm (91.9 vs. 87.6%), ejaculates were not affected by the semen collection method (Morrell et al. 1996).

SEMEN ANALYSIS AND CHARACTERISTICS

Semen analysis of South American primates is performed by simple methods, using techniques that assess samples' characteristics in an accurate and affordable manner, due to their simplicity, cost, and portability (Arakaki et al. 2019b).

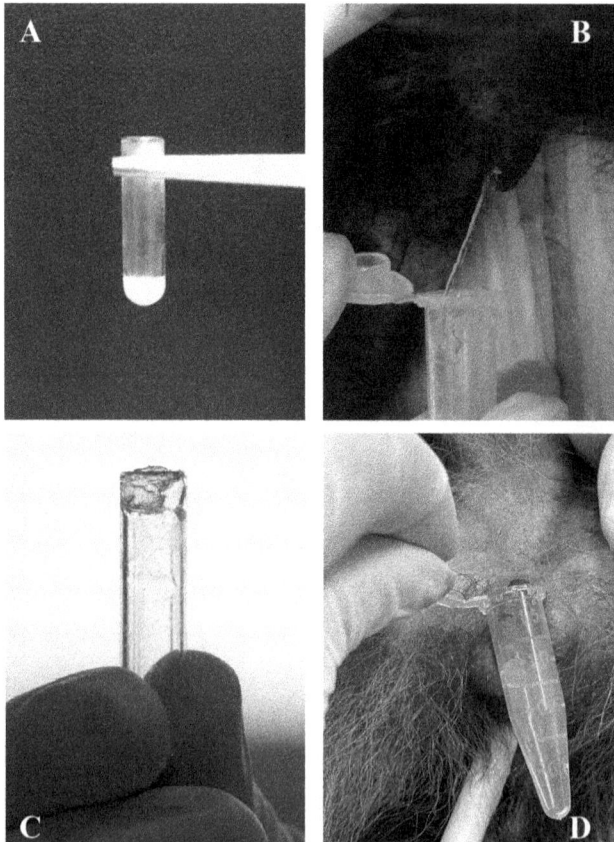

Figure 11.2   Seminal coagulation in Callitrichids. Samples from A–*Callithrix jacchus,* B–*Callimico goeldii,* C–*Leontopithecus chrysomelas* and D–*L. crhrysopygus.* Photos: A, B and D–Paloma Arakaki/FPZSP, C–Paulo Gil/FPZSP.

Data on sperm parameters of six Callitrichidae species are summarized in Table 11.1. In this Family, regarding semen coagulum presence, all species studies to date are scored by (Dixson and Anderson 2002) as 2, in a 4-point scale, meaning a "gelatinous semen with no distinct coagulum". During semen collection in *C. goeldii*, semen coagulum was noticed during ejaculation (Figure 11.2). The same was observed for *L. chrysomelas* (Arakaki et al. 2020) and *L. chrysopygus* (Arakaki 2021). In the scale proposed by (Dixson and Anderson 2002), these 3 species should be considered under score 3—presenting semen coagulum.

Incubation of *C. goeldii* semen in fresh coconut water at 37°C for 30 min was performed in an attempt to dissolve the seminal coagulum, and partial or complete liquefaction of the samples were obtained (Arakaki et al. 2017). Different attempts to dissolve the seminal coagulum of other Neotropical primates have been tested. However, no effective protocol is available.

As mentioned in the chapter "The Atelids", the method of semen collection may also influence the ejaculates characteristics, besides samples' quality. In *L. chrysomelas*, seminal samples obtained by RPE showed both liquid and coagulated fraction (Vidal et al. 2007). When PVS was used, all ejaculates were formed only by the coagulum, without the liquid fraction (Arakaki et al. 2020). When devising research or assisted reproduction program, the choice of semen collection method must consider aspects such as the target species, the expertise of the technical team, and the local conditions where the procedures will be performed.

SEMEN CRYOPRESERVATION

Semen cryopreservation from South American primates is usually performed using the slow freezing method, with egg-yolk based extenders being widely used, and glycerol, the cryoprotectant of choice. As for all other non-human primates, there is no standard protocol for semen cryopreservation. Table 11.2 presents the extender and percentages of glycerol used in each Callitrichid species as well as semen collection method, and motility for fresh and frozen thawed sperm.

TESTICULAR MORPHOMETRY/TESTES SIZE

Testicles are an important part of the male genital system, producing spermatozoa, hormones, and secretions that compose seminal plasma (see the chapter "The Atelids").

Total testicular mass of Callitrichidae primates is shown in Table 11.3. In South American primates, the reports of testicular volume intended to describe the testes morpho-physiology, their correlation with mating systems, the presence or absence of reproductive seasonality, and the reproductive strategies of males (Araújo and Sousa 2008, Møller 1988, Harcourt et al. 1995, Garber et al. 1996, Guimarães et al. 2003, Arakaki et al. 2019d, Cardoso et al. 2021).

Table 11.2 Semen cryopreservation from Callitrichidae Family primate's species

Common name Scientific name	Extender	Cryoprotectant	Semen collection method	Fresh sperm motility (%)	Thawed sperm motility (%)	References
Common marmoset (*Callithrix jacchus*)	TES–TRIS	5% GLY and EY	VW or RFE	65 (at least)	33 (at least)	(Morrell et al. 1998)
	TES–TRIS	3% GLY and EY	RFE	–	43.5 ± 21.5	(O'Brien et al. 2003)
	TES–TRIS	4% GLY and EY	PVS	63.93 ± 7.21	7.38 ± 2.14	(Valle 2007)
	TES–TRIS	4% and 6% GLY and EY	PVS	51.11 ± 4.19	3.11 ± 0.58 (4% GLY) 2.00 ± 0.40 (6% GLY)	(Arakaki et al. 2019a)
Black–tufted–ear marmoset (*Callithrix penicilatta*)	TES–TRIS	4% and 6% GLY and EY	PVS	56.67 ± 2.11	2.83 ± 0.87 (4% GLY) 0.83 ± 0.31 (6% GLY)	(Arakaki et al. 2019a)
Red–bellied tamarin (*Saguinus labiatus*)	TES–TRIS	5% GLY and EY	RFE	80 to 90	50 to 70	(Sankai et al. 1997)
Golden–headed lion tamarin (*Leontopithecus chrysomelas*)	TYB→	6% GLY and EY	PVS	89.38 ± 1.18	60.00 ± 2.32	
	BotuBOV→	(?) GLY and EY	PVS		53.08 ± 2.73	(Arakaki et al. 2019c)

TES, 2-[(2-hydroxy-1,1-bis(hydroxymethyl)ethyl) amino]ethanesulfonic acid; TRIS, tris(hydroxymethyl)aminomethane; GLY, glycerol; EY, egg yolk; VW: vaginal washing; RFE, recovery from epididymis; PVS, penile vibrostimulation.

Table 11.3 Mean values for total testicular mass of Callitrichidae primates

Common name Scientific name	Total weight (g)	References
Common marmoset (*Callithrix jacchus*)	1.30	(Harcourt et al. 1981)
Western pygmy marmoset (*Cebuella pygmaea*)	0.33	(Soini 1993)
Golden headed lion tamarin (*Leontopithecus chrysomelas*)	1.02	This chapter
Black lion tamarin (*Leontopithecus chrysopygus*)	0.87	This chapter
Golden lion tamarin (*Leontopithecus rosalia*)	0.88	This chapter
Saddleback tamarin (*Leontocebus fuscicollis*) (*Saguinus fuscicollis*)	1.53	(Dixson and Anderson 2004)
Cotton-top tamarin (*Saguinus oedipus*)	3.4	(Harcourt et al. 1981)
Mean Callitrichidae Family	1.33	

TESTICULAR TISSUE CRYOPRESERVATION AND CULTURE

Cryopreservation of testicular tissue is an option to recover male germplasm in cases of unexpected deaths. This technology is not widely applied, but new reports have become more frequent. Testicular tissue cryopreservation is particularly challenging when compared to cell cryopreservation and requires better permeation of the cryoprotectant because of the structural composition and presence of different cell types that differ in size and membrane permeability (Pothana et al. 2016a). Adult testicular tissue from *C. geoffroyi* and immature testicular tissue from *C. jacchus* were cryopreserved during studies to analyze cryoprotectants and testicular germ cell transplantation, known as testicular grafting (Luetjens et al. 2008, Pothana et al. 2016b).

Culture of testicular tissue to provide conditions for tissues to resume spermatogenesis can be achieved by *in vivo* or *in vitro* culture (Silva et al. 2020). The transplant of *C. jacchus* testicular fragments into immunodeficient nude mice (xenografting) promoted spermatogenesis until spermatocytes (Schlatt et al. 2002).

MONITORING AND CONTROL OF THE REPRODUCTION

Ovarian Cycle Synchronization and Stimulation

Female reproduction control pursuing the application of assisted reproductive technologies involves ovarian cycle synchronization, between oocyte donors and embryo recipients, for example, and ovarian stimulation, to increase the population of oocytes prior to retrieval (Kropp et al. 2017). In Table 11.4, ovarian cycle lengths for Callitrichidae primates are summarized.

In *C. jacchus*, ovarian cycle can be regulated by the administration of prostaglandin $F_{2\alpha}$(cloprostenol) to induce luteolysis (Summers et al. 1985), and the induction of ovulation can be achieved by intramuscular administration of human chorionic gonadotropin (hCG) after prostaglandin-induced luteal regression (Hodges et al. 1987). *L. rosalia* presents reduced sensitivity to the cloprostenol compared to *C. jacchus*, but a single midcycle injection of 1.6 µg effectively induced luteolysis and synchronized ovulation (Monfort et al. 1996).

Controlled ovarian stimulation integrates recombinant human follicle-stimulating hormone (r-hFSH) into ovarian synchronization protocols to stimulate follicular development, followed by timed hCG, at the end of the protocol to promote oocyte maturation. For Callitrichids, follicular aspiration is usually performed by laparotomy or after removal of the ovaries in biomedical research (Kropp et al. 2017).

Table 11.4   Ovarian cycle lengths for Callitrichidae primates

Common name Scientific name	Days (mean ± SD)	Range	References
Goeldi's monkey (*Callimico goeldii*)	23.9 ± 0.4	23–26	(Dettling 2002)
Common marmoset (*Callithrix jacchus*)	30.1 ± 3.8	24–41	(Harding et al. 1982)
Wied's marmoset (*Callithrix kuhlii*)	24.9 ± 0.6	-	(Ziegler et al. 2009)
Pygmy marmoset (*Cebuella pygmea*)	33.3 ± 5.5	26–37	(Ziegler et al. 2009)
Golden lion tamarin (*Leontopithecus rosalia*)	18.5 ± 0.3	16–21	(Monfort et al. 1996)
Golden-headed lion tamarin (*Leontopithecus chrysomelas*)	21.5 ± 2.5	18–25	(De Vleeschouwer et al. 2000)
Saddleback tamarin (*Saguinus fuscicollis*)	25.5 ± 1.0	19–31	(Ziegler et al. 2009)
Cotton-headed tamarin (*Saguinus oedipus*)	23.2 ± 1.4	18–31	(Ziegler et al. 2009)

Oocyte *in vitro* Maturation

It is known that the length of oocyte culture needed prior to fertilization depends on the oocyte maturity status at the time of aspiration. Protocols for *in vitro* maturation of *Callithrix* oocytes varies greatly, regarding gas concentrations, media components, and supplements (Gilchrist et al. 1997, Grupen et al. 2007, Sasaki et al. 2009 , Tkachenko et al. 2010, Takahashi et al. 2014). In *Saguinus labiatus*, fifty-nine germinal vesicle stage oocytes were collected from 4 ovaries of 4 females, and after *in vitro* maturation, were inseminated by frozen-thawed sperm; however, no oocytes were fertilized (Sankai et al. 1997). The scarcity in the number of research groups, the lack of financial investment in research, and the difficulty in accessing animals are factors that hinder the advance in this area.

OVARIAN CORTEX CRYOPRESERVATION

In *C. jacchus*, ovarian tissue was cryopreserved using a slow freezing protocol with dimethylsulfoxide (DMSO), 1,2-propanediol (PrOH), or ethylene glycol (EG). DMSO provided a higher percentage of morphologically normal primordial ($26.2 \pm 2.5\%$) and primary follicles ($28.1 \pm 5.4\%$) compared with PrOH (12.2 ± 3.0 and $5.4 \pm 2.1\%$, respectively) (von Schönfeldt et al. 2011a). The same research group grafted frozen-thawed ovarian cortex fragments into immunodeficient nude mice, with development of follicles up to secondary and preantral stages (von Schönfeldt et al. 2011b).

Vitrification of whole neonatal ovaries from *C. jacchus* resulted in no differences in morphology between fresh (control) and frozen-thawed ovaries, no significant difference in the number of oocytes with normal morphology evaluated by histology was found, and no significant difference in DNA damage analyzed by comet assay. However, there was a decrease in the viability of vitrified follicles (Motohashi and Ishibashi 2016).

It was also in *C. jacchus* that, for the first time, full-sized maturable oocytes from primary and early secondary follicles were produced from fragments of ovarian cortex and isolated follicles, of a non-rodent species (Nayudu et al. 2003).

ARTIFICIAL INSEMINATION

The first successful AI, resulting in pregnancy and production of offspring in *C. jacchus,* was obtained by utilizing epididymal sperm, which was deposited in the cervix around the time of expected ovulation, using either 3, 2, or 1 inseminations (Morrell et al. 1997). One year later, the same research group also successfully used fresh and cryopreserved sperm for AI (Morrell et al. 1998). Only fifteen years later, in another report on AI in *C. jacchus*, researchers used semen collected by PVS. Instead of using a medical grade vibrator, they used an electric toothbrush with a frequency of 117 or 100 Hz, fitted with a silicone tube (Ishibashi and Motohashi 2013). Successful AI was reported in the same species combining PVS and the swim-up method (Takabayashi et al. 2015).

In Vitro Fertilization and Intracytoplasmic Sperm Injection (ICSI)

In vitro fertilization (IVF, the co-incubation of male and female gametes) and intracytoplasmic sperm injection (ICSI) are two techniques that have already been successfully performed in *C. jacchus*. For IVF, addition of sperm to a final concentration ranges from 0.5 to 5.0×10^6 sperm/mL, and the length of gamete co-incubation varies between 14 and 24 h (Kropp et al. 2017). Birth of healthy offspring following ICSI as a fertilization method were reported for the species (Takahashi et al. 2014). In this study, oocytes were matured *in vitro* and ICSI was performed at various time points, after extrusion of the first polar body. To investigate *in vivo* development of embryos followed by ICSI, 6-cell- to 8-cell-stage embryos and blastocysts were non-surgically transferred into recipient marmosets. Four healthy offspring were produced and grew normally.

EMBRYO TECHNOLOGY

Embryo Culture and Development

Embryonic development is comparatively slower in *C. jacchus* when compared to other nonhuman primates and humans. Embryo culture conditions vary greatly across studies, which generally utilize a sequential or two-step medium system. The medium used in the initial and later culture period also varies, and includes TL to CMRL-1066, G1.2 to G2.2 and ISM1 to ISM2. Embryos are cultured in the first medium for about 48 h (days 1–3 of culture) and then are transferred to the second medium, being replaced approximately every 2 days (Kropp et al. 2017).

Embryo Transfer

Transfer of marmoset embryos to recipient females has been achieved by both surgical and non-surgical methods (Ishibashi et al. 2013, Lopata et al. 1988, Marshall et al. 1997, Summers et al. 1987). In a study conducted 35 years ago, surgically transferred embryos by delivery of fresh morula and blastocysts to the uterine lumen resulted in a pregnancy rate of 66.6% at day 40, with ten young born from 11 transferred embryos (Summers et al. 1987). In the same study, nonsurgical embryo transfer (ET) was also performed—17 embryos to nine recipients. One pregnancy was established but the conceptus was aborted on Day 46.

In the first successful non-surgical ET in *C. jacchus*, vaginal dilation was achieved by the introduction and removal of a lubricated glass speculum, then the cervix was visualized by illumination provided by an otoscope and a cannula and stylet were placed at cervix and gently guided into the uterus by transabdominal palpation. A needle containing 1 or 2 embryos was passed through the cannula, which were delivered into the uterine lumen by gentle pressure on the plunger (Marshall et al. 1997).

Currently, in marmosets non-surgical ET, a glass tube is inserted into the vagina and fitted to the uterine cervix of the recipient female, which is anesthetized. A long cannula, combined with a stainless-steel stylet, is inserted into the uterus via the glass tube. Then, the inner stainless-steel stylet is removed and replaced with a dummy catheter. Abdominal ultrasound is performed to confirm that the dummy catheter entered the uterus. The dummy catheter is removed and replaced by the catheter containing the embryos (one to three embryos), with a Hamilton syringe. Embryos are simultaneously injected into the uterus, and ultrasound is used to confirm the location of the embryos by the air accompanying the pre-implantation embryos (Kurotaki and Sasaki 2017).

GENETIC ENGINEERING

The first transgenic non-human primate showing not only the transgene expression in somatic tissues, but also germline transmission of the transgene with the full, normal development of the embryo was generated in *C. jacchus*, by virus infection (Sasaki et al. 2009). Another study using the same principle—lentiviral vectors—produced symptomatic transgenic marmosets of the polyQ diseases, a human age-associated neurodegenerative diseases, one more example demonstrating *C. jacchus* as models for investigations of regenerative medicine and gene therapy (Tomioka et al. 2017).

Generation of target gene knockout or knockin marmosets were successfully achieved by innovative genome editing technologies, such as zinc-finger nucleases (ZFNs), transcription activator-like effector nucleases (TALENs), and clustered regularly interspaced short palindromic repeat/CRISPR-associated protein 9 (CRISPR/Cas9) (Abe et al. 2021, Sato et al. 2016). Currently, genetic engineering of *C. jacchus* and non-human primates in general is being used only in biomedical research, to generate ideal animal models for human diseases. However, genetic engineering tools have the potential to make significant contributions to the fields of evolutionary biology, ecology, and conservation (Phelps et al. 2020).

FINAL CONSIDERATIONS

South American primate species conservation faces an ever-growing number of dangers, such as habitat loss, hunting, populational bottlenecks, among others. The use of ARTs in order to preserve those species' genetic diversity is even more urgent. *C. jacchus* has been one of the most common model species of Neotropical primate for biomedical studies and development of ARTs, and therefore, it is the species among the Callitrichids that has the most available information on reproductive biology. This knowledge should be used in favor of the other species of this Family, acting as a starting point for studies in other Genera.

ACKNOWLEDGMENTS

I thank São Paulo Zoo Foundation for providing the pictures for this chapter and Dr. Barbara R.A. Lindsey for English proofreading.

REFERENCES

Abe, Y., H. Nakao, M. Goto, M. Tamano, M. Koebis, K. Nakao, et al. 2021. Efficient marmoset genome engineering by autologous embryo transfer and CRISPR/Cas9 technology. Sci. Rep. 11: 20234. https://doi.org/10.1038/s41598-021-99656-4

Arakaki, P.R., F.M Carvalho, P.H.G. Castro, J.A.P.C. Muniz and R.R. Valle. 2017. Collection, evaluation, and coagulum dissolution of semen from Goeldi's Monkey, *Callimico goeldii*. Folia Primatol. 88: 334–343. https://doi.org/10.1159/000480501

Arakaki, P.R., M. Nichi, F.O.B. Monteiro, J.A.P.C. Muniz, M.A.B.V. Guimarães and R.R. Valle, 2019a. Comparison of semen characteristics and sperm cryopreservation in common marmoset (*Callithrix jacchus*) and black-tufted-ear marmoset (*Callithrix penicillata*). J. Med. Primatol. 48: 32–42. https://doi.org/10.1111/jmp.12388

Arakaki, P.R., P.A.B. Salgado, J.D.A. Losano, M.H. Blank, M. Nichi and R.J.G. Pereira. 2019b. Assessment of different sperm functional tests in golden-headed lion tamarins (*Leontopithecus chrysomelas*). Am. J. Primatol. 81: 1–9. https://doi.org/10.1002/ajp.23034

Arakaki, P.R., P.A.B. Salgado, J.D.A. Losano, D.R. Gonçalves, R.R. Valle, R.J.G. Pereira, et al. 2019c. Semen cryopreservation in golden-headed lion tamarin, *Leontopithecus chrysomelas*. Am. J. Primatol. 81: e23071. https://doi.org/10.1002/ajp.23071

Arakaki, P.R., P.A.B. Salgado, R.H.F. Teixeira, F.B. Rassy, M.A.B.V. Guimarães and R.R. Valle. 2019d. Testicular volume and semen characteristics in the endangered southern muriqui (*Brachyteles arachnoides*). J. Med. Primatol. 48: 244–250. https://doi.org/10.1111/jmp.12418

Arakaki, P.R., J.D.A. Losano, P.A.B. Salgado and R.J.G. Pereira. 2020. Seasonal effects on testes size and sustained semen quality in captive golden-headed lion tamarins, *Leontopithecus chrysomelas*. Anim. Reprod. Sci. 218: 106472. https://doi.org/10.1016/j.anireprosci.2020.106472

Araújo, A. and M.B.C. Sousa. 2008. Testicular Volume and Reproductive Status of Wild Callithrix jacchus. Int. J. Primatol. 29: 1355–1364. https://doi.org/10.1007/s10764-008-9291-4

Cardoso, D.L., D.A.A. Guimarães, P. Mayor, M.A.P. Ferreira, H.L.T. Dias, R.F. Espinheiro, et al. 2021. Reproductive biology of owl (*Aotus* spp.) and capuchin (*Sapajus* spp.) monkeys. Anim. Reprod. Sci. 227, 106732. https://doi.org/10.1016/j.anireprosci.2021.106732

Cui, K.-H, S.P. Flaherty, C.D. Newble, M.V. Guerin, A.J. Napier and C.D. Matthews. 1991. Collection and analysis of semen from the common marmoset (*Callithrix jacchus*). J. Androl. 12, 214–220. https://doi.org/10.1002/j.1939-4640.1991.tb00253.x

Cui, K.-H. 1996. The effect of stress on semen reduction in the marmoset monkey (*Callithrix jacchus*). Hum. Reprod. 11, 568–573. https://doi.org/10.1093/HUMREP/11.3.568

de Melo, F.R., R.R. Hilário, D.S. Ferraz, D.G. Pereira, J.C. Bicca-Marques, L. Jerusalinsky, et al. 2021a. *Callithrix flaviceps* [WWW Document]. IUCN Red List Threat. Species.

de Melo, F.R., M. Port-Carvalho, D.G. Pereira, C.R. Ruiz-Miranda, D.S. Ferraz, J.C. Bicca-Marques, et al. 2021b. *Callithrix aurita* [WWW Document]. IUCN Red List Threat. Species.

De Vleeschouwer, K., M. Heistermann, L. Van Elsacker and R.F. Verheyen, 2000. Signaling of reproductive status in captive female golden-headed lion tamarins (*Leontopithecus chrysomelas*). Int. J. Primatol. 21: 445–465. https://doi.org/10.1023/A:1005439919150

Dettling, A.C. 2002. Reproduction and development in Goeldi's Monkey (*Callimico goeldii*). Evol. Anthropol. 11: 207–210. https://doi.org/10.1002/evan.10093

Digby, L.J. 1999. Sexual behavior and extragroup copulations in a wild population of common marmosets (*Callithrix jacchus*). Folia Primatol. 70: 136–145. https://doi.org/10.1159/000021686

Dixson, A.F. and M.J. Anderson. 2002. Sexual selection, seminal coagulation and copulatory plug formation in Primates. Folia Primatol. 73: 63–69. https://doi.org/10.1159/000064784

Dixson, A.F. and M.J. Anderson. 2004. Sexual behavior, reproductive physiology and sperm competition in male mammals. Physiol. Behav. 83: 361–371. https://doi.org/10.1016/j.physbeh.2004.08.022

Fernandez-Duque, E., A. Di Fiore and M. Huck. 2012. The behavior, ecology, and social evolution of New World monkeys. pp. 43–64. *In*: J.C. Mitani, J. Call, P.M. Kappeler, R.A. Palombit and J.B. Silk (eds). The Evolution of Primate Societies. The University of Chicago Press, Chicago, IL, USA.

Ferrari, S.F. and M.A. Lopes Ferrari. 1989. A re-evaluation of the social organisation of the callitrichidae, with reference to the ecological differences between genera. Folia Primatol. 52: 132–147. https://doi.org/10.1159/000156392

Garber, P.A, L. Moya, J.D. Pruetz and C. Ique. 1996. Social and seasonal influences on reproductive biology in male moustached tamarins (*Saguinus mystax*). Am. J. Primatol. 38: 29–46. https://doi.org/10.1002/(SICI)1098-2345(1996)38:1<29::AID-AJP4>3.0.CO;2-V

Gilchrist, R.B., P.L. Nayudu and J.K. Hodges. 1997. Maturation, fertilization, and development of marmoset monkey oocytes *in vitro*. Biol. Reprod. 56: 238–246. https://doi.org/10.1095/biolreprod56.1.238

Gordo, M., F. Röhe, M.D. Vidal, R. Subirá, J.P. Boubli, R.A. Mittermeier, et al. 2021. *Saguinus bicolor* [WWW Document]. IUCN Red List Threat. Species. URL https://www.iucnredlist.org/species/40644/192551696

Grupen, C.G., R.B. Gilchrist, P.L. Nayudu, M.F. Barry, S.J. Schulz, L.J. Ritter, et al. 2007. Effects of ovarian stimulation, with and without human chorionic gonadotrophin, on oocyte meiotic and developmental competence in the marmoset monkey (*Callithrix jacchus*). Theriogenology 68: 861–872. https://doi.org/10.1016/j.theriogenology.2007.07.009

Guimarães, M.A. de B.V., C.A. De Oliveira and R.C. Barnabe, 2003. Seasonal variation in the testicular volume of capuchin monkeys (*Cebus apella*) in captivity. Folia Primatol. 74: 54–56. https://doi.org/10.1159/000068394

Harcourt, A.H., P.H. Harvey, S.G. Larson and R.V. Short. 1981. Testis weight, body weight and breeding system in primates. Nature 293: 55–56.

Harcourt, A.H., A. Purvis and L. Liles. 1995. Sperm competition: mating system, not breeding season, affects testes size of primates. Funct. Ecol. 9: 468. https://doi.org/10.2307/2390011

Harding, R.D., R.D. Hulme, S.F. Lunn, C. Henderson and R.J. Aitken. 1982. Plasma Progesterone Levels throughout the Ovarian Cycle of the Common Marmoset (Callithrix jacchus). J. Med. Primatol. 11: 43–51. https://doi.org/10.1159/000460023

Harrison, R.M. and R.H. Wolf. 1985. Sperm parameters and testicular volumes in Saguinus mystax. J. Med. Primatol. 14: 281–4.

Hodges, J.K., P.G. Cottingham, P.M. Summers and L. Yingnan. 1987. Controlled ovulation in the marmoset monkey (*Callithrix jacchus*) with human chorionic gonadotropin following prostaglandin-induced luteal regression. Fertil. Steril. 48: 299–305. https://doi.org/10.1016/s0015-0282(16)59360-1

Ishibashi, H. and H.H. Motohashi. 2013. Artificial insemination in common marmosets using sperm collected by penile vibratory stimulation. Neotrop. Primates 20: 54–56. https://doi.org/10.1896/044.020.0110

Ishibashi, H., H.H. Motohashi, M. Kumon, K. Yamamoto, H. Okada, T. Okada, et al. 2013. Efficient embryo transfer in the common marmoset monkey (*Callithrix jacchus*) with a reduced transfer volume: a non-surgical approach with cryopreserved late-stage embryos. Biol. Reprod. 88: 1–5. https://doi.org/10.1095/biolreprod.113.109165

IUCN. 2022. The IUCN Red List of Threatened Species. Version 2022-1 [WWW Document]. URL https://www.iucnredlist.org/search/stats?query=PRIMATES&searchType=species (accessed 7.15.22).

Kropp, J., A. Di Marzo and T. Golos, 2017. Assisted reproductive technologies in the common marmoset: an integral species for developing nonhuman primate models of human diseases. Biol. Reprod. 96: 277–287. https://doi.org/10.1095/biolreprod.116.146514

Kuederling, I., J.M. Morrell and P.L. Nayudu. 1996. Collection of semen from marmoset monkeys (*Callithrix jacchus*) for experimental use by vaginal washing. Lab. Anim. 30: 260–266. https://doi.org/10.1258/002367796780684845

Kuederling, I., A. Schneiders, J. Sønksen, P.L. Nayudu, and J.K. Hodges. 2000. Non-invasive collection of ejaculates from the common marmoset (*Callithrix jacchus*) using penile vibrostimulation. Am. J. Primatol. 52: 149–154. https://doi.org/10.1002/1098-2345(200011)52:3<149::AID-AJP4>3.0.CO;2-B

Kurotaki, Y. and E. Sasaki. 2017. Practical reproductive techniques for the common marmoset practical reproductive techniques for the common marmoset. J. Mamm. Ova Res. 34: 3–12.

Lopata, A., P.M. Summers and J.P. Hearn. 1988. Births following the transfer of cultured embryos obtained by *in vitro* and *in vivo* fertilization in the marmoset monkey (*Callithrix jacchus*). Fertil. Steril. 50: 503–509. https://doi.org/10.1016/S0015-0282(16)60141-3

Ludwig, G., A.T.A. Nascimento, J.M.D. Miranda, M. Martins, L. Jerusalinsky, R.A. Mittermeier, et al. 2021. *Leontopithecus caissara* [WWW Document]. IUCN Red List Threat. Species. URL https://www.iucnredlist.org/species/11503/206547044

Luetjens, C.M., J.-B. Stukenborg, E. Nieschlag, M. Simoni and J. Wistuba. 2008. Complete spermatogenesis in orthotopic but not in ectopic transplants of autologously grafted marmoset testicular tissue. Endocrinology 149: 1736–1747. https://doi.org/10.1210/en.2007-1325

Marshall, V., J. Kalishman and J. Thomson. 1997. Nonsurgical embryo transfer in the common marmoset monkey. J. Med. Primatol. 26: 241–247.

Møller, A.P. 1988. Ejaculate quality, testes size and sperm competition in primates. J. Hum. Evol. 17: 479–488. https://doi.org/10.1016/0047-2484(88)90037-1

Monfort, S.L., M, Bush and D.E. Wildt. 1996. Natural and induced ovarian synchrony in golden lion tamarins (*Leontopithecus rosalia*). Biol. Reprod. 55: 875–882. https://doi.org/10.1095/biolreprod55.4.875

Moore, H.D., T.D. Hartman and W.V. Holt. 1984. The structure and epididymal maturation of the spermatozoon of the common marmoset (*Callithrix jacchus*). J. Anat. 138(Pt. 2): 227–235.

Morrell, J., R. Nubbemeyer, M. Heistermann, J. Rosenbusch, I. Küderling, W. Holt, et al. 1998. Artificial insemination in *Callithrix jacchus* using fresh or cryopreserved sperm. Anim. Reprod. Sci. 52: 165–174. https://doi.org/10.1016/S0378-4320(97)00092-4

Morrell, J.M., I. Küderling and J.K. Hodges. 1996. Influence of semen collection method on ejaculate characteristics in the common marmoset, *Callithrix jacchus*. J. Androl. 17: 164–172. https://doi.org/10.1002/j.1939-4640.1996.tb01766.x

Morrell, J.M., M. Nowshari, J. Rosenbusch, P.L. Nayudu and J.K. Hodges. 1997. Birth of offspring following artificial insemination in the common marmoset, *Callithrix jacchus*. Am. J. Primatol. 41: 37–43. https://doi.org/10.1002/(SICI)1098-2345(1997)41:1<37::AID-AJP3>3.0.CO;2-0

Motohashi, H.H. and H. Ishibashi. 2016. Cryopreservation of ovaries from neonatal marmoset monkeys. Exp. Anim. 65: 189–196. https://doi.org/10.1538/expanim.15-0097

Nayudu, P., J. Wu and H. Michelmann. 2003. *In vitro* development of marmoset monkey oocytes by pre-antral follicle culture. Reprod. Domest. Anim. 38: 90–96. https://doi. org/10.1046/j.1439-0531.2003.00398.x

O'Brien, J.K., F.K. Hollinshead, K.M. Evans, G. Evans and W.M.C. Maxwell. 2003. Flow cytometric sorting of frozen—thawed spermatozoa in sheep and non-human primates. Reprod. Fertil. Dev. 15: 367. https://doi.org/10.1071/RD03065

Phelps, M.P., L.W. Seeb, and J.E. Seeb. 2020. Transforming ecology and conservation biology through genome editing. Conserv. Biol. 34: 54–65. https://doi.org/10.1111/cobi.13292

Poches, R.A., C.I. Brieva and C. Jiménez. 2013. Características seminales del tití gris (*Saguinus leucopus*) bajo condiciones de cautiverio, obtenidas por estimulación vibratoria del pene (evp). Rev. la Fac. Med. Vet. y Zootec. 60: 11–22.

Pothana, L., N.K. Venna, L. Devi, A. Singh, I. Chatterjee and S. Goel. 2016a. Cryopreservation of adult primate testes. Eur. J. Wildl. Res. 62: 619–626. https://doi.org/10.1007/s10344-016-1024-y

Rezende, G., C. Knogge, F. Passos, G. Ludwig, L.C. Oliveira, L. Jerusalinsky, et al. 2020. *Leontopithecus chrysopygus* [WWW Document]. IUCN Red List Threat. Species. URL https://www.iucnredlist.org/species/11505/17935400

Rodríguez, V., A. Link, D. Guzman-Caro, T.R. Defler, E. Palacios, P.R. Stevenson, et al. 2021. *Saguinus oedipus* [WWW Document]. IUCN Red List Threat. Species. URL https://www.iucnredlist.org/species/19823/192551067 (accessed 9.20.22).

Rylands, A.B. and R.A. Mittermeier. 2008. The diversity of the New World primates—an annotated taxonomy. pp. 23–54. *In*: P.A. Garber, A. Estrada, J.C. Bicca-Marques, E.W. Heymann and K.B. Strier (eds). South American Primates—Comparative Perspectives in the Study of Behavior, Ecology and Conservation. Springer New York, NY, USA.

Rylands, A.B., E.W. Heymann, J.L. Alfaro, J.C. Buckner, C. Roos, C. Matauschek, et al. 2016. Taxonomic review of the New World tamarins (Primates: Callitrichidae). Zool. J. Linn. Soc. 177: 1003–1028. https://doi.org/10.1111/zoj.12386

Sasaki, E., H. Suemizu, A. Shimada, K. Hanazawa, R. Oiwa, M. Kamioka, et al. 2009. Generation of transgenic non-human primates with germline transmission. Nature 459: 523–527. https://doi.org/10.1038/nature08090

Sankai, T., H. Tsuchiya, N. Ogonuki, F. Cho and Y. Yoshikawa. 1997. A trial of oocyte maturation and *in vitro* fertilization by frozen-thawed spermatozoa in the red-bellied tamarin (*Saguinus labiatus*). J. Mamm. Ova Res. 14: 205–208. https://doi.org/10.1274/jmor.14.205

Sato, K., R. Oiwa, W. Kumita, R. Henry, T. Sakuma, R. Ito, et al. 2016. Generation of a nonhuman primate model of severe combined immunodeficiency using highly efficient genome editing. Cell Stem Cell 19: 127–138. https://doi.org/10.1016/j.stem.2016.06.003

Schiel, N. and A. Souto. 2017. The common marmoset: An overview of its natural history, ecology and behavior. Dev. Neurobiol. 77: 244–262. https://doi.org/10.1002/dneu.22458

Schlatt, S., S. Kim and R. Gosden. 2002. Spermatogenesis and steroidogenesis in mouse, hamster and monkey testicular tissue after cryopreservation and heterotopic grafting to castrated hosts. Reproduction 124: 339–346. https://doi.org/10.1530/rep.0.1240339

Schneiders, A., J. Sonksen and J.K. Hodges. 2004. Penile vibratory stimulation in the marmoset monkey: a practical alternative to electro-ejaculation, yielding ejaculates of enhanced quality. J. Med. Primatol. 33: 98–104. https://doi.org/10.1111/j.1600-0684.2004.00058.x

Silva, A.M. da, A.F. Pereira, P. Comizzoli and A.R. Silva. 2020. Cryopreservation and culture of testicular tissues: an essential tool for biodiversity preservation. Biopreserv. Biobank. 18: 235–243. https://doi.org/10.1089/bio.2020.0010

Soini, P. 1993. The ecology of the pygmy marmoset, Cebuella pygmaea: some comparisons with two sympatric tamarins. pp. 252–272. *In*: A.B. Rylands (ed.). Marmosets and Tamarins: Systematics, Behaviour, and Ecology. Oxford University Press, Oxford, England. https://doi.org/0.1002/ajpa.1330970413

Sønksen, J., F. Biering-Sørensen and J.K. Kristensen. 1994. Ejaculation induced by penile vibratory stimulation in men with spinal cord injuries. The importance of the vibratory amplitude. Paraplegia 32: 651–660. https://doi.org/10.1038/sc.1994.105

Sønksen, J. and D.A. Ohl. 2002. Penile vibratory stimulation and electroejaculation in the treatment of ejaculatory dysfunction. Int. J. Androl. 25: 324–332. https://doi.org/10.1046/j.1365-2605.2002.00378.x

Summers, P.M., A.M. Shephard, C.T. Taylor and J.P. Hearn. 1987. The effects of cryopreservation and transfer on embryonic development in the common marmoset monkey, Callithrix jacchus. J. Reprod. Fertil. 79: 241–250.

Summers, P.M., C.J. Wennink and J.K. Hodges. 1985. Cloprostenol-induced luteolysis in the marmoset monkey (*Callithrix jacchus*). Reproduction 73: 133–138. https://doi.org/10.1530/jrf.0.0730133

Takabayashi, S., Y. Suzuki and H. Katoh. 2015. Development of a modified artificial insemination technique combining penile vibration stimulation and the swim-up method in the common marmoset. Theriogenology 83: 1304–1309. https://doi.org/10.1016/j.theriogenology.2015.01.017

Takahashi, T., K. Hanazawa, T. Inoue, K. Sato, A. Sedohara, J. Okahara, et al. 2014. Birth of healthy offspring following ICSI in *in vitro*-matured common marmoset (*Callithrix jacchus*) oocytes. PLoSOne 9: 1–11. https://doi.org/10.1371/journal.pone.0095560

Tardif, S.D., D.A. Smucny, D.H. Abbott, K. Mansfield, N. Schultz-Darken and M.E. Yamamoto. 2003. Reproduction in captive common marmosets (*Callithrix jacchus*). Comp. Med. 53: 364–8.

Tkachenko, O.Y., S. Delimitreva, E. Isachenko, R.R. Valle, H.W. Michelmann, A. Berenson, et al. 2010. Epidermal growth factor effects on marmoset monkey (*Callithrix jacchus*) oocyte *in vitro* maturation, IVF and embryo development are altered by gonadotrophin concentration during oocyte maturation. Hum. Reprod. 25: 2047–2058. https://doi.org/10.1093/humrep/deq148

Tomioka, I., H. Ishibashi, E.N. Minakawa, H.H. Motohashi, O. Takayama, Y. Saito, et al., 2017. Transgenic monkey model of the polyglutamine diseases recapitulating progressive neurological symptoms. eNeuro 4: 1–16. https://doi.org/10.1523/ENEURO.0250-16.2017

Valle, R.R. 2007. Collection, analysis and cryopreservation of semen from a model species, the common marmoset (*Callithrix jacchus*) [dissertation]. Universidade de São Paulo. https://doi.org/10.11606/T.10.2007.tde-21052007-103859

Valle, R.R., C.M.R. Valle, M. Nichi, J.A.P.C. Muniz, P.L. Nayudu and M.A.B.V. Guimarães. 2014. Semen characteristics of captive common marmoset (*Callithrix jacchus*): A comparison of a German with a Brazilian colony. J. Med. Primatol. 43: 225–230. https://doi.org/10.1111/jmp.12111

Vidal, F.D., M.S. Luz, T.G. De Pinho and A. Pissinatti. 2007. Coleta de sêmen em mico-leão-de-cara-dourada (*Leontopithecus chrysomelas*) (Kuhl, 1820) através da eletroejaculação Callitrichidae primates. Rev. Bras. Ciência Veterinária 14: 67–71. https://doi.org/10.4322/rbcv.2014.235

von Schönfeldt, V., R. Chandolia, L. Kiesel, E. Nieschlag, S. Schlatt and B. Sonntag. 2011a. Assessment of follicular development in cryopreserved primate ovarian tissue by xenografting: prepubertal tissues are less sensitive to the choice of cryoprotectant. Reproduction 141: 481–490. https://doi.org/10.1530/REP-10-0454

von Schönfeldt, V., R. Chandolia, L. Kiesel, E. Nieschlag, S. Schlatt and B. Sonntag. 2011b. Advanced follicle development in xenografted prepubertal ovarian tissue: the common marmoset as a nonhuman primate model for ovarian tissue transplantation. Fertil. Steril. 95: 1428–1434. https://doi.org/10.1016/j.fertnstert.2010.11.003

Wilton, L.J., V.S. Marshall, E.C Piercy and H.D.M. Moore. 1993. *In vitro* fertilization and embryo development in the marmoset monkey (*Callithrix jacchus*). Reproduction 97: 481–486. https://doi.org/10.1530/jrf.0.0970481

Yeoman, R.R., R.B. Ricker, L.E. Williams, J. Sonksen and C.R. Abee. 1997. Vibratory stimulation of ejaculation yields increased motile spermatozoa, compared with electroejaculation, in squirrel monkeys (*Saimiri boliviensis*). Contemp. Top. Lab. Anim. Sci. 36(1): 62–64.

Yeoman, R.R., J. Sonksen, S.V. Gibson, B.M. Rizk and C.R. Abee. 1998. Penile vibratory stimulation yields increased spermatozoa and accessory gland production compared with rectal electroejaculation in a neurologically intact primate (*Saimiri boliviensis*). Hum. Reprod. 13: 2527–2531. https://doi.org/10.1093/humrep/13.9.2527

Yeung, C.H., J.M. Morrell, T.G. Cooper, G.F. Weinbauer, J.K. Hodges and E. Nieschlag. 1996. Maturation of sperm motility in the epididymis of the common marmoset (*Callithrix jacchus*) and the cynomolgus monkey (*Macaca fascicularis*). Int. J. Androl 19: 113–121.

Ziegler, T.E., K.B. Strier and S. Van Belle. 2009. The reproductive ecology of south american primates: ecological adaptations in ovulation and conception. pp. 191–210. *In*: P.A. Garber, A. Estrada, J.C. Bicca-Marques, E.W. Heymann and K.B. Strier (eds). South American Primates, Developments in Primatology: Progress and Prospects. Springer New York, N.Y. https://doi.org/10.1007/978-0-387-78705-3

Chapter 12

The Cebidae: *Saimiri* spp.

Danuza Leite Leão[1,2], Wlaisa Vasconcelos Sampaio[2,3],
Karol Guimarães Oliveira[4], Daniele Cristina Calado de Brito[5]
and Sheyla Farhaydes Souza Domingues[2,6]*

[1]Grupo de Pesquisa em Ecologia de Primatas Amazônicos do
Instituto Mamirauá, Mamirauá Institute for Sustainable Development,
Tefé, Amazonas, Brazil.
Email: danuzalleao@gmail.com

[2]Laboratory of Wild Animal Biotechnology and Medicine,
Faculty of Veterinary Medicine, Federal University of Pará, Castanhal, Pará, Brazil.
Email: wlaisa.sampaio@ufra.edu.br or wlaisa.sampaio@gmail.com;

[3]Federal Rural University of the Amazon, Parauapebas, Pará, Brazil.

[4]National Primate Center, Ananindeua, Pará, Brazil.
Email: karolgoliveira83@gmail.com

[5]School of Veterinary Medicine and Animal Science,
University of São Paulo, São Paulo, Brazil.
Email: dcaladobrito@gmail.com

[6]Federal University of Pará, Castanhal, Pará, Brazil.
Email: shfarha@ufpa.br or shfarha@gmail.com

INTRODUCTION

The genus *Saimiri* is commonly known as the squirrel monkeys. The most recent taxonomic classification, which covers the entire genus (Rylands et al. 2012), considers 11 taxa distributed among species and subspecies. They are *Saimiri oerstedii oerstedii, Saimiri o. citronellus, Saimiri cassiquiarensis cassiquiarensis, Saimiri c. albigena, Saimiri macrodon, Saimiri ustus, Saimiri sciureus sciureus, Saimiri s. collinsi, Saimiri boliviensis boliviensis, Saimiri b. peruviensis,* and

*Corresponding author: Email: shfarha@ufpa.br or shfarha@gmail.com

Saimiri vanzolinii. Notably, Mercês et al. (2015) had reclassified the subspecies *S. s. collinsi* to *S. collinsi* (Figure 12.1). According to the International Union for Conservation of Nature's (IUCN) Red List, *S. oerstedii* and *S. vanzolinii* are currently endangered (Lynch et al. 2021, Solano-Rojas 2021), while *Saimiri ustus* is near threatened (Alves et al. 2021).

Studies on the reproductive biology of *Saimiri* are restricted to those concerning *S. sciureus* and *S. boliviensis* (Chen et al. 1981, Kuehl and Dukelow 1982, Dukelow 1983). Recently, *S. collinsi* was adopted as an experimental model for the development of reproduction biotechniques (Oliveira et al. 2015, Scalercio et al. 2015, Oliveira et al., 2016a, b, Sampaio et al. 2017, Almeida et al. 2018, Leão et al. 2020, 2021, Sousa et al. 2021). However, there are significant field limitations due to insufficient knowledge about fundamental reproductive aspects such as anatomy, physiology, and seasonal reproduction patterns. This insufficient information on the most varied aspects is an obstacle to implementing *ex situ* conservation measures. In light of this, this chapter is dedicated to the Neotropical primates of the genus *Saimiri*, with emphasis on their anatomy and reproductive biology, and on the development of reproductive biotechnologies aimed at the *ex situ* conservation of these primates.

Figure 12.1   Free-living male squirrel monkey (*Saimiri collinsi*). Picture: Sampaio, W.V.

THE GENUS *SAIMIRI*

The genus *Saimiri* is of great interest to biomedical research centers due to their small size, easy handling, availability in nature, and similarity with humans in the development of some diseases (Dukelow 1983, 1985, Mehlotra et al. 2017, Vanchiere et al. 2018).

Concerning morphological characteristics, squirrel monkeys have a greenish-black coloration that varies according to the species, with shades ranging from black to yellow-orange (Groves 2005). Another striking feature is the mask of white eyebrow hairs, which form an arch with a pattern that divides them into two groups: Gothic and Roman (Groves 2005, Ingbermann et al. 2008). The Gothic group is characterized by a white facemask that forms a high arch above the eyes, and a tail with very thick tufts of fur. In the Roman group, the white band over the eyebrow is narrower, and the tufts of hair on the tail are less thick. As for sexual dimorphism, the adult males are more noticeable than females. The most notable features are the blackened pre-auricular spots, which is an exclusive feature of females (Muniz 2005).

Squirrel monkeys have great manual agility (Laska et al. 2007); they are arboreal, occupy the middle stratum of the forest (Stone 2006), move predominantly in a quadrupedal way (Boinski 1989), and have daytime habits (Ingbermann et al. 2008). They are omnivores whose diet in the wild is composed of fruits, small vertebrates, flowers, nuts, gum, and insects (Mittermeier and van Roosmalen 1981, Terborgh and Winter 1983, Stone 2007).

The Reproductive System

The reproductive organs of the male squirrel monkey are characterized by a semi-pendulous and asymmetrical scrotum. The testis and epididymis form a relatively small and globular mass. The penis is cylindrical, approximately 35 mm long, and is covered by a retractable foreskin. It also has a penile bone and some vestigial keratinized spicules lateral to its body (Steinberg et al. 2005). The prostate is located in the proximal part of the urethra, just distal to the bladder (Hill 1960).

Lopes et al. (2017) described the macroscopic aspects of the female genital organs and external female genitalia of *Saimiri macrodon*, *Saimiri cassiquiarensis*, and *Saimiri vanzolinii*. Regarding the external genitalia, the labia majora shows a rough cutaneous aspect and is sparsely covered with hair. It is also a yellowish color with the presence of black dotted pigmentation in amounts that vary between animals. Furthermore, a hood-shaped prepuce has been identified, and the clitoris has a conical shape. The clitoral glands have a smooth surface, callous aspect, and dark color. The vagina consists of a long channel with a dorsoventrally compressed lumen. The vaginal vestibule is located in the caudal portion of the vagina, while the cranial region is connected to the uterine cervix. The cervix is located in the abdominal cavity, arranged ventrally to the rectum and dorsally to the urinary bladder. Its caudal portion is delimited by the vagina, and the cranial portion is delimited by the uterus.

The uterus in the abdominal cavity is simple, with an inverted pear shape. The uterine tubes consist of two tubular structures located on the upper margin and between the broad ligament's folds called the mesosalpinx. In *S. collinsi* (previously called *S. sciureus*), Almeida et al. (2012) characterized the population of normal and atretic preantral follicles in ovaries from senile squirrel monkeys. The mean ovarian population (± standard deviation) was estimated to be 915.04 ± 78.83, 230.46 ± 20.82, and 115.88 ± 15.72 primordial, primary, and secondary follicles per ovary, respectively.

REPRODUCTIVE BIOLOGY

Sexual maturity occurs at around age three in females (Coe et al. 1985), while in males it occurs around two and a half and three and a half years of age (Richter et al. 1984). Squirrel monkeys are seasonal animals with well-defined breeding seasons (Taub et al. 1978, Baldwin and Baldwin 1981). In the Amazon region, mating occurs between June and September (dry season), whereas the final third stage of gestation, parturition, and lactation correspond to the wettest season, between December and February (Dukelow 1983). Before and during the breeding season, morphological, behavioral, and physiological changes occur in both males and females (Chen et al. 1981, Mendonza 1999, Schiml et al. 1999, Moorman et al. 2002). In males, the phenomenon of weight gain or 'fattening' is remarkable (Mendonza et al. 1978a, b, Mendonza 1999), and will be discussed later.

Squirrel monkeys live in social groups composed of several males and females (Goodall and Mittermeier 1999), with group sizes varying from tens to hundreds of individuals (Baldwin and Baldwin 1981). The known life span for this animal is over 20 years (Mendoza 1999), but it is estimated that they can reach up to 30 years (Williams 2008). *Saimiri* is among four Neotropical primate genera universally distributed throughout the Amazon (Ferrari 2004), predominantly in South America, but can also be found in southern Mexico, Central America, and even northern Argentina (Goodall and Mittermeier 1999, Alfaro et al. 2015).

Squirrel monkeys have developed a unique attention-grabbing phenomenon due to the morphophysiological changes that affect males during the mating season; this is known as fattening or seasonal fattening (Dumond and Hutchinson 1967, Stone 2014). Over the last few decades, this phenomenon has been studied from different perspectives such as behavioral (Boinski 1987, 1992, Mendonza et al. 1978a,b, Stone 2014), endocrinological (Schiml et al. 1996, Nadler and Ronsemblum 1972), and physiological (Coe et al. 1985, Dumond and Hutchinson 1967, Belt and Cavazos 1971). However, there is still much speculation about the role of fattening and its relationship with the reproduction of these animals.

In the wild, the fattening condition varies between individuals (Stone 2014), and appears to be advantageous as the fattest males participate in 70% of the mating that occurs during the mating season (Boinski 1987, 1992). This suggests a possible selection mechanism of pre-copulatory sexual intercourse (Stone 2014). In semi-captivity, the fatted condition is related to increased testicular volume, improved seminal quality (increased sperm concentration and motility), and seasonal spermatogenesis (Dumond and Hutchinson 1967, Belt and Cavazos 1971, Nadler and Rosemblum 1972), which may all be related to post-copulatory sexual selection. However, in captivity, reproductive seasonality varies widely (Stone and Williams 2021), and quality semen samples are obtained throughout the year regardless of the fattening condition (Chen et al. 1981, Dukelow 1983, 1985, Oliveira et al. 2015, 2016a, b), which generates disagreement about the function of this phenomenon. To try to answer these questions, Sampaio et al. (2022) recently tested the fatted condition and advantageous seminal characteristics in squirrel monkeys, and evaluated its implications for sperm competition. In their study, they found that the fattening condition is associated with larger testicular volume, higher seminal quality in the form of higher seminal volume, neutral seminal pH, and firmer ejaculate coagulation. Because these characteristics are related to sperm competition, the results suggest that fatted males have important advantages in the production and survival of sperm, with increased fertilization probability.

The gestational period is approximately 150 d (Coe and Rosenblum 1978), and a single calf (Baldwin 1985) is born with approximately 15% of the mother's body weight. For this reason, it is considered one of the most expensive pregnancies and deliveries among primates (Dukelow 1983, Mitchell 1990).

SEMEN COLLECTION AND SEMINAL CHARACTERISTICS

The semen collection methods used for the genus *Saimiri* are penile vibrostimulation (Kugelmeier 2011, Laverde-Corrêa et al. 2001, Viana 2013, Yeoman et al. 1998) and electrostimulation (Ackerman and Roussel 1968, Almeida et al. 2018, Bennett 1967a,b, Denis et al. 1976, Dukelow 1983, Leão et al. 2020, 2021, Oliveira et al. 2015, 2016a,b, Sampaio et al. 2017, 2021, Sousa et al. 2021, Yeoman et al. 1998). The protocols defined for the application of electrical stimuli vary between species. In *S. sciureus*, the best seminal collection results were achieved using the voltage range of 0.95–1.15 volts, achieving between 2 and 3 min as the average time for ejaculation (Lang 1967). The induction of ejaculation is performed using stimuli levels ranging from 12.5 mA to 100 mA, with 30-s intervals between each series of stimuli, in *S. collinsi* (Oliveira et al. 2015, 2016a,b, Leão et al. 2021, Sampaio et al. 2021, Sousa et al. 2021), *S. vanzolinii*, *S. cassiquiarensis*, and *S. macrodon* (Oliveira et al. 2016a).

Bennett (1967a) was the first researcher to collect and describe semen from *Saimiri* species, and characterized it as containing two fractions: the liquid, which is rich in sperm and the other fraction, which is coagulated. However, subsequent studies observed the presence of sperm in both fractions (liquid and coagulated – Figure 12.2) (*Saimiri collinsi*: Oliveira et al. 2015, 2016a, b, Sampaio et al. 2017, 2022, Almeida et al. 2018, Leão et al. 2020, 2021, Sousa et al. 2021). The seminal volume varies according to the species and collection method, corresponding to 5–500 µL. The semen color varies from whitish to yellowish, and also varies in opacity (transparent to opaque) (Oliveira et al. 2016a). Data on the seminal parameters of *Saimiri* spp. are summarized in Table 12.1.

Table 12.1 Seminal characteristics of *Saimiri* spp. based on the semen collection method

	Collection method	Volume (µL)	Concentration (10⁶ sptz/mL)	Motility (%)	PMI (%)	References
		50–81	—	—	—	Bennett 1967a,b
	EEJ	—	153 ± 96	35 ± 11	37 ± 19	Ackerman and Roussel 1968
Saimiri sciureus		160 ± 57	427 ± 160	66 ± 15	79 ± 2	Denis et al. 1976
		125 (g)	80–300	40–80	—	Dukelow 1983
	VP	106 ± 80	362 ± 344	70 ± 22	74 ± 17	Laverde-Corrêa et al. 2001
		454	15 ± 3	83 ± 2.1	61.8 ± 8.5	Kugelmeier 2011
		478.8 ± 53.8	39.5 ± 4.9	86.3 ± 1.7	82.3 ± 1.9	Viana 2013
Saimiri boliviensis	EEJ	205 ± 25	4.7 ± 2.7	44.1 ± 11.4	—	Yeoman et al. 1998
	VP	436 ± 90	77.1 ± 20.4	80.6 ± 4.3	—	Yeoman et al. 1998
		*49.2 ± 68.9 **65.4 ± 142.1	88.31 ± 36.64	~45	~56	Oliveira et al. 2016a
		*51.8 ± 49.5 **304 ± 286.6	13–29	~80–100	~75–95	Oliveira et al. 2015
		*100 ± 106.14 **287 ± 184.28	—	~75–95	~58–75	Oliveira et al. 2016b
Saimiri collinsi	EEJ	*203 ± 88**177 ± 89	—	79 ± 13	77 ± 9	Sampaio et al. 2017
		*85 ± 10 **280 ± 51	24 ± 19	44 ± 8.4	61 ± 3.1	Almeida et al. 2018
		339 ± 61	171 ± 610	67 ± 11	43 ± 8	Leão et al. 2020
		32.2 ± 15* 339 ± 61**	—	—	—	Sousa et al. 2021
		179 ± 18.9	127 ± 164	60.6 ± 4	60.4 ± 3.5	Leão et al. 2021
		165.9 ± 148.8*** (non-fatted) 227.3 ± 178.5****(fatted)	—	—	—	Sampaio et al. 2021

(Contd.)

Table 12.1 Seminal characteristics of *Saimiri* spp. based on the semen collection method (*Contd.*)

	Collection method	Volume (μL)	Concentration (10⁶sptz/mL)	Motility (%)	PMI (%)	References
Saimiri vanzolinii	EEJ	*28.3 ± 59.8 **126 ± 142	—	~77	~78	Oliveira et al. 2016a
		*85 ± 92 ** 90 ± 85	—	~95	~75	Oliveira et al. 2016b
		*83 ± 6 **93 ± 58	—	95 ± 5	71 ± 11	Sampaio et al. 2017
Saimiri cassiquiarensis	EEJ	*5 ± 7 ** 175 ± 177	—	~43	~ 90	Oliveira et al. 2016a
		*10**5	—	80	62	Oliveira et al. 2016b
Saimiri macrodon	EEJ	*0** 500	—	90	98	Oliveira et al. 2016b

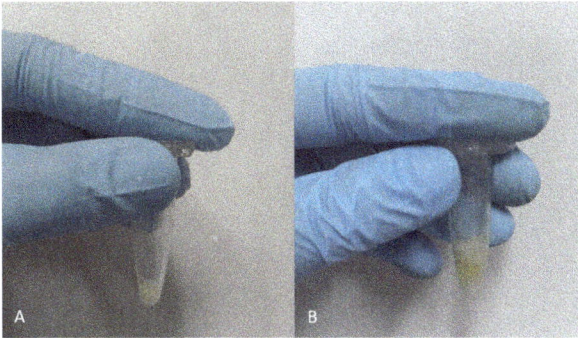

Figure 12.2   Liquid (A) and coagulated (B) fractions of *Saimiri collinsi* semen. Picture: Leão, D.L.

Concerning morphological characteristics, *Saimiri* sperm have a narrow head with a reduced surface area and apical protrusion. The posterior margin of the acrosome has a serrated or microvilli appearance. As in lemurids and other primate species (Gould and Martins 1978), the midpiece may be found laterally to the back of the head (Dukelow 1983, Gage 1998, Sampaio et al. 2017, Leão et al. 2020). The mitochondria in the cross-section are very thin and go around the midpiece in approximately 53 turns (Dukelow 1983).

Sperm morphometry is another characteristic that varies between different species and levels of sperm quality (Sampaio et al. 2017). Sampaio et al. (2017) observed that in low-quality *S. collinsi* semen samples, even when measuring morphologically normal gametes, there is a trend of decreases in sperm head size, which may be a reflex caused by a defect of chromatin. The abovementioned study is the only one proposed to evaluate sperm morphometry and semen quality in squirrel monkeys. Sperm morphometry data for *Saimiri* spp. are summarized in Table 12.2.

SEMINAL CRYOPRESERVATION

In the genus *Saimiri*, seminal cryopreservation has already been studied in *S. sciureus* (Denis et al. 1976), *S. boliviensis* (Yeoman et al. 1997), *S. collinsi* (Oliveira et al. 2015, 2016b), *S. vanzolinii*, and *S. cassiquiarensis* (Oliveira et al. 2016b). The extender used was lactose for *S. sciureus* (Denis et al. 1976); TES-TRIS for *S. boliviensis* (Yeoman et al. 1997); and powdered coconut water for *S. collinsi*, *S. vanzolinii*, and *S. cassiquiarensis* (ACP-118®; Oliveira et al. 2015, 2016a, b). It is important to note that, in *Saimiri* sp., conventional cryopreservation is slow, with cooling rates from –0.3°C/min (Oliveira et al. 2015, 2016a, b). This is usually done by applying a freezing curve with an average duration of 2 h until reaching 4°C.

Among the non-penetrating cryoprotectants, egg yolk is most commonly used (Oliveira et al. 2015, 2016b), while glycerol is the penetrating cryoprotectant used in concentrations ranging from 1.5% to 8% for squirrel monkeys (Denis et al. 1976, Oliveira et al. 2016b). Notably, in *Saimiri* spp., a decrease in post-thawing

Table 12.2 Sperm morphometry of the genus *Saimiri* including linear dimensions of sperm (µm) – Mean ± SD

	Head width	Head length	IP length	Total tail length	Total length	References
Saimiri sciureus	3.51 ± 0.01	5.11 ± 0.02	9.03 ± 0.12	55.34 ± 0.13	69.24 ± 0.15	Dukelow 1983
	3.6 ± 0.3	5.5 ± 0.03	10 ± 0.5	55.1 ± 2.7	70.6 ± 2.8	Laverde-Corrêa et al. 2001
	–	5.1	9.0		69.2	Gage and Freckleton 2003
Saimiri boliviensis	3.76 ± 0.45[4]	5.71 ± 0.45	12.2 ± 0.45	65.68 ± 0.45	71.39 ± 0.45	Steinberg et al. 2007
Saimiri collinsi	4.3 ± 0.01	6.2 ± 0.01	–	70.5 ± 0.19	76.7 ± 0.19	Sampaio et al. 2017
	4.1 ± 0.01	6.2 ± 0.02	–	70.6 ± 0.19	76.9 ± 0.19	Sampaio et al. 2017
Saimiri vanzolinii	3.36 ± 0.03	5.14 ± 0.08	9.49 ± 0.38	–	71.7 ± 0.7	Leão et al. 2020
	4.8 ± 0.01	6.8 ± 0.01	–	69.2±0.09	76.1 ± 0.09	Sampaio et al. 2017
Saimiri cassiquiarensis	6.95 ± 0.38	5.09 ± 0.36	–	–	–	Soares et al. 2020
Saimiri macrodon	6.79 ± 0.46	4.91 ± 0.33	–	–	–	Torres et al. 2021

*IP: Intermediate piece

sperm quality has been observed, which demonstrates the need for more research on the optimal concentration of cryoprotectant to be added to the diluent medium for each species to maintain post-thawing sperm parameters. Data from the seminal cryopreservation of *Saimiri* spp. are summarized in Table 12.3.

OOCYTE COLLECTION

Studies with females of *S. sciureus* involve ovulation induction (Kuehl and Dukelow 1975b, 1978) and ovarian hormonal stimulation. These are generally based on pretreatment with progesterone, followed by the association of follicle-stimulating hormone and human chorionic gonadotropin (FSH-hCG) (Dukelow et al. 1981, Asakawa et al. 1982, Pierce et al. 1993).

In *S. sciureus*, oocytes recovered via laparoscopic follicular aspiration (Dukelow et al. 1971, Kuehl and Dukelow 1979, Asakawa et al. 1982, Asakawa and Dukelow 1982, Dukelow 1983, Yano and Gould 1985, Pierce et al. 1993) or laparotomy (Kuehl and Dukelow 1979) have been described, where the number of oocytes obtained range from two to six per animal (Kuehl and Dukelow 1979, Asakawa et al. 1982, Asakawa and Dukelow 1982, Pierce et al. 1993).

With the oocytes obtained from *S. sciureus*, after hormonal stimulation and follicular puncture, studies aiming at *in vitro* fertilization have been carried out (Johnson et al. 1972, Cline 1972, Gould et al. 1973, Kuehl and Dukelow 1975a, 1979, Asakawa et al. 1982, Asakawa and Dukelow 1982, Pierce et al. 1993). Because *S. sciureus* shows seasonality, ovarian hormonal stimulation and the induction of ovulation are important steps in performing artificial insemination. Notably, *S. sciureus* was the first Neotropical primate species in which artificial insemination was performed (Bennett 1967b).

In Vitro Oocyte Maturation

In vitro oocyte maturation was studied in *S. boliviensis* by Yeoman et al. (1994) from unstimulated females during the non-breeding season. This was done to compare their *in vitro* development with that of oocytes obtained from monkeys who received a low level of FSH stimulation but did not have a preovulatory gonadotropin peak. These authors demonstrated that, in squirrel monkeys, oocytes retrieved during the non-breeding season need prior FSH stimulation for maturational competence.

ARTIFICIAL INSEMINATION

The only successful report of artificial insemination in squirrel monkeys was that of Bennett (1967b) in *S. sciureus* using semen recovered via rectal electroejaculation. The induction of ovulation in the female was performed over nine consecutive days using injections of pregnant mare serum (PMS), and during the last 4 d, human chorionic gonadotrophin (HCG) was also administered. The female was artificially inseminated on the first or second day after the gonadotropin treatment.

Table 12.3	Seminal cryopreservation data of the genus *Saimiri*

Species	Extender	Cryoprotectant	Frozen semen					References
			Motility (%)	Vigor	PMI (%)	PMF (%)	NSM (%)	
Saimiri sciureus	Lactose	4% Glycerol and egg yolk	58.8 ± 22.9 / 53.8 ± 17.1	NI	NI	NI	NI	Denis et al. 1976
Saimiri boliviensis	TES-TRIS	8% Glycerol and egg yolk	41%	NI	NI	NI	NI	Yeoman et al. 1997
Saimiri collinsi	PCW	1.5% Glycerol and egg yolk	≅10	1	≅15	≅20	≅90	Oliveira et al. 2015
		3% Glycerol and egg yolk	≅10	1	≅20	≅25	≅95	
	PCW	3% Glycerol and egg yolk	≅15	2	≅30	≅35	100	
Saimiri vanzolinii	PCW	3% Glycerol and egg yolk	30 / 60	2 / 4	2 / 55	NI / 24	95 ± 1.41	Oliveira et al. 2016a
Saimiri cassiquiarensis	PCW	3% Glycerol and egg yolk	6	2	7	20	83	
Saimiri macrodon	PCW	3% Glycerol and egg yolk	10	5	38	49	87	

NI: Not informed
Abbreviations—PMI: plasma membrane integrity; PMF: plasma membrane functionality; NSM: normal sperm morphology;
TES: 2-[(2-hydroxy-1,1-bis(hydroxymethyl)ethyl) aminol]ethanesulfonic acid; TRIS, Tris(hydroxymethyl)aminomethane;
PCW: powdered coconut water

Fertilized eggs were recovered at the two-cell stage from the Fallopian tubes, and at the two-cell and four-cell stages from the uterus.

PARTHENOGENETIC ACTIVATION

In *S. collinsi* (previously called *S. sciureus*), a protocol of *in vitro* embryo production (IVEP) was developed, performing *in vitro* maturation (IVM) for 44 h at 37°C in an atmosphere of 5% CO_2, and was subjected to the AP protocol using 5 µM of ionomycin for 5 min in association with 100 µM of 6-DMAP for 3 h. In total, 89% of oocytes reached metaphase II, observing an 85% cleavage rate and 33% blastocist rate (Lima et al. 2012).

IN VITRO FERTILIZATION AND EMBRYO TRANSFER

In vitro fertilization has only been described in *S. sciureus*. Generally, in *in vitro* fertilization, females undergo ovulation induction using gonadotropins, the oocytes are collected via laparotomy, and sperm are collected via electroejaculation using a rectal probe. The medium used for *in vitro* fertilization is TC199 (DeMayo et al. 1985, Gould et al. 1973, Kuehl and Dukelow 1975a, 1979, 1982).

In the study performed by Gould et al. (1973), of the 22 mature oocytes utilized in the *in vitro* fertilization system, 11 showed changes that could be indicative of the initial stages of fertilization, such as sperm in the perivitelline space, extrusion of the second polar body, or the formation of pronuclei. Among these oocytes, six were observed to cleave into two cells. Extrusion of the second polar body occurred 20–24 h after insemination, and cleavage occurred between 36 and 42 h after insemination.

Kuehl and Dukelow (1975a) evaluated the effect of the medium used (TC199 or Ham's F 10, both supplemented to 20% with heat-inactivated agamma calf serum and containing 50 units/mL of penicillin G) on *in vitro* fertilization cultures, and verified that there was no difference between the media in terms of oocyte maturation and fertilization. Twenty-one of the 32 fertilized oocytes reached the two-polar body, two-pronuclear stage, some developed to two or more cells, and seven embryos reached the four-cell stage. Years later, in 1979, these same authors evaluated five media, which were compared for their effectiveness in *in vitro* fertilization. Their results showed that medium 3 (80% TC199, 20% serum, 72 µg Pyruvate, 100 u Pen-strep, and 1 U Heparin) gave the best *in vitro* maturation (56.5%) and fertilization (74%) results.

In 1982, Kuehl and Dukelow conducted a study to determine the times taken for the expulsion of the first and second polar bodies in *S. sciureus*. The time taken for the expulsion of the first polar body after hCG injection is the time required for ovum maturation and the time needed for sperm capacitation. Of the 28 eggs, 24 (85–7%) exhibited extrusion of the second polar body within one standard deviation of the mean (4–7 h). After the addition of sperm to the system, the shortest time required for an egg to extrude the second polar body is 2–0 h for an egg with a single polar body, 1–3 h after insemination, and 0–7 h for an egg with two polar bodies. Furthermore, cleavage to the two-cell stage occurs at an average time of 16–2 ± 7–6 h after insemination. The intervals for the earliest and latest possible times of this first cleavage for three ovules were 6-2-10-8, 10-9-22-3, and 22-4-24-8 h.

DeMayo et al. (1985) evaluated the *in vitro* fertilization using the cryopreserved oocytes of squirrel monkeys. They found no difference in maturation or fertilization rate between the control and frozen oocytes.

OVARIAN TISSUE CULTURE AND AUTOGRAFTING

Ovarian tissue culture and autografting in *S. collinsi* were described by Scalercio et al. (2015), who evaluated ovarian tissue pretreatment with 50 mM Trolox antioxidant, followed by heterotopic transplantation. The results verified that Trolox prevented massive follicular activation and kept the percentages of morphologically normal follicles higher than in untreated grafts. However, Trolox increased the quality of follicular tissue and collagen synthesis. Although it increased the rate of ovarian tissue, stromal fibrosis was increased in grafts treated with Trolox, mainly due to increased collagen synthesis.

FINAL CONSIDERATIONS

In recent decades, there has been considerable progress in the strategies for reproducing the genus *Saimiri*, partly due to its use as a biomedical model of animal experimentation. The first studies on the reproductive biotechniques applied to squirrel monkeys began in the 1970s, in which protocols for semen collection, cryopreservation, artificial insemination, and *in vitro* fertilization were described. At that time, those protocols represented a daring feat because little was known about the reproductive biology of the genus. In subsequent years, studies failed to efficiently repeat these protocols, partly due to the difficulty in obtaining gametes. In the case of females, superovulation protocols can trigger polycystic ovary syndrome. In males, the manipulation of semen in its solid state (clotted) hinders the recovery of spermatozoa, which is an obstacle to cryopreservation. This demonstrates the need to develop research aimed at the dissolution of the clot, which is an important goal in optimizing seminal cryopreservation techniques.

There is a demand for improving the collection, handling, maturation, and cryopreservation protocols for semen and oocytes that are crucial for forming animal germplasm banks. In the future, this tool will be essential for the management of metapopulations, thereby enabling genetic exchange. We can also highlight the need for developing protocols for the *in vitro* production and cryopreservation of embryos, embryo transfer, and collecting physiological data on embryonic development, as there is a significant gap for these between the sexes. Intracytoplasmic sperm injection, a technique preferably used in primates as it does not depend on high sperm motility rates, has potential for use with *Saimiri* due to the difficulty in handling their semen. Although reproductive biotechniques facilitate the fertilization process and improve reproductive efficiency, it is imperative to invest in basic research that elucidates the mechanisms behind the reproductive physiology of these primates to obtain consistent results with high repeatability and satisfactory success rates.

REFERENCES

Ackerman, D.R. and J.D. Roussel. 1968. Fructose, lactic acid and citric acid content of the semen of eleven subhuman primate species and of man. J. Reprod. Fertil. 17: 563–566.

Asakawa, W. and R. Dukelow. 1982. Chromosomal analyses after *in vitro* fertilization of squirrel monkey (*Saimiri sciureus*) oocytes. Biol. Reprod. 26: 579–583.

Asakawa, T., P.J. Chan and W.R. Dukelow. 1982. Time Sequence of *in vitro* maturation and chromosomal normality in metaphase i and metaphase ii of the squirrel monkey (*Saimiri sciureus*) Oocyte. Biol. Reprod. 27: 118–124.

Alfaro, J.W.L., J.P. Boubli, F.P. Paim, C.C. Ribas, M.N.F. Silva, M.R. Messias, et al. 2015. Biogeography of squirrel monkeys (genus *Saimiri*): South-central Amazon origin and rapid pan-Amazonian diversification of a lowland primate. Mol. Phylogenet. Evol. 82: 436–454.

Almeida, D.V., R.R. Santos, S.R. Scalercio, D.L. Leão, A. Haritova, I.C. Oskam, et al. 2012. Morphological and morphometrical characterization, and estimation of population of preantral ovarian follicles from senile common squirrel monkey (*Saimiri sciureus*). Anim. Reprod. Sci. 134: 210–215.

Almeida, D.V.C., J.S. Lima, D.L. Leão, K.G. Oliveira, R.R. Santos, S.F.S. Domingues, et al. 2018. The effects of Trolox on the quality of sperm from captive squirrel monkey during liquefaction in the extender ACP-118. Zygote. 26(4): 333–335.

Alves, S.L., J.S. Silva Júnior, A.L. Ravetta, M.R. Messias and J.W. Lynch Alfaro. 2021. *Saimiri ustus* (amended version of 2019 assessment). The IUCN Red List of Threatened Species. e.T41538A192584351.

Baldwin, J.D. and J.I. Baldwin. 1981. Primates. The squirrel monkeys, genus *Saimiri*. pp. 277–330. *In*: Coimbra-Filho, A.F. and R.A. Mittermeier. Ecology and Behavior Neotropical. Academia Brasileira de Ciências, Rio de Janeiro.

Baldwin, J.D. 1985. The behavior of squirrel monkeys (*Saimiri*) in natural environments. pp. 35–53. *In*: L.A. Rosenblum and C.L. Coe (eds). Handbook of Squirrel Monkey Research. New York: Plenum Press.

Belt, W.D. and L.F. Cavazos. 1971. Fine structure of the interstitial cells of leydig in the squirrel monkey during seasonal regression. Anat. Rec. 169: 115–127.

Bennett, J.P. 1967a. Semen collection in the squirrel monkey. J. Reprod. Fertil. 13: 353–355.

Bennett, J.P. 1967b, Artificial insemination of the squirrel monkey. J. Endocrinal. 37: 473–474.

Boinski, S. 1987. Mating patterns in squirrel monkeys (*Saimiri oerstedii*). Behav. Ecol. Sociobiol. 21: 13–21.

Boinski, S. 1989. The positional behavior and substrate use of squirrel monkey: Ecological implications. J. Hum. Evol. 18: 659–678.

Boinski, S. 1992. Monkeys with inflated sex appeal. Nat. Hist. 101: 42–49.

Chen, J.J., E.R. Smith, G.D. Gray and J.M. Davidson. 1981. Seasonal changes in plasma testosterone and ejaculatory capacity in squirrel monkeys (*Saimiri sciureus*). Primates. 22: 253–260.

Cline, E.M. 1972. Investigations on *in vitro* fertilization for three mammalian species: golden hamster, squirrel monkey and New Zealand White rabbit. Thesis (MSc)—University of Georgia, Department of Biochemistry, Athens, Georgia.

Coe, C.L. and L.A. Rosenblum. 1978. Annual reproductive strategy of the squirrel monkey (*Saimiri sciureus*). Folia Primatol. 29: 19–42.

Coe, C.L., E.R. Smith and S. Levine. 1985. The endocrine system of the squirrel monkey. pp. 191–218. *In*: LA. Rosenblum and C.L. Coe (eds). Handbook of Squirrel Monkey Research. Springer, US.

DeMayo, F.J., R.G. Rawlins and W.R. Dukelow. 1985. Xenogenous and *in vitro* fertilization of frozen/thawed primate oocytes and blastomere separation of embryos. Fertil. Steril. 43: 295–300.

Denis, L.T., A.N. Poindexter, M.B. Ritter, S.W. Seager and R.L. Deter. 1976. Freeze preservation of squirrel monkey sperm for use in timed fertilization studies. Fertil. Steril. 27: 723–729.

Dukelow, W.R., S.J. Jarosz, D.A. Jewett and R.M. Harrison. 1971. Laparoscopic examination of the ovaries in goats and primates. Lab. Anim. Sci. 21: 594–597.

Dukelow, W.R., C.G. Theodoran, J. Howe-Baughman and W.T. Magee. 1981. Ovulatory patterns in the squirrel monkey (*Saimiri sciureus*). Anim. Reprod. Sci. 4: 55–63.

Dukelow, W.R. 1983. The squirrel monkey (*Saimiri sciureus*). pp. 149–180. *In:* J.P. Hearn (ed.). Reproduction in New World Primates: New Models in Medical Science. Lancaster: MTP Press Limited.

Dukelow, W.R. 1985. Reproductive cyclicity and breeding in the Squirrel Monkey. pp. 169–190. *In*: L.A. Rosenblum and C.L. Coe (eds). Handbook of Squirrel Monkey Research. New York: Plenum Press.

Dumond, F.V. and T.C. Hutchinson. 1967. Squirrel monkey reproduction: the "*fatted*" male phenomenon and seasonal spermatogenesis. Science. 158: 1067–1070.

Ferrari, S.F. 2004. Biogeography of Amazonian primates. A primatologia no Brasil. 8: 101–122.

Gage, M.J.G. 1998. Mammalian sperm morphometry. Proc. R. Soc. Lond. B. 263: 97–103.

Gage, M.J.G. and R.P. Freckleton. 2003. Relative testis size and sperm morphometry across mammals: no evidence for an association between sperm competition and sperm length. Proc. Royal Soc. B. Biol. Sci. 270: 625–632.

Goodall, J. and R. Mittermeier. 1999. The Pictorial Guide to the Living Primates. Hong Kong: Pogonias Press.

Gould, K.G., E.M. Cline and W.L. Williams. 1973. Observations on the induction of ovulation and fertilization *in vitro* in the squirrel monkey (*Saimiri sciureus*). Fertil. Steril. 24: 260–268.

Gould, K.G. and D.E. Martin. 1978. Comparative morphology of primate spermatozoa using scanning electron microscopy. II. Families cercopithecidae, Lorisidae, Lemuridae. J. Hum. Evol. 7: 637–642.

Groves, C.P. 2005. Order primates. pp. 111–184. *In*: D.E. Wilson and D.M. Reeder (eds). Mammal Species of the World: A Taxonomic and Geographic Reference, 3rd Ed. Baltimore. Johns Hopkins University Press.

Hill, W.C.O. 1960. Genus *Saimiri* Voigt, 1831. pp. 251–319. *In*: W.C.O. Hill (ed.). Primates Comparative Anatomy and Taxonomy, Cebidae, Part A. Edinburgh: R & R Clark, Ltda.

Ingbermann, B., A.I. Stone and C.C. Cheida. 2008. Gênero *Saimiri* (Voigt, 1831). pp. 41–46. *In*: N.R. Revis, A.L. Peroech and F.R. Andrade (orgs). Primatas do Brasil. Londrina: Technical Book Editora.

Johnson, M.P, R.M. Harrison and W.R. Dukelow. 1972. Studies on oviductal fluid and *in vitro* fertilization in rabbits and nonhuman primates. Fed. Proc. 31: 369.

Kuehl, T.J and W.R. Dukelow. 1975a. Fertilization *in vitro* of *Saimiri sciureus* follicular oocytes. J. Med. Primatol. 4: 209–216.

Kuehl, T.J and W.R. Dukelow. 1975b. Ovulation induction during the anovulatory season in *Saimiri sciureus*. J. Med. Primatol. 4: 23–31.

Kuehl, T.J. and W.R. Dukelow. 1978. The effect of a synthetic polypeptide threonyl-prolyl-arginyl-lysine, on ovulation in the squirrel monkey (*Saimiri sciureus*) and hamster. J. Reprod. Fertil. 2: 25–28.

Kuehl, T.J. and W.R. Dukelow. 1979. Maturation and *in vitro* fertilization of follicular oocytes of the squirrel monkey (*Saimiri sciureus*). Biol. Reprod. 21: 545–556.

Kuehl, T.J. and W.R. Dukelow. 1982, Time relations of squirrel monkey (*Saimiri sciureus*) sperm capacitation and ovum maturation in an *in vitro* fertilization system. J. Reprod. Ferti. 64: 135–137.

Kugelmeier, T. 2011. Colheita e análise do sêmen de macacos de cheiro (*Saimiri sciureus*) por vibroestimulação: do condicionamento ao coágulo seminal. Ph.D. Thesis, University of São Paulo, São Paulo, BR.

Lang, C.M. 1967. A Technique for the collection of semen from squirrel monkeys (*Saimiri sciureus*) by electro-ejaculation. Lab. Anim. Care. 17: 218–221.

Laska, M., P. Freist and S. Krause. 2007. Which senses play a role in nonhuman primate food selection? A comparison between squirrel monkeys and spider monkeys. Am. J. Primatol. 69: 282–294.

Laverde-Corrêa, H.J., V.M. Robles-Medina and P.E. Cruz-Casallas. 2001. Contribucion al conocimento de las caracteristicas seminales del mono ardilla (*Saimiri sciureus*) en cautiverio, obtenido por el metodo de estimulacion vibratoria del pene. Universidade de los Lianos-Escuela de Medicina Veterinaria y Zootecnia Villaavivencio-Colômbia.

Leão, D.L., W.V. Sampaio, P.C. Sousa, A.A. Moura, I.C. Oskam, R.R. Santos, et al. 2020. Micromorphological and ultrastructural description of sperm from squirrel monkeys (*Saimiri collinsi* Osgood, 1916). Zygote. 28: 203–207.

Leão, D.L., W.V. Sampaio, P.C. Sousa, I.C. Oskam, R.R. Santos and S.F.S. Domingues. 2021. The use of anogenital distance as a non-invasive predictor of seminal quality in captive squirrel monkey (*Saimiri collinsi* Osgood 1961). J. Med. Primatol. 502: 99–305.

Lima, J.S., S.R.R.A. Scalercio, N.N. Costa, A.B. Brito, E. Nikolak and S.F.S. Domingues. 2012. Ciência Animal—Suplemento. Trabalho apresentado no VI Congresso Norte Nordeste de Reprodução Animal. 14–16.

Lopes, G.P., A.B. Brito, F.P. Paim, R.R. Santos, H.L. Queiroz and S.F.S. Domingues. 2017. Comparative characterization of the external genitalia and reproductive tubular organs of three species of the genus *Saimiri* Voigt, 1831 (Primates: Cebidae). Anat. Histol. Embryol. 46: 143–161.

Lynch, J.W., F.P. Paim, R.M. Rabelo, J.S. Silva Júnior and H.L. Queiroz. 2021. *Saimiri vanzolinii*. The IUCN Red List of Threatened Species. e.T19839A17940474.

Mehlotra, R.K., R.E. Howes, T.A. Rakotomanga, B. Ramiranirina, S. Ramboarina, T Franchard, et al. 2017. Long-term in vitro culture of *Plasmodium vivax* isolates from Madagascar maintained in *Saimiri boliviensis* blood. Malaria J. 16: 442.

Mendoza, S.P., E.L. Lowe, J.M. Davidson and S. Levine. 1978a. Annual cyclicity in the squirrel moneky (*Saimiri sciureus*): The relationship between testosterone, fatting and sexual behavior. Horm. Behav. 11: 295–303.

Mendoza, S.P., E.L. Lowe, J.A. Resko and S. Levine. 1978b. Seasonal variations in gonadal hormones and social behavior in squirrel monkeys. Physiol. Behav. 20: 515–522.

Mendoza, S.P. 1999. Squirrel monkey. pp. 591–600. *In*: T. Poole (ed.). The Ufaw Handbook on the Care and Management of Laboratory Animals, 7th Ed. Terrestrial Vertebrades. Oxford Black. Science.

Mercês, M.P., J.W.L. Alfaro, W.A. Ferreira, M.L. Harada and J.S.S. Júnior. 2015. Morphology and mitochondrial phylogenetics reveal that the Amazon River separates two eastern squirrel monkey species: *Saimiri sciureus* and *S. collinsi*. Mol. Phylogenetics Evol. 82: 426–435.

Mitchell, C.L. 1990. The ecological basis of female dominance: a behavioral study of the squirrel monkey (*Saimiri sciureus*) in the wild. Ph.D. thesis, Princeton University.

Mittermeier, R.A. and G.M.V. Roosmalen. 1981. Preliminary observations of habitat utilization and diet in eight Surinam monkeys. Folia Primatol. 36: 1–39.

Moorman, E.A., S.P. Mendoza, S.E. Shideler and B.L. Lasley. 2002. Excretion and measurement of estradiol and progesterone metabolites in the feces and urine of female squirrel monkeys (*Saimiri sciureus*). Am. J. Primatol. 57: 79–90.

Muniz, I.C.M. 2005. Desenvolvimento do dimorfismo sexual nos macacos-de-cheiro (*Saimiri* Voigt, 1831). Tese de Doutorado apresentada ao programa de Pós Graduação em Zoologia-Museu Emílo Goeldi, Belém.

Nadler, R.D. and L.A. Rosenblum. 1972. Hormonal regulation of the "fatted" phenomenon in male squirrel monkeys. Anat. Rec. 173: 181–187.

Oliveira, K.G., D.L. Leão, D.V.C. Almeida, R.R. Santos and S.F.S. Domingues. 2015. Seminal characteristics and cryopreservation of sperm from the squirrel monkey, *Saimiri collinsi*. Theriogenology 84: 743–749.

Oliveira, K.G., R.R. Santos, D.L. Leão, H.L. Queiroz, F.P. Paim, J.L.D.S.G. Viamez-Júnior, et al. 2016a. Testicular biometry and semen characteristics in captive and wild squirrel monkey species (*Saimiri* spp.). Theriogenology 86: 879–887.

Oliveira, K.G., R.R. Santos, D.L. Leão, A.B. Brito, J.S. Lima, W.V. Sampaio, et al. 2016b. Cooling and freezing of sperm from captive, free-living and endangered squirrel monkey species. Cryobiology. 72: 283–289.

Pierce, D.L., M.P. Jonson and J.B. Kaneene. 1993. *In vitro* fertilization analysis of squirrel monkey oocytes produced by various follicular Induction regimens and the Incidence of triploidy. Am. J. Primatol. 16: 321–330.

Richte, C.B., N.D.M. Lehner and R.V. Hendrickson. 1984. Primates. pp. 298-383. *In*: J. Fox, B. Coehen and F. Loew (eds). Laboratory Animal Medicine. Academic Press, Inc., San Diego, Calif.

Rylands, A.B., R.A. Mittermeier and J.S. Silva. 2012. Neotropical primates: taxonomy and recently described species and subspecies. Int. Zoo. Yearb. 46: 11–24.

Sampaio, W.V., K.G. Oliveira, D.L. Leão, M.C. Caldas-Bussiere, H.L. Queiroz, F.P. Paim, et al. 2017. Morphologic analysis of sperm from two neotropical primate species: comparisons between the squirrel monkeys *Saimiri collinsi* and *Saimiri vanzolinii*. Zygote. 25: 141–148.

Sampaio, W.S, D.L. Leão, P.C. Sousa, H.L. Queiroz and S.F.D. Fararhayldes. 2021. Male fattening is related to increased seminal quality of squirrel monkeys *(Saimiri collinsi)*: Implications for sperm competition. Am. J. Primatol. 84: e23353-e23353.

Scalercio, S.R.A., C.A. Amorim, D.C. Brito, S. Percário, I.C. Oskam, S.F.S. Domingues, et al. 2015. Trolox enhances follicular survival after ovarian tissue autograft in squirrel monkey (*Saimiri collinsi*). Reprod. Fertil. Dev. 21: 1–15.

Schiml, P.A., S.P. Mendonza, W. Saltman, D.M. Lyons and W.A. Mason. 1996. Seasonality in squirrel monkeys (*Saimiri sciureus*). Physiol. Behav. 60: 1105–1113.

Schiml, P.A., S.P. Mendoza, W. Saltzman, D.M. Lyons and W.A Mason. 1999. Annual physiological changes in individually housed squirrel monkeys (*Saimiri sciureus*). Am. J. Primatol. 47: 93–103.

Soares, A.R.B., E.C.B. Torres, W.V. Sampaio, D.L. Leão, K.G. Oliveira, H.L. Queiroz and S.F.S. Domingues. 2020. Análise morfométrica da cabeça espermática em *Saimiri cassiquiarensis*. Cienc. Anim. 30: 161–165.

Solano-Rojas, D. 2021. *Saimiri oerstedii*. The IUCN Red List of Threatened Species. e.T19836A17940807.

Sousa, P.C., D.L. Leão, W.V. Sampaio, F.R. Vasconcelos, S.K.P. Pinheiro, E.M. Castro et al. 2021. Morphological and ultrastructural changes in seminal coagulum of the squirrel monkey (*Saimiri collinsi* Osgood, 1916) before and after liquefaction. Anim. Reprod. Sci. 226: 106710.

Steinberg, E.R., A.M. Palermo, M. Nieves, A. Burna, G. Solis, G. Zunino, et al. 2005. Sex determination and sperm morphology in cebidae. Anais do XI Congresso brasileiro de primatologia. Porto Alegre.

Steinberg, E.R., M. Nieves and M.D. Mudry. 2007. Meiotic characterization and sex determination system of neotropical primates: Bolivian squirrel monkey *Saimiri boliviensis* (Primates: Cebidae). Am. J. Primatol. 69: 1236–1241.

Stone, A.I. 2006. Foraging ontogeny is not linked to delayed maturation in squirrel monkeys. Ethology. 112: 105–115.

Stone, A.I. 2007. Responses of squirrel monkeys to seasonal changes in food availability in na Eastern Amazonian rainforest. Am. J. Primatol. 69: 142–157.

Stone, A.I. 2014. Is fatter sexier? Male reproductive strategies squirrel monkeys, *Saimiri sciureus*. *In*: The 82nd annual meeting of the American association of physical anthropologists.

Stone, A.I. and L. Williams. 2021. Behavioral biology of squirrel monkeys. pp. 395–407. *In*: K. Coleman and S.J. Schapiro (eds). Behavioral Biology of Laboratory Animals. CRC Press, Boca Raton.

Taub, O.M., M.R. Adams and K.G. Auerbach. 1978. Reproductive performance in a breeding colony of Brazilian squirrel monkeys (*Saimiri sciureus*). Lab. Anim. Sci. v. 28: 562–566.

Terborgh, J. and B. Winter. 1983. A method for siting parks and reserves with special reference to Columbia and Ecuador. Biol. Conserv. 27: 45–58.

Torres, E.C.B., A.R.B. Soares, W.V. Sampaio, D.L. Leão and S.F.S. Domingues. 2021. Análise da morfometria das cabeças de espermatozoides *in natura* e após a liquefação em ACP-118® do sêmen de macaco-de-cheiro (*Saimiri collinsi* Osgood, 1916). Rev. Bras. Reprod. Anim. Brazil 45: 703.

Vanchiere, J.A., J.C. Ruiz, A.G. Brady, T.J. Kuehl, L.E. Williams, W.B. Baze, et al. 2018. Experimental Zika virus infection of Neotropical primates. Am. J. Trop. Med. Hyg. 98: 173–177.

Viana, C.F. 2013. Características do sêmen, perfil da concentração de testosterona no extrato fecal, variação da massa corporal e volume testicular de micos-de-cheiro (*Saimiri sciureus*, Linnaeus, 1758) mantidos em cativeiro sob condições ambientais controladas. Dissertations, Universidade Estadual do Norte Fluminense Darcy Ribeiro, Rio de Janeiro, BR.

Williams, L. 2008. Aging Cebidae. pp. 49–61. *In*: S. Atsalis, S.W. Magulis, S.W. Primate (eds). Reproductive Aging: Cross-taxon Perspectives. Switzerland: Ed Karger.

Yano, J. and K.G. Gould. 1985. Induction of follicular growth in the squirrel monkey (*Saimiri sciureus*) with human urinary follicle stimulation hormone (Metrodin). Fertil. Steril. 43: 799–803.

Yeoman, R.R., A. Helvacioglu, L.E. Williams, S. Aksel and C.R. Abee. 1994. Restoration of oocyte maturational competency during the nonbreeding season with follicle-stimulating hormone stimulation in squirrel monkeys (*Saimiri boliviensis boliviensis*). Biol. Reprod. 50: 329–335.

Yeoman, R.R., R.B. Ricker, A.M. Hossain and C.R. Abee. 1997. Cryopreservation of sperm from squirrel monkeys. Am. J. Primatol. 42: 157.

Yeoman, R.R., J. Sonksen, S.V. Gibson, B.M. Rizk and C.R. Abee. 1998. Penile vibratory stimulation yields increased spermatozoa and accessory gland production compared with rectal electroejaculation in a neurologically intact primate (*Saimiri boliviensis*). Hum. Reprod. 13: 2527–2531.

Chapter 13

The Cebidae: *Sapajus apella*

Danuza Leite Leão[1,2], Karol Guimarães Oliveira[3],
Wlaisa Vasconcelos Sampaio[2,4], Daniele Cristina Calado de Brito[5],
and Sheyla Farhayldes Souza Domingues[2,6]*

[1]Grupo de Pesquisa em Ecologia de Primatas Amazônicos do Instituto Mamirauá, Mamirauá Institute for Sustainable Development, Tefé, Amazonas, Brazil.
Email: danuzalleao@gmail.com

[2]Laboratory of Wild Animal Biotechnology and Medicine, Faculty of Veterinary Medicine, Federal University of Pará, Castanhal, Pará, Brazil.

[3]National Primate Center, Ananindeua, Pará, Brazil.
Email: karolgoliveira83@gmail.com

[4]Federal Rural University of the Amazon, Parauapebas, Pará, Brazil.
Email: wlaisa.sampaio@ufra.edu.br or Email: wlaisa.sampaio@gmail.com

[5]School of Veterinary Medicine and Animal Science, University of São Paulo, São Paulo, Brazil.
Email: dcaladobrito@gmail.com

[6]Federal University of Pará, Castanhal, Pará, Brazil.
Email: shfarha@ufpa.br or shfarha@gmail.com

INTRODUCTION

Non-human primates are one of the most species-rich groups of mammals and are found in the following four regions: the Neotropics, mainland Africa, Madagascar, and Asia. They are divided into two groups: Old World primates and New World primates. Of the 183 species from the New World and the Neotropics, 42.2% are listed as threatened, according to the International Union for Conservation of Nature—IUCN (Mittermeier et al. 2022). According to the red list of the IUCN,

*Corresponding author: shfarha@ufpa.br or shfarha@gmail.com

regarding the genus Sapajus: *S. robustus* and *S. flavius* are endangered (Martins et al. 2021, Valença-Montenegro et al. 2021), *S. nigritus* and *S. libidinosus* are near threatened (Martins et al. 2021, Ludwig et al. 2022), *S. xanthosternos* is critically endangered (Canale et al. 2021), and *S. cay* is vulnerable (Rímoli et al. 2022).

Efforts to conserve biodiversity are increasingly necessary as an alternative to *in situ* conservation measures (conservation of ecosystems and natural habitats). Maintenance outside the natural habitat (*ex situ* conservation) is a tool that can help assure the survival of these species, increase genetic exchange, and restore free-living populations (Comizzoli 2015). In this proposal, *Sapajus apella* has been used as an experimental model for reproduction biotechnologies (Oliveira et al. 2011, Lima et al. 2013, Leão et al. 2015, 2017). So this chapter is dedicated to the Neotropical primates *Sapajus apella*, emphasizing reproductive anatomy, reproductive physiology, and the development of reproductive biotechnologies aimed at *ex situ* conservation.

Sapajus apella

Sapajus apella occurs in east-central Colombia, southern Venezuela, Guyana, and northern Brazil (Alfaro et al. 2012). They are medium-sized animals, between 0.32 and 0.56 cm long, and the tail is between 0.30 and 0.55 cm (Kindlovits and Kindlovits 2009). Males reach sexual maturity at seven years (Nagle and Denari 1983), and females at around five years (Nagle and Denari 1982, Phillips et al. 1994, Di Bitetti and Janson 2001). They have a polygamous mating system, and females reproduce preferentially with the dominant male (Rowe 1996). The most common morphological pattern of the robust capuchin monkey includes the following characteristics: a blackened hood with two small, erect tufts in the form of "horns"; a reddish-brown coloring to the throat and ventral parts of the trunk; blackened extremities and lateral surfaces of the limbs, which contrasts with the coloration of the proximal parts; and a blackish-brown tail, with a gradual darkening toward the tip (Silva 2001, 2002, Silva et al. 2008).

The male reproductive system of *Sapajus* sp. consists of two testes, two epididymides, two *vas deferens*, two seminal vesicles, a prostate, two bulbourethral glands (Teixeira 2005), a penis with a vestige of penile bone (baculum), and a very developed glans in the shape of a nail (Napier and Napier 1985).

Regarding the females of *S. apella*, the reproductive system consists of two ovaries characterized as compact and symmetrical with a smooth surface. The uterine tubes are bilaterally present; they are convoluted in adult animals and straight in young animals. The uterus is simple, located in the pelvic region, and characterized as slender in young females and thicker in adults. The vagina is a long structure due to the position of the uterus, with longitudinal folds in the mucosa of the vaginal canal. The external genitalia is located in the urogenital perineum and consists of a vulva with labia majora characterized by few and sparse hairs, labia minora, and a vaginal vestibule. The clitoris is highly developed in this species and consists of a body and a glans (Lima et al. 2015).

Domingues et al. (2004) described general aspects of oogenesis and folliculogenesis in the preantral and antral phases through the stereological study of the ovarian follicular population. According to the study, adult females of *S. apella* presented an average (± standard deviation) population of 56.938 ± 21.888 preantral follicles (without antral cavity) for the right ovary and 49.133 ± 26.896 for the left ovary, with 80 ± 4.95% normal follicles. The mean percentage of primordial, transition, primary, and secondary follicles were 30 ± 4.3, 60 ± 4.8, 6 ± 0.96, and 4 ± 0.67, respectively, presenting with follicular diameters ranging from 22 ± 0.5 μm to 61.2 ± 4.0 μm. Regarding the small antral follicles, the number of normal follicles per ovary was 60.0 ± 19.0, presenting with diameter of 514.4 ± 56.6 μm. The mean duration of the ovarian cycle in *S. apella* was 21.7 days, ranging from 19–26 days. According to plasma hormonal dosage, the ovarian cycle of *S. apella* had a duration of 22.7 days (Lima et al. 2019).

SEMEN COLLECTION AND CHARACTERISTICS

The semen collection method in *Sapajus apella* is electrostimulation (Martinez and Garcia 2020). The protocols defined for the application of electrical stimuli vary between species. Barnabe et al. (2002) described a protocol with five series of 20 stimuli and an intensity progression from 50 to 300 mA. However, Oliveira et al. (2011) and Leão et al. (2015, 2017) demonstrated the induction of ejaculation with stimuli ranging from 12.5 mA to 100 mA, with 30-second intervals between each series of stimuli, making it possible to obtain liquid and clotted semen fractions.

The semen of *Sapajus apella* has two fractions—one liquid (Grade I) and the other coagulated (Grade II to IV) (Figure 13.1). The semen varies in color, from whitish to yellowish, and opacity, from transparent to opaque (Oliveira et al. 2011, Leão et al. 2015, 2017, Lima et al. 2017). Lima et al. (2017) evaluated the

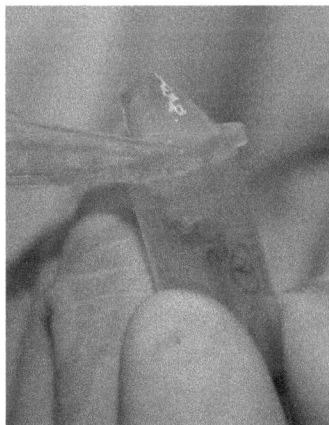

Figure 13.1 Coagulated fraction of *Sapajus apella* semen. Picture: Domingues, S.F.S.

influence of different social contexts on seminal coagulation and sperm quality in captive *S. apella,* and while no statistical differences in the degrees of coagulation were found in males housed individually, animals housed in group cages (male-only groups and mixed-sex groups) showed a significantly higher percentage of ejaculates at degree IV. Animals housed individually showed the highest seminal volume (543 µl) when compared with those animals from the male (273 µl) and mixed-sex (318 µl) groups. Coagula degree IV samples from animals housed individually showed the highest (72%) sperm motility percentages, while sperm plasma membrane integrity was lower with coagula degree II and III samples collected from male—(17%) and mixed-sex (23%) groups.

For sperm to be released from the seminal coagulum and to be used in reproductive biotechnologies such as seminal cryopreservation and *in vitro* fertilization, the semen generally needs to undergo an incubation period in an extender at 37°C for 30 to 60 min (Oliveira et al. 2011, Leão 2015, 2017). In *S. apella*, protocols aimed at dissolving the seminal coagulum have already been developed using 0.9% saline solution (Nagle and Denari 1983), medium 199 plus hyaluronidase and trypsin (Paz et al. 2006), in natura coconut water (CWS; Araújo et al. 2009), powdered coconut water (ACP®-118; Oliveira et al. 2010, 2011, Lima et al. 2013, Leão et al. 2015, 2017), and TES-TRIS (Oliveira et al. 2011, Leão et al. 2015).

Paz et al. (2006) evaluated the effects of hyaluronidase (1 mg/ml) and trypsin (1 mg/ml) enzymes on seminal coagulum liquefaction (after 5 and 15 min), motility, vigor, and acrosome integrity in capuchin monkey semen. They verified that both enzymes could not completely liquefy the seminal coagulum. However, there was a significant difference in motility and vigor between the liquid and coagulated semen fractions after 15 min. On the other hand, Araújo et al. (2009) obtained total liquefaction of the seminal coagulum after mechanical fragmentation and an incubation period of one to two hours in a water-based coconut extender at 37°C; however, only one sample of the liquid fraction of the semen with the extender was verified as having a 20% motility and vigor of 2 during the entire period of analysis. In addition, after seven hours of incubation in a water bath at 37°C, the total average percentage of live sperm was 72.0 + 3.0%. Years later, Oliveira et al. (2011) observed that the supplementation of CWS with caffeine (6 or 10 mM) resulted in recovered sperm motility.

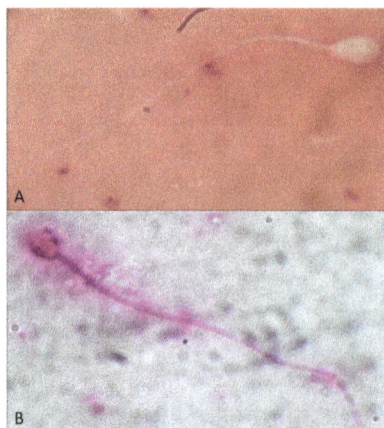

Figure 13.2   *Sapajus apella* sperm with an intact plasma membrane (A) and non-intact plasma membrane (B). Picture: Leão, D.L.

Coconut water powder (ACP) was also evaluated for the liquefaction of the seminal coagulum of *Sapajus apella*, after incubation in a water bath at 33, 35, and 37°C, for 24 hours. Only in a sample of the liquid fraction was sperm verified as having motility of 20% and vigor of 4, which lasted for 40 minutes. Most of the coagulum liquefied in ACP-118® after 12 hours of incubation; moreover, after 24 hours of incubation at 33°C, it was possible to obtain up to $47 \pm 12.8\%$ of sperm with an intact plasmatic membrane (Oliveira et al. 2010; Figure 13.2). Data on seminal parameters of *S. apella* are summarized in Tables 13.1 and 13.2.

Table 13.1 Seminal parameters of the capuchin monkey (*Sapajus apella*) semen liquid fraction

Seminal parameters	Barnabe et al. 2002	Paz et al. 2006	Araújo et al. 2009	Oliveira et al. 2010
Volume	0.2 ± 0.02 ml	0.7 ± 0.1	0.26 ± 0.06	0.20 ± 0.1
Motility	68.4 ± 3.1	71.7 ± 7.9	20	20*
Vigor	2.6 ± 0.1	3.5 ± 0.6	2	4*
Intact membrane	UN	UN	$83.0 + 6.0$	51.5 ± 5
Normal sperm	UN	62.8 ± 7.2	UN	UN
Major defect	33 ± 3.5	37.3 ± 7.2*	UN	UN
Minor defect	28 ± 3.8		UN	UN

*Abnormal sperm (%); UN: uninformed

Seminal Cryopreservation

Semen cryopreservation from *S. apella* was performed using the slow freezing method, with TES-TRIS, *in natura* coconut water (Oliveira et al. 2011), and powdered coconut water extenders (ACP-118®; Oliveira et al. 2011, Leão et al. 2015, 2017). The cooling rates were from −0.3°C/min (Oliveira et al. 2011, Leão et al. 2015, 2017), applying a freezing curve with an average duration of 2 hours until reaching 4°C.

Another critical step in the seminal cryopreservation process is the addition of cryoprotectants, either penetrating (intracellular) or non-penetrating (extracellular). Among the non-penetrating cryoprotectants, egg yolk was used (Oliveira et al. 2011, Leão et al. 2015, 2017), while glycerol was the penetrating cryoprotectant used in concentrations ranging from 2.5% to 7% for capuchin monkeys (Oliveira et al. 2011, Leão et al. 2015), being added before (Oliveira et al. 2011) or after (Leão et al. 2015, 2017) cooling.

In *S. apella*, a decrease in post-thawing sperm quality has been observed by Oliveira et al. (2011), as sperm motility after thawing was null in a TES–TRIS solution plus 3.5% glycerol and CWS plus 2.5% glycerol. However, the live sperm percentages in TES–TRIS and CWS were 26.2% and 13.2%, respectively. On the other hand, in the study carried out by Leão et al. (2015), it was possible to

Table 13.2 Seminal parameters of the capuchin monkey (*Sapajus apella*) coagulated semen fraction

Authors	Volume	Protocol*	Sperm concen.	Motility	Vigor	Intact membrane	Normal sperm	Major defect	Minor defect
Ackerman and Roussel 1968	UN	Trypsin-digested	48 ± 15	28 ± 28	UN	45 ± 12	UN	41 ± 10**	UN
Paz et al. 2006	UN	Hyaluronidase (H) and trypsin (T) digested	UN	24 ± 12.9 (H) 26 ± 14.7 (T)	1.6 ± 0.8 (H) 1.8 ± 0.98 (T)	UN	UN	UN	UN
Araújo et al. 2009	0.98 + 0.02	CWS	1.600 ± 900	0	0	82.0 + 8.0	UN	UN	UN
Oliveira et al. 2010	0.20 ± 0.02	ACP-118®	$1.1 ± 0.3 × 10^8$	0	0	45.8 ± 8	UN	UN	UN
Oliveira et al. 2011	0.6 ± 0.2	TEST-TRIS and CWS	$1806 ± 367 × 10^6$	38.0 ± 22.0 (TES-TRIS) 22.0 ± 16.0 (CWS)	3.0 ± 05 (TES-TRIS) 3.0 ± 01 (CWS)	≈50 (TES-TRIS and CWS)	78.2 ± 2.0 (TES-TRIS) 83.7 ± 1.0 (CWS)	09.1 ± 0.7 (TES-TRIS) 12.0 ± 0.6 (CWS)	6.06 ± 0.4 (TES-TRIS) 10.6 ± 0.5 (CWS)
Leão et al. 2015	0.6 ± 0.1	ACP-118®	$1199 ± 608 × 10^6$	80.6 ± 8	3	70.3 ± 5.5	81.7 ± 10.9	UN	UN
Leão et al. 2017	410 ± 129	ACP-118®	UN	77 ± 19	5 ± 1	60 ± 22	77 ± 4	5 ± 1	11 ± 3 7 ± 1
Lima et al. 2017	377.83 ± 57.18	ACP-118®	$608 ± 98 × 10^6$ and $416 ± 162 × 10^6$	39.17 ±	3 ± 0.40	36.4 ± 8.76	UN	UN	UN

*Protocol: Seminal coagulum liquefaction protocol; **Abnormal sperm (%); UN: uninformed; CWS: coconut water solution

obtain motile sperm after cryopreservation of *S. apella* semen in ACP-118® at 3%, 5%, and 7% glycerol. Sperm motility and maintenance of motility were higher when 3% (34.0 ± 10.2% and 43.3 ± 12.8%, respectively) or 5% (30.0 ± 16.1% and 36.3 ± 16.6%, respectively) glycerol was added to the extender (ACP-118®). In comparison, the highest percentages of plasma membrane integrity (37.7 ± 13.8%) were observed when sperm was frozen in the presence of 3% glycerol. There was no statistical difference between the percentages of normal sperm using the different glycerol concentrations analyzed (92.2 ± 5.1 in 3% glycerol, 96.3 ± 1.8 in 5% glycerol, and 96 ± 2.3 in 7% glycerol). Leão et al. (2017) also verified that extender supplementation (ACP-118®) with catalase (50 µg/ml) maintains the integrity of sperm plasma membrane after frozen–thawed (39 ± 12% diluted, 39 ± 12% cooled, and 15 ± 29% after frozen–thawed).

OOCYTE COLLECTION

In female non-human primates, obtaining viable oocytes is one of the main limiting factors of biotechniques applied to reproduction. However, it is possible to recover oocytes of *S. apella* by laparotomy, obtaining 2–5 oocytes per animal by puncturing the antral follicles measuring 2–9 mm in diameter, just by following the cycle using colpocytology and follicular growth using ultrasound, without the need for hormonal stimulation (Domingues et al. 2007). In *S. apella*, the acquisition of immature oocytes has also been described by puncturing the antral follicles to be studied regarding the protein profile of the oocyte and its surrounding cells (cumulus cells). The meiotic competence it was achieved after 36 h of *in vitro* culture. The protein profile found in oocytes and cumulus cells of *S. apella* was similar to the molecular weight of proteins involved in the control of oocyte maturation (Domingues et al. 2010). Furthermore, it was possible to use a mechanical procedure as a tissue chopper for the successful isolation of a great number of ovarian preantral follicles from *Cebus apella* ovaries (Domingues et al. 2003). This procedure allows the use of these follicles for *in vitro* cultures, allowing for the future possibility of obtaining viable oocytes for assisted reproduction programs.

OVARIAN TISSUE CRYOPRESERVATION AND CULTURE

Ovarian tissue cryopreservation has shown great receptivity in the scientific community, as it is an option to preserve the fertility of prepubertal individuals through ovarian fragments (Donnez and Dolmans 2017). In addition, this technique can be performed at any time during the reproductive cycle, allowing the acquisition of a pool of oocytes included in follicles (Campos and Silva 2011), which are considered more cryoresistant, as they are immature (oocytes arrested in prophase I) (Oktay et al. 1997, Varghese et al. 2008). The obtaining of these ovarian fragments has been reported in Neotropical primates by using laparotomy, ultrasound monitoring (Domingues et al. 2010), and the trap door technique, a modified method developed for gingival tissue (Santana et al. 2013).

The ovary is still considered a limited source of reproductive cells that form a reserve pool represented by the preantral follicles. In this way, the preservation of ovarian tissue becomes relevant and possible.

In *S. apella*, a 24-hour *in vitro* culture of ovarian tissue in a medium supplemented with pregnant mare serum gonadotropin, b-mercaptoethanol, and bone morphogenetic protein 4, favors follicular transition from primary to secondary stages, with an expression of genes encoding the bone morphogenetic protein 15 (BMP15), connective tissue growth factor (CTGF), and kit ligand (KL) (Brito et al. 2013). In these same Neotropical primates, the conventional freezing of ovarian fragments was performed in solution with 0.5 M sucrose and 1 M ethylene glycol with or without the antioxidant Trolox (50 μM), resulting in higher rates of viable follicles (67%), even after *in vitro* culture (61%) (Brito et al. 2014). This antioxidant prevented endoplasmic reticulum-related vacuolization, which is commonly observed in oocytes and stromal tissue after exposure to the freezing solution, and resulted in higher rates of viable follicles (67%), even after *in vitro* culture (61%) (Brito et al. 2014).

IN VITRO PRODUCTION OF EMBRYOS

In vitro oocyte maturation was studied for the first time in *S. apella* by Domingues et al. (2010), which examined the protein profile of oocytes and cumulus cells recovered from different-sized follicles throughout the follicular phase, and assessed the ability of the oocytes to progress to the MII stage. The authors verified a change in the protein profile between the follicular phases, and the oocyte MII competence was achieved only after 36 hours *in vitro*, with a rate of maturation of 55%. Maturation rates were statistically significantly higher in oocytes recovered from dominant follicles than those from subordinate follicles. Subsequently, Lima et al. (2013), using the same *in vitro* maturation protocol, observed the effect of follicle-stimulating hormone (FSH) and luteinizing hormone (LH) on parthenogenetic activation and IVF of oocytes collected from unstimulated females, obtaining 87% metaphase II oocytes after 40 IVM and four-cell embryos after IVF and pathogenesis.

FINAL CONSIDERATIONS

Despite using the Neotropical primate *Sapajus apella* in biomedical research and studies aimed at *ex situ* conservation of endangered species, the successful development of assisted reproduction biotechniques is still in its early stages. Knowledge of these species' physiology and reproductive biology will be of great relevance in optimizing and improving the manipulation and preservation of gametes, as well as the *in vitro* production of embryos.

REFERENCES

Alfaro, J.W.L., J.P. Boubli, L.E. Olson, A.D. Fiore, B. Wilson, G.A. Gutiérrez-Espeleta, et al. 2012. Explosive Pleistocene range expansion leads to widespread Amazonian sympatry between robust and gracile capuchin monkeys. J. Biogeogr. 39: 272–288.

Araújo, L.L., J.S. Lima, K.G. Oliveira, J.A.P.C. Muniz, R.R. Valle and S.F.S. Domingues. 2009. Uso de solução à base de água de coco a 37°C como diluidor de sêmen de *Cebus apella* (macaco-prego) mantido em cativeiro. Cienc. Anim. Bras. (UFG). 10: 588–594.

Barnabe, R.C., M.A.B.V. Guimarães, C.A. Oliveira and A.H. Barnabe. 2002. Analysis of some normal parameters of the spermiogram of captive capuchin monkeys (*Cebus paella* Linnaeus, 1758). Braz. J. Vet. Res. Anim. Sci. 39: 331–333.

Brito, A.B., R.R. Santos, R. van den Hurk, J.S. Lima, M.S. Miranda, O.M. Ohashi, et al. 2013. Short-term culture of ovarian cortical strips from capuchin monkeys (*Sapajus apella*): a morphological, viability, and molecular study of preantral follicular development *in vitro*. Reprod. Sci. 20: 990–997.

Brito, D.C., A.B. Brito, S.R.R.A. Scalercio, S. Percário, M.S. Miranda, R.M. Rocha, et al. 2014. Vitamin e-analog trolox prevents endoplasmic reticulum stress in frozen-thawed ovarian tissue of capuchin monkey (*Sapajus apella*). Cell Tissue Res. 355: 471–480.

Campos, J.R. and A.C.J.S.R. Silva. 2011. Cryopreservation and fertility: current and prospective possibilities for female cancer patients. Obstet Gynecol. 2011; 2011: 350813.

Canale, G.R., A.C. Alonso, W.P. Martins, L. Jerusalinsky, F.R. de Melo, M.C.M. Kierulff et al. 2021. *Sapajus xanthosternos* (amended version of 2020 assessment). The IUCN Red List of Threatened Species 2021: e.T4074A192592138.

Comizzoli, P. 2015. Biotechnologies for wildlife fertility preservation. Anim. Front. 5: 73–78.

Di Bitetti, M.S. and C.H. Janson. 2001. Reproductive socioecology of tufted capuchins (*Cebus apella nigritus*) in northeastern Argentina. Int. J. Primatol. 22: 127–142.

Domingues, S.F.S., H.S. Ferreira, J.A.P.C. Muniz, A.K.F. Lima, O.M. Ohashi, J.R. Figueiredo, et al. 2003. Mechanical isolation of capuchin monkey (*Cebus apella*) preantral ovarian follicles. Arq. Bras. Med. Vet. Zootec. 55: 301–308.

Domingues, S.F.S., L.V. Diniz, S.H.C. Furtado, O.M. Ohashi, D. Rondina and L.D.M. Silva. 2004. Histological study of capuchin monkey (*Cebus apella*) ovarian follicles. Acta Amaz. 34: 495–501.

Domingues, S.F.S., M. Caldas-Bussiere, N. Martins and R. Carvalho. 2007. Ultrasonographic imaging of the reproductive tract and surgical recovery of oocytes in *Cebus apella* (capuchin monkeys). Theriogenology. 68: 1251–1259.

Domingues, S.F.S., M.C. Caldas-Bussiere, M.D.A. Petretski, O.M. Ohashi, J.S. Lima, R.R. Santos, et al. 2010. Effects of follicular phase and oocyte–cumulus complexes quality on the protein profile and *in vitro* oocyte meiosis competence in *Cebus apella*. Fertil Steril. 93: 1662–1667.

Donnez, J. and M.M. Dolmans. 2017. Fertility preservation in women. N. Engl. J. Med. 377:1657–1665.

IUCN. 2022. The IUCN Red List of Threatened Species. Version 2022-2. https://www. iucnredlist.org. Accessed on 20/11/2022.

Kindlovits, A. and L.M. Kindlovits. 2009. Primatas em Cativeiro: Classificação, Descrição, Biologia, Comportamento e Distribuição Geográfica. pp. 8–27. *In*: A. Kindlovits and L.M. Kindlovits (eds). Clínica e Terapêutica em Primatas Neotropicais. L.F. Livros, Rio de Janeiro, Brazil.

Leão, D.L., S.A. Miranda, A.B. Brito, J.S. Lima, R.R. Santos and S.F.S. Domingues. 2015. Efficacious long-term cooling and freezing of *Sapajus apella* semen in ACP-118®. Anim. Reprod. Sci. 159: 118–123.

Leão, D.L., A.B. Brito, S.A. Miranda, K.G. Oliveira, D.V. Almeida, R.R. Santos, et al. 2017. Extender supplementation with catalase maintains the integrity of sperm plasma membrane after freezing-thawing of semen from capuchin monkey. Zygote. 25: 231–234.

Lima, J.S., D.L. Leão, R.V. Sampaio, A.B. Brito, R.R. Santos, M.S. Miranda, et al. 2013. Embryo production by parthenogenetic activation and fertilization of *in vitro* matured oocytes from *Cebus apella*. Zygote. 21: 162–166.

Lima, A.R., S.B. Guimarães, E. Branco, E.G. Giese, J.A.P.C. Muniz, W.L.A. Pereira et al. 2015. Morphological and morphometric description of female reproductive tract of *Sapajus apella* (Capuchin monkey). Anat. Histol. Embryol. 44: 262–268.

Lima, M.C.M., S.R.R.A. Scalercio, C.T.A. Lopes, N.D. Martins, K.G. Oliveira, M.C. Caldas-Bussiere, et al. 2019. Monitoring sexual steroids and cortisol at different stages of the ovarian cycle from two capuchin monkey species: use of non- or less invasive methods than blood sampling. Heliyon. 5: e02166.

Lima, J.S., D.L. Leão, K.G. Oliveira, A.B. Brito, W.V. Sampaio, R.R. Santos, et al. 2017. Seminal coagulation and sperm quality in different social contexts in captive tufted capuchin monkeys (*Sapajus apella*). Am. J. Primatol. 9999: e"22643".

Ludwig, G., F.R. de Melo, W.P. Martins, J.M.D. Miranda, J.W. Lynch Alfaro, A.C. Alonso, et al. 2022. *Sapajus nigritus* (amended version of 2021 assessment). The IUCN Red List of Threatened Species 2022: e.T136717A210336199.

Martinez, G. and C. Garcia. 2020. Sexual selection and sperm diversity in primates. Mol. Cell. Endocrinol. Mol Cell. 18: 1–25.

Martins, A.B., M.S. Fialho, L. Jerusalinsky, M.M. Valença-Montenegro, B.M. Bezerra, P.O. Laroque, et al. 2021. *Sapajus libidinosus* (amended version of 2019 assessment). The IUCN Red List of Threatened Species 2021: e.T136346A192593226.

Mittermeier, R.A., K.E. Reuter, A.B. Rylands, L. Jerusalinsky, C. Schwitzer, K.B. Strier, et al. 2022. Primates in Peril: The World's 25 Most Endangered Primates 2022–2023. IUCN SSC Primate Specialist Group, International Primatological Society, Re: wild, Washington, DC.

Nagle, C.A. and J.H. Denari. 1982. The reproductive biology of capuchin monkeys Cebus spp. Int. Zoo. Yearbook. 22: 143–150.

Nagle, C.A. and J.H. Denari. 1983. The Cebus monkey (*Cebus apella*). pp. 39–67. *In*: J. Hearn (ed.). Reproduction in the New World Primates. New Model in Medical Science. MTP Press, Lancaster.

Napier, J.R. and P.H. Napier. 1985. Structure and Function. pp. 30–59. *In*: J.R. Naiper and P.H. Naiper (eds). The Natural History of the Primates. The MITT Press, Cambridge, Mass.

Oktay, K., D. Nugent, H. Newton, O. Salha, P. Chatterjee and R.G. Gosden. 1997. Isolation and characterization of primordial follicles from fresh and cryopreserved human ovarian tissue. Fertil. Steril. 67: 481–486.

Oliveira, K.G., P.H.G. Castro, J.A.P.C Muniz and S.F.S. Domingues. 2010. Conservação do sêmen e liquefação do coágulo seminal macaco-prego (*Cebus apella*) em água de coco em pó (ACP-118®), em diferentes temperaturas. Ciência Rural. 4: 617–621.

Oliveira, K.G., S.A. Miranda, D.L. Leão, A.B. Brito, R.R. Santos and S.F.S. Domigues. 2011. Semen coagulum liquefaction, sperm activation and cryopreservation of capuchin monkey (*Cebus apella*) semen in coconut water solution (CWS) and TES–TRIS. Anim. Reprod. Sci. 123: 75–80.

Paz, R.C.R., R.L. Zacariotti, R.H.F. Teixeira and M.A.B.V. Guimarães. 2006. O efeito das enzimas hialuroni¬dase e tripsina na liquefação do sêmen de macacos-pregos (*Cebus apella*). Braz. J. Vet. Res. Anim. Sci. 43: 196–201.

Phillips, K.A., I.S. Bernstein, E.L. Dettmer, H. Devermann and M. Powers. 1994. Sexual behavior in brown capuchins (*Cebus apella*). Int. J. Primatol. 15: 907–917.

Rímoli, J., L. Smith, G. Ludwig, M. Martinez, M. Kowalewski, R. Melo, et al. 2022. *Sapajus cay*. The IUCN Red List of Threatened Species 2022: e.T136366A215548623.

Rowe, N. 1996. The Pictorial Guide to the Living Primates. East Hampton: Pogonias Press.

Santana, L.N.N., A.B. Brito, D.C. Brito, J.S. Lima, S.F.S. Domingues and R.R. Santos. 2013. Adaptation of a trap door technique for the recovery of ovarian cortical biopsies from *Cebus apella* (capuchin monkey). Zygote (Cambridge. Print). 21: 158–161.

Silva, J.S. Jr. 2001. Especiação em macacos-prego e caiararas, gênero Cebus Erxleben, 1777 (Primates, Cebidae). Rio de Janeiro (Brazil): Universidade Federal do Rio de Janeiro.

Silva, J.S. Jr. 2002. Taxonomy of capuchin monkey, Cebus Erxleben, 1777. Neotrop. Primates. 10: 29.

Silva, J.S. Jr., C.R. Silva and T.P. Kasecker. 2008. Primatas do Amapá: Guia de identificação de Bolso. Panamérica Editorial Ltda, Bogotá, Colombia.

Texeira, D.G. 2005. Estudo anatômico descritivo dos órgãos genitais masculinos do macaco-prego (*Cebus apella* Linnaeus, 1758). São Paulo, Tese (Doutorado em Ciências)-Universidade de São Paulo, São Paulo.

Valença-Montenegro, M.M., B.M. Bezerra, A.B. Martins, L. Jerusalinsky, M.S. Fialho and J.W. Lynch Alfaro. 2021. *Sapajus flavius* (amended version of 2020 assessment). The IUCN Red List of Threatened Species 2021: e.T136253A192592928.

Varghese, A.C., S.S. Plessis, T. Falcone and A. Agarwal. 2008. Cryopreservation/transplantation of ovarian tissue and *in vitro* maturation of follicles and oocytes: challenges for fertility preservation. Reprod. Biol. Endocrinol. 6: 47.

Part IV

Rodentia

Part Co-Editor: Alexsandra F. Pereira
Laboratory of Animal Biotechnology
Federal Rural University of Semi-Arid (UFERSA)
Mossoró, RN, Brazil
alexsandra.pereira@ufersa.edu.br

Chapter 14

The Agoutis

Alexsandra Fernandes Pereira[1]*, Érika Almeida Praxedes[1],
Leonardo Vitorino Costa de Aquino[1],
Lhara Ricarliany Medeiros de Oliveira[1],
Maria Valéria de Oliveira Santos[1] and Alexandre Rodrigues Silva[2]

[1]Laboratory of Animal Biotechnology, Federal Rural University of Semi-Arid (UFERSA),
Av. Francisco Mota, 572, Costa e Silva, Mossoró, RN, Brazil, 59625900.
Email: alexsandra.pereira@ufersa.edu.br; erikaalmeida-@hotmail.com
leonardovt@live.com; lharagirs@hotmail.com; valeriasnts07@gmail.com

[2]Laboratory on Animal Germplasm Conservation,
Federal Rural University of Semi-Arid (UFERSA),
Av. Francisco Mota, 572, Costa e Silva, Mossoró, RN, Brazil, 59625900
Email: alexrs@ufersa.edu.br

INTRODUCTION

The success of assisted reproduction techniques (ARTs) in wild mammals relies on anatomical knowledge as well as the general mechanisms that regulate the reproductive cycle of the species (Silva et al. 2017). This approach does not differ when related to wild rodent species, which has been explored to develop protocols associated to their conservation, involving the understanding of their reproductive biology, making it possible to assist in the storage and exchange of genetic material among populations (Lall et al. 2022).

Among the species of wild rodents, the agoutis—the name popularly given to the species of the genus *Dasyprocta*—stand out as excellent seed dispersers, being important in the recovery of tropical forests (Mittelman et al. 2020). In general, the genus *Dasyprocta* comprises 13 species, of which 10 are in South America (Table 14.1; Patton and Emmons 2015), while *D. coibae, D. mexicana,* and

*Corresponding author: alexsandra.pereira@ufersa.edu.br

Table 14.1   Geographical distribution of specimens of the genus *Dasyprocta* from South America

Specimens	Conservation status	Population location	Fur color	Specimen weight (kg)	ARTs developed
D. azarae	Data Deficient	Argentina, Bolivia, Brazil, and Paraguay	Black and orange	3.0	Collection and cryopreservation/ conservation of sperm
D. croconota	Data Deficient	Brazil	Orange	2.0	–
D. fuliginosa	Least Concern	Brazil, Colombia, Ecuador, Peru, and Venezuela	Black	2.0–3.0	Reproductive monitoring and estrus control methods
D. guamara	Near Threatened	Venezuela	Dark coloring	2.7–3.3	–
D. iacki	Data Deficient	Brazil	Orange back	1.1–1.5	–
D. kalinowski	Data Deficient	Peru	Dark and lighter regions	2.0	–
D. leporina	Least Concern	Brazil, French Guyana, Guyana, Suriname, and Venezuela	Reddish	2.0–3.0	Collection and cryopreservation/ conservation of sperm, reproductive monitoring, estrus control methods, embryo production and somatic resources banks
D. prymnolopha	Least Concern	Brazil	Orange Coffee/dark	1.0–2.6	Estrus control methods, embryo production, and somatic resources banks
D. punctata	Least Concern	Colombia, Ecuador, and Panama	Uniform reddish Brown	2.0–4.0	–
D. variegata	Data Deficient	Brazil, Bolivia, and Peru	Brown	1.0–2.5	–

Source: International Union for Conservation of Nature (IUCN 2022). *kg: weight in kilograms of specimens of the genus *Dasyprocta*

D. ruatanica are located between the extreme South of Mexico and the islands of Panama. Nevertheless, human actions that involve deforestation, loss of habitat, and consumption of its meat have caused a significant reduction in its population, especially of the quantitatively captive species (Vanegas et al. 2016).

Given the importance of the species, *ex situ* conservation efforts by ARTs are alternatives to allow the maintenance of species. For ARTs related to the genus, we highlight studies developed on males and females of *D. azarae*, *D. fuliginosa*, *D. prymnolopha* and *D. leporina*, with *D. leporina* (Figure 14.1) having a greater number of studies among the species. These studies cover from reproductive monitoring to *in vitro* embryo production aiming at cloning.

Figure 14.1 Representative image of the *Dasyprocta* genus: a male red-rumped agouti (*Dasyprocta leporina*) specimen.

Therefore, according to benefits promoted using ARTs, as well as the importance of agoutis in terms of ecological, populational and reproductive aspects, this chapter describes the scientific advances achieved in males and females of the *Dasyprocta* genus using ARTs. Additionally, it emphasizes the difficulties observed in the development of efficient protocols and how they can contribute to the preservation of these small rodents.

MONITORING AND CONTROL OF THE MALE REPRODUCTION

For the development of ARTs in agouti males, morphology and histology techniques have been used to understand the reproductive physiology of these animals, especially in *D. leporina*. From birth to the fifth month of life, *D. leporina* are considered prepubertal and the testes show little growth, as well as the absence of tubular lumen in the testicular cords (Arroyo et al. 2017). In the sixth month of life the prepubertal phase begins, in which testicular cords are in the process of tubular lumination and the first signs of spermatogenesis are observed with the differentiation of first primary spermatocytes and rounded spermatids (Costa et al. 2010). This phase extends until the eighth month, and at nine months the male agoutis have puberty established, making it possible to observe the presence of all cell types and spermatozoa in the testicular lumen (Assis Neto et al. 2003).

Using histological tools, it was observed that the mean duration of the seminiferous epithelium cycle in *D. leporina* is 9.5 ± 0.03 days, and since approximately 4.5 cycles are necessary for the spermatogenic process to be completed, the total length of spermatogenesis was 42.8 ± 0.16 days (Costa et al. 2010). Dubost et al. (2005) observed that males are generally in reproductive activity throughout the year; nevertheless, fruit production was linked to the percentage of individuals with spermatozoa. Moreover, in semiarid regions during dry season, wind speed and solar radiation have a strong negative effect on the quality of sperm parameters of *D. leporina* (Dantas et al. 2021).

According to morphological aspects, the reproductive system in the agouti is composed of the testes, epididymis, accessory glands, and penis. The tests of agoutis are oval and are located intra-abdominally. According to Mollineau et al. (2006), on average the testes are 3.67 ± 0.12 cm in length and 1.67 ± 0.04 cm in diameter, which can be seen as two protuberances next to the inner region of the hind legs. The epididymis runs the entire length of the testes with the coiling intensity increasing towards the cauda epididymis and straightening after leaving the cauda epididymis in paired ducts (Mollineau et al. 2006). The accessory sex glands include seminal vesicles, coagulating glands, prostate, and bulbourethral glands (Menezes et al. 2010). The penis is cylindrical, mean length is 9.90 ± 0.43 cm, caudally directed, and U-shaped, containing paired ventral keratinous spicules (Menezes et al. 2003).

SPERM COLLECTION AND CONSERVATION

Sperm collection is a key step in the application of artificial insemination (AI) and *in vitro* fertilization (IVF). Most of the related studies were carried out on *D. leporina*, but advances were also made in *D. azarae*. Spermatozoa have been recovered using the electroejaculation technique (*D. leporina*, and *D. azarae*) and by retrograde epididymal washing (*D. leporina*). Semen collection by electroejaculation has variable efficiency in relation to the presence of sperm in the ejaculate, ranging from 30 to 100% (Table 14.2).

The anesthetic protocol used in the preparation of the animal for electroejaculation has great influence on the result. Since the first report of electroejaculation in *D. leporina*, the protocol has been improved by the inclusion of low doses of xylazine (40 mg/kg) in combination with ketamine hydrochloride intramuscularly five minutes before electroejaculation, resulting in an increase from 30 to 75% of ejaculate samples containing sperm (Mollineau et al. 2008, 2010a).

Table 14.2 Efficiency of agouti sperm collection and conservation

Species	Collection efficiency (%)	Volume (mL)	Concentration × 10⁶/mL	Motility (%)	Conservation method	Motility frozen thawed (%)	References
Electro-ejaculation							
D. azarae	100	NA	NA	50	NA	NA	Martinez et al. (2013)
D. leporina	30	0.47	106.7	50.3	NA	NA	Mollineau et al. (2008)
D. leporina	75	0.35	431.0	47.2	NA	NA	Mollineau et al. (2010a)
D. leporina	71.4	0.26	635.9	≥50	Cooling at 5°C	59.5	Mollineau et al. (2010b)
D. leporina	57	0.15	75.0	63.8	NA	NA	Castelo et al. (2015)
Epididymis–Retrograde flush							
	100	0.03	748.0	86.5	NA	NA	Ferraz et al. (2011)
D. leporina	100	0.30	140.0	91.5	Cryopreservation	26.5	Silva et al. (2012)
	100	1.65	104.0	95.9	Cryopreservation	47.5	Castelo et al. (2015)

NA: Not assessed.

A different anesthetic protocol used in the species *D. azarae* achieved an efficiency of 100%. In the present study, pre-anesthesia intramuscular was performed with azaperone (4 mg/kg) and meperidine (4 mg/kg); then, 10 min later anesthesia was induced intramuscularly with xylazine hydrochloride (0.4 mg/kg) and ketamine hydrochloride (20 mg/kg), and 5 min later a lumbosacral application of 5 mg/kg lidocaine was performed (Martinez et al. 2013). Furthermore, Martinez et al. (2013) used a different stimulus sequence with a wild-type device. For *D. azarae*, ejaculation occurred when stimuli of 6 V were used, while for *D. leporina*, an average of 9.33 V was needed (Mollineau et al. 2008). The use of a device with ring electrodes and emitting sine waves, associated with a protocol of serial stimuli, also improved the electroejaculation efficiency in *D. leporina*, increasing from approximately 41.3% (Mollineau et al. 2010a) to 57% of samples containing sperm (Castelo et al. 2015).

Retrograde epididymal washing has been shown to be very useful and efficient for obtaining *D. leporina* sperm. In general, the efficiency is 100%, and it is possible to obtain a good volume of quality samples (Table 14.2). This method can be performed after orchidectomy (Ferraz et al. 2011) or euthanasia (Silva et al. 2011), and the highest sperm concentration and motility is found in the caudal region of the epididymis (Dantas et al. 2022). Different solutions can be used for retrograde washing, including physiologic salt solution (Ferraz et al. 2011), powdered coconut water (ACP®-109c; ACP Biotecnologia, Brazil) and Tris extenders (Silva et al. 2011).

After sperm collection, conservation tools can be applied to preserve genotypes and prolong the lifetime of these gametes. Cryopreservation allows for long-term conservation, with the ACP® –109c being a better extender than Tris for *D. leporina* sperm processing and cryopreservation maintaining 26.5% of motile sperm compared to 9.7% (Silva et al. 2011). On the other hand, ultra-high-temperature (UHT) milk showed a better motility rate after freeze-thaw compared to unpasteurized or pasteurized coconut water, but the rate of 12.5% appears to be lower than the 26.5% reported by Silva et al. (2011) using ACP®-109c.

Castelo et al. (2015) recommend, for *D. leporina,* sperm cryopreservation by glycerol (3 or 6%) or dimethyl sulfoxide (DMSO; 3%) and 20% egg yolk, compared to ethylene glycol (EG; 6%), dimethylformamide (6%), and DMSO 6%. Furthermore, epididymal sperm of agouti can be efficiently cryopreserved in 0.25 mL or 0.50 mL straws and thawing should be conducted at 37°C for 60 s, preserving 26.6% sperm motility (Silva et al. 2012). For short-term conservation, refrigeration at 5°C using UHT milk achieved better sperm motility after 24 h of storage with value of 59.5%, compared to unpasteurized or pasteurized coconut water (Mollineau et al. 2010b).

TESTICULAR TISSUE CRYOPRESERVATION AND CULTURE

In addition to sperm collection and cryopreservation, another possibility has arisen as a promising technology to safeguard the genetic material of males, which is the cryopreservation and culture of testicular tissue. For the genus *Dasyprocta*, only one testicular cryopreservation protocol was performed in *D. leporina* (Silva et al. 2022). Fragments (3.0 mm³) were recovered by *post-mortem* collection and compared among three cryopreservation techniques. For slow freezing, the use of 1.5 M of DMSO, 1.5 M of EG, or the association of these cryoprotectants at 0.75 M were used. In both vitrification techniques, direct vitrification in cryovials or solid-surface vitrification (SSV), the use of cryoprotectants at a concentration of 3.0 M DMSO, 3.0 M EG, or the combination of these solutions at 1.5 M were employed. As a result, DMSO and EG in SSV were considered the best testicular cryopreservation technique in *D. leporina*, ensuring greater maintenance of the tissue structural morphology, as well as the proliferative potential of spermatogonia, Leydig and Sertoli cells. Therefore, the first testicular cryopreservation protocol was effectively obtained for the species and represented an initial step for the development of methodologies aimed at the conservation of the species through gonadal tissues.

Finally, in agoutis, the physiological and reproductive behavior becomes unique when compared to other rodents, and characteristics such as these are still challenging when they are related to sperm processing for species of the genus *Dasyprocta*. In addition to sperm processing, another technology that has reached initial steps is testicular cryopreservation and culture. Although the results from said technology are promising, only one protocol has been described so far, making it necessary to obtain more information about the technique, since later samples show how it can be used in the development of protocols involving AI and IVF.

MONITORING AND CONTROL OF THE FEMALE REPRODUCTION

With the monitoring of reproductive parameters in females, it is possible to achieve greater reproductive control in captive animals, as well as application and advances in ARTs (Guimarães et al. 2009, Lall et al. 2022). Studies have sought to understand the basic reproductive biology of species of the genus *Dasyprocta* through the elucidation of anatomical, morphological, and physiological parameters (Almeida et al. 2003, Campos et al. 2015, Singh et al. 2016). Notwithstanding strategies aimed at the reproductive control, there is still much to be elucidated for agoutis, where only initial steps have been reported, with one study directed at inducing estrus in *D. leporina* (Peixoto et al. 2018).

Thus, as general reproductive aspects, *D. prymnolopha* showed continuous polyestral not influenced by the seasons, giving rise to one to two offspring per pregnancy, with a gestation ranging from 103 to 120 days (Guimarães et al. 1997, Sousa et al. 2012). Puberty begins early, after about nine months, and its onset is influenced by the presence of males in the same environment (Guimarães et al. 2009). Using histological tools, the anatomical and morphological characterization of ovaries and uterine tubes in agouti showed an ovarian pattern macro and microscopically similar to that observed in sexually active rodents, and uterine tubes similar to mammals in general (Almeida et al. 2003, Fortes et al. 2005).

Additionally, the functional anatomy of genital organs in *D. fuliginosa* and *D. leporina* during estrous cycle and pregnancy shows changes in the uterine and vaginal epithelium features in accordance with the reproductive state of the female, presenting a thicker vaginal and uterine epithelium in the follicular phase and a reduction in thickness in the luteal phase (Mayor et al. 2011, Oliveira et al. 2019).

Regarding the monitoring of the estrous cycle, studies conducted in both *D. leporina* and *D. prymnolopha* using vaginal cytology report a high presence of superficial cells at proestrus and estrus phases, followed by the predominance of intermediate cells in the metestrus and parabasal cells in diestrus, highlighting small variations between species in terms of cycle duration 28.2 ± 0.7 days and 30.69 ± 4.65 days, respectively (Guimarães et al. 1997, Campos et al. 2015). With the use of ultrasonography in the same species, the absence of ovarian morphological changes was noted throughout the different phases of the estrous cycle (Campos et al. 2015, Carreiro et al. 2018). Nevertheless, follicles were observed during the estrogenic phases (proestrus and estrus), with an average diameter of 1.0 ± 0.5 mm and in 12.5% of luteal phases, corpora lutea measuring 1.4 ± 0.9 mm (Campos et al. 2015). Furthermore, by monitoring hormonal parameters by saphenous venipuncture, with 17β estradiol (E2) and progesterone (P4) concentrations being measured via enzyme linked immunosorbent assay (ELISA), E2 peaks during proestrus (2998 pg/mL) at estrus (3500 pg/mL) and in early metestrus (2363 pg/mL) were observed along the phases of the estrous cycle referring to concentrations, and P4 reached a peak concentration of 4.26 ng/mL in diestrus in captive *D. leporina* (Singh et al. 2016).

Furthermore, the use of ultrasound proved to be effective in the diagnosis of early pregnancy in *D. prymnolopha*, helping to elucidate information on embryonic and fetal structures that could be used to predict gestational age and birth, thereby contributing to their reproductive management in captivity (Sousa et al. 2012). The gestational age and biometry of embryos and fetuses at different days of development were performed in *D. prymnolopha* (Fortes et al. 2013) and *D. leporina* (Oliveira et al. 2017). Furthermore, regarding gestational parameters, monitoring the hormonal dynamics of the progesterone profile throughout pregnancy using radioimmunoassay showed an increase in progesterone concentration between the 1st and 5th week of gestation, with a peak of 6.88 ng/mL from the 5th week of pregnancy. After the 6th week, a decrease in this concentration begins (Guimarães et al. 2016). Thus, such information provides a basis for knowledge that can assist in the application of ARTs, such as embryo transfer and embryonic stem cell isolation (Oliveira et al. 2017).

As for the control of reproduction *in vivo*, initial steps aimed at inducing estrus were taken in *D. leporina*, reporting the use of prostaglandin alone or in association with a gonadotropin-releasing hormone (GnRH) analogue (Peixoto et al. 2018). Nevertheless, the evaluated protocols showed a limited efficiency in females that are in the luteal phase at the beginning of the treatment. Thus, it is noted that for the elucidation of reproductive parameters in female agoutis, a greater range of studies are concentrated in the *D. leporina* and *D. prymnolopha* species, with only some studies on *D. fuliginosa*. Additionally, there are also few studies on the control of *in vivo* reproduction using *D. leporina*.

ANTRAL AND PREANTRAL FOLLICLE MANIPULATION

The elucidation obtained by the reproductive monitoring of the agouti allows the practical application of its material, such as studies with the preantral and antral follicles. Preantral ovarian follicles (PAFs) are defined as those presenting an oocyte surrounded by one or more layers or granulosa cells, with absence of the antrum, classified as primordial, primary, and secondary. These structures represent about 90–95% of the ovarian follicles, being an important source of female germplasm for use in many ARTs, such as manipulation of oocytes enclosed in preantral follicles (MOEPF).

Initial steps with MOEPF were established in agoutis by describing the follicular population of these animals. Santos et al. (2018) furthered the characterization of the follicular population in *D. leporina*, where the authors reported a mean per pair of ovaries at 4419.8 ± 532.26 and 5397.52 ± 574.91 for right and left ovaries, respectively. Most of the follicles found belonged to the primordial category (86.63%), followed by primary (13.01%) and secondary (0.35%). Moreover, with studies of ovarian tissue cryopreservation of *D. leporina,* it was possible to observe the preservation of these follicles using a conventional freezing method with propanediol at 1.5 M, where up to 64% PAFs were preserved (Wanderley et al. 2012), as well as the application of an open and closed cryopreservation system, which were equally efficient in preserving agouti PAF morphology and DNA integrity (Praxedes et al. 2021b).

Moreover, Praxedes et al. (2018) was able to evaluate the development of fresh and vitrified *D. leporina* ovarian tissue after xenografting C57Bl/6 in female mice with severe combined immunodeficiency (SCID). The authors reported that according to the characteristics of the external genitalia and the predominance of cornified cells on vaginal cytology, ovarian activity returned in 80.0% of the xeno-fresh group and 16.7% in the xeno-vitrified group, a mean of 20.6 ± 8.6 days after xenotransplantation. After graft removal, a predominance of primordial and primary follicles was observed in all grafts, demonstrating the technique efficiency in *D. leporina.*

Regarding antral follicles, these structures are formed when the follicle gains a fluid-filled cavity known as the antrum. They are 200 times bigger than a primordial follicle and are widely used as they contain immature oocytes for ARTs. These structures have already been measured in *D. fuliginosa*, in which the authors observed the follicular distribution of antral follicles according to the reproductive state of the female, where the smallest and largest antral follicles were of a diameter of 0.25 mm and 2.34 mm, respectively (Mayor et al. 2011).

Likewise, the application of antral follicles is highlighted in studies aimed at the *in vitro* maturation of oocytes in agoutis (Table 14.3). Through such

Table 14.3   *In vitro* maturation and *in vitro* embryo production data in agouti

Species	Methods for oocyte recovery	Oocyte recovery rate (% ± SE)	Oocyte viability, %	IVM duration	1PB (% ± SE)	References
D. prymnolopha	Slicing	17.0 ± 1.9	9.7	–	–	Ferraz et al. (2016)
D. prymnolopha	Slicing	–	–	24 h	27.4 ± 2.0 (17/43)	Ferraz et al. (2020)
D. leporina	Slicing	18.7 ± 7.2	92.8	24 h	37.3 ± 7.7 (19/51)	Praxedes et al. (2020b)

Species	IVEP protocol	Chemical activation/IVF parameters	Embryo culture medium	Cleavage rate D2 (%)	Cleavage rate D7 (%)	References
D. prymnolopha	AA	5 µM ionomycin + 2 mM 6–DMAP	SOF	63.6 (21/33)	15.1 (5/33)	Ferraz et al. (2020)
	IVF	25 × 10^6 sptz/mL 15 h	SOF	8.6 (3/35)	2.9 (1/35)	
D. leporina	AA	10 mM SrCl$_2$ + 5 µg/mL cytochalasin B	SOF	43.2 (19/44)	6.8 (3/44)	Praxedes et al. (2020b)

SE: standard error. IVM: *in vitro* maturation. 1PB: first polar body. AA: artificial activation. IVF: *in vitro* fertilization. 6-DMAP: 6-dimethylaminopurine. SrCl$_2$: strontium chloride. SOF: synthetic oviductal fluid.

research, it is possible to obtain the rate of oocyte recovery of these follicles and their subsequent development. In *D. prymnolopha*, the morphological parameters were evaluated of cumulus-oocyte complexes (COCs) obtained by the ovarian slicing technique. This species presented an average of 8.5 COCs per ovary, or 17.0 COCs per agouti. As for their quality parameters, about 3.5% were grade I, 6.2% grade II, 68.2% grade III, 13.5% grade IV and 8.6% degenerate. In *D. prymnolopha*, different *in vitro* maturation protocols were tested with immature oocytes obtained by slicing, where maturation for 24 h provided the best results, with about 27.4% presenting the first polar body (1PB) (Ferraz et al. 2020). Similar studies were carried out on *D. leporina*, where it was observed that applying the slicing technique it was possible to obtain an average of 18.7 oocytes per female, with 92.8% oocyte viability. As for the *in vitro* maturation protocol, after 24 h it was observed that 37.3% of oocytes presented the 1PB (Praxedes et al. 2020b).

These findings reinforce the importance of studying all ovarian material of the species, whether antral or preantral follicles, in addition to highlighting the lack of research on different species of agouti, since only three out of the thirteen have been explored for ovarian material so far. Nevertheless, with these data, it was possible to make advancements with ARTs, generating the production of embryos in agoutis.

EMBRYO TECHNOLOGY

In vitro embryo production (IVEP), despite being well developed in laboratory rodents such as rats and mice, is still in its early stages in agouti. The two main lines of research on the *Dasyprocta* genus are the establishment of IVF and the optimization of protocols for artificial activation (AA) of *in vitro* matured oocytes.

For AA in *D. prymnolopha*, a protocol adapted from bovines was used, where the authors used 5 µM of ionomycin as primary activator, and 2 mM of 6-dimethylaminopurine (6-DMAP) as secondary activator. The structures were cultured in synthetic oviductal fluid (SOF) medium for seven days. The cleavage rate in D2 was 63.6% and embryonic development in D4 was 15.1%. Nevertheless, embryo development did not go beyond the morula stage, and at D7 no blastocyst was observed.

In *D. leporina*, the authors chose to follow a protocol based on rodents, using a combination of 10 mM of $SrCl_2$ and 5 µg/mL of cytochalasin B for 6 h in Tyrode Albumin Lactate-Free Calcium Pyruvate (TALP) medium. They also chose to culture in SOF medium for seven days. Regarding embryonic development, 43.2% (19/44) of cleaved structures was observed, 6.8% (3/44) of morulae in relation to the number of oocytes, and 18.8% (3/16) of morulae in relation to the number of cleaved structures. Like D. *prymnolopha* at D7, no blastocyst was observed (Praxedes et al. 2020b). These data are thoroughly presented in Table 14.3.

Finally, as for IVF, an initial protocol was used similar to that of bovines, where the matured oocytes were transferred to drops of IVF medium with epididymal sperm. The final concentration was adjusted to 25×10^6 spermatozoa/mL, with IVF medium, which were added to the microdroplets containing the

COCs. The co-incubation period of COCs and sperm was 15 h. In D2, there was observation of embryo with two cells in 8.6% (3/35) of the COCs, and in D4 the presence of morula was observed in 2.9% (1/35) of the COCs. None of the oocytes submitted to IVF reached the blastocyst stage on D7, stopping their development at the morula stage (Ferraz et al. 2020).

Considering these data, it is possible to conclude that several steps still need to be established to increase the efficiency in the technique of producing embryos in agoutis.

SOMATIC RESOURCE BANKS AND CLONING

The creation of somatic biobanks aiming at the storage of either tissue or cells has been an explored approach with great potential for the multiplication and conservation of wild species. This is due to the possibility of using these cells in biotechnologies such as somatic cell nuclear transfer (SCNT) or in the induction of pluripotency cells (iPSCs) (Praxedes et al. 2020a). To explore this source, steps such as harvesting, transport, processing, immediate *in vitro* culture, or cryopreservation aiming at a later cell isolation (Figure 14.2) must be carried out taking into account aseptic conditions. For the genus *Dasyprocta*, this approach has been, observed mainly in animals kept in captivity, especially in *D. leporina* and *D. prymnolopha* (Carvalho et al. 2019, Praxedes et al. 2021a). Furthermore, among the advantages of using somatic resources for conservation purposes, the recovery of samples from live or dead animals of both sexes can be cited (Mattos et al. 2022) such as easier tissue collection, recovery of a high number of cells from a few tissue explants, and the possibility of recovering several cell types (Praxedes et al. 2021a).

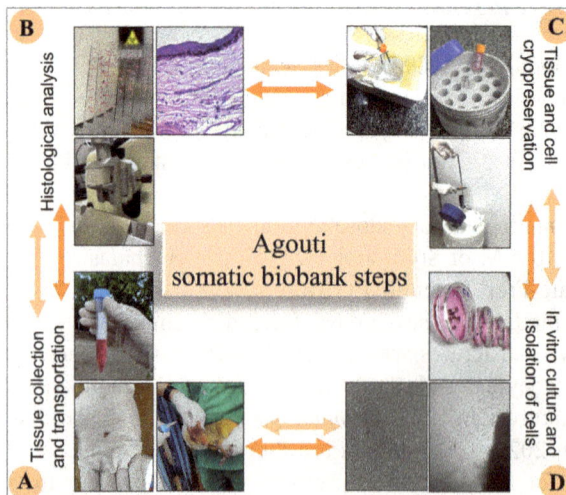

Figure 14.2 Steps in the establishment of somatic biobanks for the *Dasyprocta* genus. (a) Tissue collection and transportation. (b) Histological analysis. (c) Tissue and cell cryopreservation. (d) *In vitro* culture and isolation of cells.

For the effective implementation of somatic biobanks, it is important to ensure the subsequent quality of the stored sample (Costa et al. 2020). Initially, the histological knowledge of the region to be stored is a relevant approach helping to define ideal conditions for cryopreservation of tissue samples, such as composition of solution for cryopreservation and the best technique to be used for storage in somatic banks (Pereira et al. 2021). In *D. leporina*, evaluating skin and cartilage from the ear region, some species-specific particularities were observed such as thickness of skin layers as well as an elevated number of epidermal cells and chondrocytes (Pereira et al. 2021).

Additionally, with regard to the effective storage of biological samples in somatic banks, SSV showed advantageous results compared to direct vitrification in cryotubes, and this technique is highlighted as advantageous for its easy execution and can be performed in the field (Costa et al. 2020). Furthermore, the use of 1.5 M EG, 1.5 M DMSO and 0.25 M of sucrose was found to be the optimal cryoprotectant solution in terms of maintenance of skin thickness, number of perinuclear halos, proliferative potential, number of empty lacunae and degenerated chondrocytes (Rodrigues et al. 2021).

Regarding the isolation and characterization of cell lines, different cell sources can be explored with a view to their use as a nucleus for cloning or source for iPSCs (Praxedes et al. 2021a). In *D. prymnolopha*, the isolation of hematopoietic progenitor cells from bone marrow as well as progenitor cells from dental pulp and umbilical cord are reported (Rocha et al. 2012, Carvalho et al. 2015, 2019). Additionally, cell lines obtained from adult fibroblasts isolated from *D. leporina* were established showing high proliferative and metabolic quality up to the 8th passage as well as minimal quality changes after cell cryopreservation (Praxedes et al. 2021a). Thus, it is noted that there are still few results from the use of somatic resources for conservation in agoutis, and that initial steps have been taken in the species *D. leporina*.

FINAL CONSIDERATIONS

Advances related to ARTs in agoutis have been achieved in several lines of research, considering reproductive aspects of males, females, as well as somatic cells. Nevertheless, essential information is lacking so that techniques such as embryo transfer, AI, IVEP, SCNT, and iPSC can be successfully applied to these animals. It is possible to observe that most studies focus on four species of agoutis, *D. leporina, D. prymnolopha, D. fuliginosa*, and *D. azarae*. Among these agoutis, *D. leporina* stands out in a plenty of studies, likely due to the greater availability of individuals, and may be a greatly advantageous model for the conservation of threatened species of the same genus. Although similarities within the genus are expected, differences have also been found that reinforce the need for studies on different species.

Therefore, we emphasize the importance of gaining knowledge on deeper morphological and physiological characteristics of male and female reproduction, as well as *in vitro* manipulation of gametes, embryos, and somatic cells. Such knowledge will surely favor future applications of ARTs and the development of sustainable breeding of these animals, aiming at multiplication and conservation of the biodiversity of agouti species.

REFERENCES

Almeida, M.M.D., M.A.M.D. Carvalho, M.F. Cavalcante Filho, M.A. Miglino and D.J.A.D. Menezes. 2003. Morfological and morfometric study of the ovary in agoutis (*Dasyprocta aguti*, Linnaeus, 1766). Braz. J. Vet. Res. Anim. Sci. 40: 55–62.

Arroyo, M.A.M., F.F.S. Silva, P.R.S. Santos, A.R. Silva, M.F. Oliveira and A.C. Assis Neto. 2017. Ultrastructure of spermatogenesis and spermatozoa in agoutis during sexual development. Reprod. Fertil. Dev. 29: 383–393.

Assis Neto, A.C.D., M.I.V.D. Melo, M.A.M.D. Carvalho, M.A. Miglino, M.F.D. Oliveira, D.J.A.D. Menezes, et al. 2003. Qualitative analysis of the established spermatogenesis in agoutis (*Dasyprocta aguti*) rised in captivity. Braz. J. Vet. Res. Anim. Sci. 40: 180–184.

Campos, L.B., G.C. Peixoto, G.L. Lima, T.S. Castelo, A.L. Souza, M.F. Oliveira, et al. 2015. Monitoring of the estrous cycle of agoutis (*Dasyprocta leporina* Lichtenstein, 1823) by vaginal exfoliative cytology and ultrasonography. Pesq. Vet. Bras. 35: 188–192.

Carreiro, A.N., J.A. Diniz, J.G. Souza, D.V.F. Araújo, R.F.F. Dias, L.M.S. Azerêdo, et al. 2018. Ovary and vaginal epithelium dynamics during the estrous cycle in *Dasyprocta prymnolopha* Wagler, 1831: ultrasound and cytological examinations. J. Vet. Sci. 19: 446–451.

Carvalho, Y.K., N.M. Argôlo-Neto, C.E. Ambrósio, L.D.J.D. Oliveira, A.R.D. Rocha, J.B.D. Silva, et al. 2015. Isolation, expansion and differentiation of cellular progenitors obtained from dental pulp of agouti (*Dasyprocta prymnolopha* Wagler, 1831). Pesq. Vet. Bras. 35: 590–598.

Carvalho, M.A.M., N.M. Argôlo-Neto, E.R.D.F. Siqueira, Y.K.C. Leite, G.T. Pessoa, D.O. Bezerra, et al. 2019. Structural plasticity and isolation of umbilical cord progenitor cells of agouti (*Dasyprocta prymnolopha*) raised in captivity. Semina: Ciênc. Agrar. 40: 225–238.

Castelo, T.S., A.M. Silva, L.G.P. Bezerra, C.Y.M. Costa, A.E.A. Lago, J.A.B. Bezerra, et al. 2015. Comparison among different cryoprotectants for cryopreservation of epididymal sperm from agouti (*Dasyprocta leporina*). Cryobiology 71: 442–447.

Costa, G.M., M.C. Leal, C.S. Ferreira, D.A. Guimarães and L.R. França. 2010. Duration of spermatogenesis and spermatogenic efficiency in 2 large neotropical rodent species: the agouti (*Dasyprocta leporina*) and paca (*Agouti paca*). J. Androl. 31: 489–499.

Costa, C.A., A.A. Borges, M.B. Nascimento, L.V.C. Aquino, A.R. Silva, M.F. Oliveira, et al. 2020. Effects of vitrification techniques on the somatic tissue preservation of agouti (*Dasyprocta leporina* Linnaeus, 1758). Biopreserv. Biobank. 18: 165–170.

Dantas, M.R.T., A.M. Silva, A.G. Pereira, C.S. Santos, J.B.F. Souza-Junior and A.R. Silva. 2021. Environmental conditions affect the sperm quality of agoutis (*Dasyprocta leporina*) during the dry period of a semiarid region. Anais—V Reunião ABRAA 5: 249–252.

Dantas, M.R., A.M. Silva, L.G. Bezerra, A.G. Pereira, N.R. Luz, J.B. Souza-Junior, et al. 2022. Morphological, morphometric, ultrastructural, and functional evaluation of red-rumped agouti (*Dasyprocta leporina*) sperm during epididymal transit. Anim. Reprod. Sci. 243: 107029.

Dubost, G., O. Henry and P. Comizzoli. 2005. Seasonality of reproduction in the three largest terrestrial rodents of French Guiana forest. Mamm. Biol. 70: 93–109.

Ferraz, M.S., D.J.A. Menezes, G.T. Pessoa, R.M. Cabral, M.J. Illera, A.R. Silva, et al. 2011. Collection and evaluation of epididymal sperm in captive agoutis (*Dasyprocta aguti*). Theriogenology 75: 459–462.

Ferraz, M.S., F.J. Morais Junior, M.L.T. Feitosa, H.M. Almeida, D.O. Bezerra, G.T. Pessoa, et al. 2016. Ovarian slicing technique to obtain oocytes from agoutis (*Dasyprocta prymnolopha*). Pesq. Vet. Bras. 36: 204–208.

Ferraz, M.S., M.A.M.D. Carvalho, F.J. Moraes Junior, M.L.T. Feitosa, M. Bertolini, H.M. Almeida, et al. 2020. *In vitro* maturation of agoutis (*Dasyprocta prymnolopha*) oocytes followed by *in vitro* fertilization and parthenogenetic activation. Arq. Bras. Med. Vet. 72: 443–451.

Fortes, E.A.M., M.A.M. Carvalho, M.M. Almeida, A.M. Conde-Júnior, N.E.A. Cruz and A.C. Assis Neto. 2005. Morphological aspects of the uterine tube in agouti (*Dasyprocta aguti*, Mammalia: Rodentia). Braz. J. Vet. Res. Anim. Sci. 42: 130–134.

Fortes, E.A.M., M.S. Ferraz, D.O. Bezerra, A.M.C. Júnior, R.M. Cabral, F.D. Sousa, et al. 2013. Prenatal development of the agouti (*Dasyprocta prymnolopha* Wagler, 1831): External features and growth curves. Anim. Reprod. Sci. 140: 195–205.

Guimarães, D.A., D. Moreira and W.G. Vale. 1997. Determination of agouti (*Dasyprocta prymnolopha*) reprodutive cycle by colpocytologyc diagnostic. Acta. Amazon. 27: 55–63.

Guimarães, D.A., O.M. Ohashi, M. Singh and W. Vale. 2016. Profile of plasmatic progesterone on pregnancy, and the postpartum estrus of *Dasyprocta prymnolopha* (Rodentia: *Dasyproctidae*). Rev. Biol. Trop. 64: 1519–1526.

IUCN Red List Categories and Criteria, Version 3.1. 2022. IUCN, Gland, Switzerland and Cambridge, UK. Available on: www.iucnredlist.org. Accessed on: September 26th.

Lall, K.R., K.R. Jones and Garcia, G.W. 2022. Reproductive technologies used in female neo-tropical hystricomorphic rodents. Animals 12: 618.

Martinez, A.C., F.S. Oliveira, C.O. Abreu, L.L. Martins, A.P. Pauloni and N. Moreira. 2013. Semen collection by electroejaculation in Azara´s agouti (*Dasyprocta azarae*). Pesq. Vet. Bras. 33: 86–88.

Mattos, L.M., F.C. Reis, A.M.C. Racanicci, I. Pivato, G.S.S.S. Tonelli, S.N. Báo, et al. 2022. Freezing for the future: obtaining fibroblast samples from deceased wild mammals for the Brazilian Cerrado germplasm bank. Biopreserv Biobank. in press.

Mayor, P., R.E. Bodmer and M. Lopez-Bejar. 2011. Functional anatomy of the female genital organs of the wild black agouti (*Dasyprocta fuliginosa*) female in the Peruvian Amazon. Anim. Reprod. Sci. 123: 249–257.

Menezes, D.J., A.C. Assis Neto, M.F. Oliveira, M.A. Miglino, G.R. Pereira, C.E. Ambrosio, et al. 2010. Morphology of the male agouti accessory genital glands (*Dasyprocta prymnolopha* Wagler, 1831). Pesq. Vet. Bras. 30: 793–797.

Menezes, D.J.A., M.A.M. Carvalho, A.C. Assis Neto, M.F. Oliveira, E.C. Farias, M.A. Miglino, et al. 2003. Morphology of the external male genital organs of agouti (*Dasyprocta aguti* Linnaeus, 1766). Braz. J. Vet. Res. Anim. Sci. 40: 148–153.

Mittelman, P., C. Kreischer, A.S. Pires and F.A.S. Fernandez. 2020. Agouti reintroduction recovers seed dispersal of a large-seeded tropical tree. Biotropica 52: 766–774.

Mollineau, W., A. Adogwa, N. Jasper, K. Young and G. Garcia. 2006. The gross anatomy of the male reproductive system of a neotropical rodent: the agouti (*Dasyprocta leporina*). Anat. Histol. Embryol. 35: 47–52.

Mollineau, W.M., A.O. Adogwa and G.W. Garcia. 2008. A preliminary technique for electro-ejaculation of agouti (*Dasyprocta leporina*). Anim. Reprod. Sci. 7: 92–108.

Mollineau, W.M., A.O. Adogwa and G.W. Garcia. 2010a. Improving the efficiency of the preliminary electroejaculation technique developed for semen collection from the agouti (*Dasyprocta leporina*). J. Zoo Wildl. Med. 41: 633–637.

Mollineau, W.M., A.O. Adogwa and G.W. Garcia. 2010b. Liquid and frozen storage of agouti (*Dasyprocta leporina*) semen extended with UHT Milk, unpasteurized coconut water, and pasteurized coco-nut water. Vet. Med. Int. 2011: 702635.

Oliveira, G.B., H.N. Araújo Júnior, H.S. Costa, A.R. Silva, C.E.B. Moura, H.A.O. Rocha, et al. 2017. Post-implantation development of red-rumped agouti (*Dasyprocta leporina* Linnaeus, 1758). Anim. Reprod. Sci. 182: 35–47.

Oliveira, G.B., H.N. Araújo Júnior, R.S. Sousa, F.V.F. Bezerra, A.C. Santos, C.E.B. Moura, et al. 2019. Morphology of the genital organs of the female red-rumped agouti (*Dasyprocta leporina*, Linnaeus, 1758) during estrous cycle phases and in advanced pregnancy. J. Morphol. 280: 1232–1245.

Patton, J.L. and L.H. Emmons. 2015. Mammals of South America. The University of Chicago Press, Chicago.

Peixoto, G.C.X., K.M. Maia, L.M. Almeida, L.B. Campos, G.B. Oliveira, M.F. Oliveira et al. 2018. Estrus induction in agoutis (*Dasyprocta leporina*) using protocols based on prostaglandin isolated or in association to GnRH. Arq. Bras. Med. Vet. Zootec. 70: 806–814.

Pereira, A.F., L.V.C. Aquino, M.B. Nascimento, F.V.F. Bezerra, A.A. Borges, É.A. Praxedes, et al. 2021. Ultrastrutural and morphometric description of the ear skin and cartilage of two south American wild histricognate rodents (*Dasyprocta leporina* and *Galea spixii*). Pesq. Vet. Bras. 41: e06775.

Praxedes, É.C., G.L. Lima, L.G. Bezerra, F.A. Santos, M.B. Bezerra, D.D. Guerreiro, et al. 2018. Development of fresh and vitrified agouti ovarian tissue after xenografting to ovariectomised severe combined immunodeficiency (SCID) mice. Reprod. Fertil. Dev. 30: 459–468.

Praxedes, É.A., F.F. Bressan and A.F. Pereira. 2020a. A Comparative approach of cellular reprogramming in the Rodentia order. Cell Reprogram. 22: 227–235.

Praxedes, É.A., L.R.M. Oliveira, M.V.O. Santos, L.V.C. Aquino, M.F.D. Oliveira and A.F. Pereira. 2020b. Production of parthenogenetic embryos from agoutis using strontium chloride and cytochalasin B. Ciênc. Anim. 30: 180–184.

Praxedes, É.A., M.B. Silva, L.R.M. Oliveira, J.V.S. Viana, A.R. Silva, M.F. Oliveira et al. 2021a. Establishment, characterization, and cryopreservation of cell lines derived from red-rumped agouti (*Dasyprocta leporina* Linnaeus, 1758)—A study in a wild rodent. Cryobiology 98: 63–72.

Praxedes, É.C.G., L.G.P. Bezerra, S.S.J. Moreira, C.S. Santos, A.V. Brasil, et al. 2021b. Microbiological load and preantral follicle preservation using different systems for ovarian tissue vitrification in the red-rumped agouti. Cryobiology 103: 123–128.

Rocha, A.R., F.R. Alves, N.M. Argôlo-Neto, L.F Santos, H.M. Almeida, Y.K.P. Carvalho, et al. 2012. Hematopoietic progenitor constituents and adherent cell progenitor morphology isolated from black-rumped agouti (*Dasyprocta prymnolopha*, Wagler 1831) bone marrow. Microsc. Res. Tech. 75: 1376–1382.

Rodrigues, L.L.V., A.A. Borges, M.B.D. Nascimento, L.V.C. Aquino, M.D.C.B. Santos, A.R. Silva, et al. 2021. Evaluation of different cryoprotectant solutions for the cryopreservation of somatic tissues of *Dasyprocta leporina* (Linnaeus, 1758). Cryoletters 42: 210–219.

Santos, E.A., G.L. Lima, E.C. Praxedes, A.M. Silva, K.M. Maia, M.F. Oliveira, et al. 2018. Estimation, morphometry and ultrastructure of ovarian preantral follicle population in agouti (*Dasyprocta leporina*). Pesq. Vet. Bras. 38: 175–182.

Silva, M.A., G.C.X. Peixoto, E.A.A. Santos, T.S. Castelo, M.F. Oliveira and A.R. Silva. 2011. Recovery and cryopreservation of epididymal sperm from agouti (*Dasiprocta aguti*) using powdered coconut water (ACP-109c) and Tris extenders. Theriogenology 76: 1084–1089.

Silva, M.A., G.C.X. Peixoto, P.C. Sousa, F.S.B. Bezerra, B.S.A. Bezerra and A.R. Silva. 2012. Interactions between straw size and thawing rates on the cryopreservation of agouti (*Dasyprocta aguti*) epididymal sperm. Reprod. Domest. Anim. 47: e4–e6.

Silva, A.R., A.M.D. Silva, E.C.G. Praxedes and A.F. Pereira. 2017. Conservation of South American wild hystricognath rodents using reproductive strategies. Rev. Bras. Reprod. Anim. 41: 231–236.

Silva, A.M., A.G. Pereira, L.G.P. Bezerra, S.S.J. Moreira, A.F. Pereira, M.F. Oliveira, et al. 2022. Cryopreservation of testicular tissue from adult red-rumped agoutis (*Dasyprocta leporine* Linnaeus, 1758). Animals 12: 738–749.

Singh, M.D., M.J. Morris, D.A. Guimarães, G. Bourne and G.W. Garcia. 2016. Serological evaluation of ovarian steroids of red-rumped agouti (*Dasyprocta leporina*) during the estrous cycle phases. Anim. Reprod. Sci. 175: 27–32.

Sousa, F.C.A., F.R. Alves, E.A.M. Fortes, M.S. Ferraz, A. Machado-Júnior, D.J.A. Menezes, et al. 2012. Pregnancy in Hystricomorpha: gestational age and embryonic-fetal development of agouti (*Dasyprocta prymnolopha*, Wagler 1831) estimated by ultrasonography. Theriogenology 78: 1278–1285.

Vanegas, L., N.V. Vliet, D. Cruz and F. Sandrin. 2016. (Contribución proteica de animales silvestres y domésticos a los menús de los contextos rurales, peri-urbanos y urbanos de varias regiones de Colombia). Protein contribution of wild and domestic animals in rural, peri-urban and urban diets in different regions of Colombia. Biota Colomb 17: 26–43.

Wanderley, L.S., H.K.M. Luz, L.R. Faustino, I.M.T. Lima, C.A.P. Lopes, A.R. Silva, et al. 2012. Ultrastructural features of agouti (*Dasyprocta aguti*) preantral follicles cryopreserved using dimethyl sulfoxide, ethylene glycol and propanediol. Theriogenology 77: 260–267.

Chapter 15

The Capybaras

Cristiane Schilbach Pizzutto* and Derek Andrew Rosenfield

Animal Reproduction Department, Faculty of Veterinary Medicine and Animal Science,
University of Sao Paulo (USP), São Paulo, SP, 05508-270, Brazil.
Email: cspizzutto@yahoo.com.br, dro@alumni.usp.br

INTRODUCTION

The capybara (*Hydrochoerus hydrochaeris,* Linnaeus 1766) is the largest rodent in the world, as well as belonging to the neotropical hystricomorphic rodents of the suborder Hystricomorpha. It has a broad distribution throughout South America, including east of the Andes, Colombia, Venezuela, Ecuador, Peru, Guyana, Brazil, Eastern Bolivia, Paraguay, Uruguay, the Northwestern and Eastern part of Argentina (Lacher et al. 1998), while the only other known species, *Hydrochoerus isthmius*, Goldman, 1912 (lesser Capybara) is concentrated in the North-Eastern and Central American Region, and some islands, including Western Colombia, Northwestern Venezuela, Panama, and Trinidad/Tobago (Moreira et al. 2013).

The most important characteristic of a typical capybara home range is a permanent water body, which serves as a safe harbor against predator attacks, as well as the principal source of drinking water, to control body temperature, and for mating. Also, eroded and collapsed riverbanks with partially flooded cavities serve as protective areas for females to give birth, or to take mud baths. In close proximity, elevated dry-areas function as a shelter against high waters during rainy seasons. Furthermore, it comprises a vast vegetation area for their extensive foraging activities, to provide shade, or temporarily as an escape for injured animals, or females with newly born pubs (Moreira and Macdonald 1996, Moreira et al. 2013, Rosenfield 2019). Their average body mass is between 50–60 kg, while synanthropic individuals can gain over 100 kg (personal observation 2017).

---

*Corresponding author: cspizzutto@yahoo.com.br

The color of their coarse and not dense fur range from light to reddish-brown, to black, with the ventral side being almost free of coat.

As semi-aquatic animals, their anatomy has several characteristics adapted to life in and around water, starting with the sturdy and barrel-shaped body, with a large proportion of adipose tissue, allowing for buoyancy and forward motion while swimming or diving, facilitated by a partially interdigital webbing, with four toes on their forefeet and three on their hind feet (Rosenfield 2019).

The head shows the most significant adaptations, with an elongated nasal bone where the nostrils, eyes, and ears are located near the top, allowing for surfacing while swimming. The ears are small and can be folded shut, to prevent water from entering.

Although gregarious animals, capybaras have a well-defined hierarchy in their polygynous society, made up of an alpha male (scent marking using nasal and perianal glands, vigilance, courtship, and agonistic actions), or principal breeder, fiercely defending its harem from any potential male intruder, or sexually maturing male, which are driven out of the group, remaining in a certain distance as satellite males. The reason is that their society is prominently female-biased, followed by dominant adult females, subordinate adult females, gender-mixed adolescents and pups. Group formation is typically from one male/two females to an average of 20–30 members, with observed extremes, reaching >100 members. These dynamics are mainly driven by carrying capacities, hunter/predator pressure, human activities, synanthropism, and seasonality. Dominant females become aggressive when it comes to feeding rights and are the principal breeders. Not all subordinate, sexually mature females breed with the alpha male. It is uncertain if this is caused by the agonistic behavior of dominant females or a potentially cyclic suppression due to their presence. On the other hand, subordinate females are observed to leave the group to mate with nearby satellite males, afterward returning to the group, or giving a start to a new group formation. A capybara survival strategy includes alloparental conduct, whereby females nurse pups indiscriminately, being their own, or from other females.

All these characteristics, regarding the biology of this species, are fundamental for reproductive management, whether intending to increase the number of individuals for conservational efforts or with the intent to control the population growth of pest species, to mitigate human-wildlife problems. The capybara is a species of variant interest, from economics to conservation, to public health. Although it still requires more research, this chapter aims to provide information about conducted reproductive studies, from reproductive physiology to applied biotechnologies.

THE CAPYBARA IN CONSERVATION AND PRODUCTION

Capybaras are listed in International Union for Conservation of Nature's (The IUCN, Red List of Threatened Species, 2021) as "Least Concern" (LC)

– (The IUCN Red List of Threatened Species), meaning IUCN has evaluated *H. hydrochaeris* as not being a focus of species conservation because the specific species is still plentiful in the wild. In Brazil, the capybara, belonging to its natural fauna, is protected by environmental laws, and therefore cannot be hunted. Any interest in breeding needs special authorization from the government. Capybara breeding activity for commercial purposes can be found, to one degree or another, in several South American countries, such as Argentina, Venezuela, Colombia, as well as Brazil (Moreira and Macdonald 1996).

Concerning reproductive productivity, the capybara is considered an exception among neo-tropical rodents. As stated in the work by Moreira and Macdonald (1996), the capybara is the largest living rodent and has a rather large litter size with an average of 4 capybara cubs/litter. Besides being highly proliferative, they have been considered an economically interesting livestock species because of their total carcass use, their muscle mass (meat consumption), fat extraction, and their hides for the leather industries. Capybaras are very resistant to diseases and environmental impacts, have long lifespans, fast growth rates, a high reproductive output, and a low-cost diet.

Using wildlife for livestock creation is believed to contribute to the reduction of the illegal use of fauna as well as to promote conservation efforts of threatened species, with important aspects such as increasing the domestic stock of wild species, positive implications for the maintenance of genetic heritage (*ex situ* conservation), domestication of new animals (new commodities), and conservation of certain species under natural conditions (*in situ*); to be an economic alternative, mainly for small and medium land owners; increase in the supply of protein and meat with less fat and less saturated composition; greater livestock diversification to attend to different properties and ecosystems, as well as providing greater sustainability in agriculture (Pinheiro 2005), a profil that Capybara perfectly fit.

When kept *in situ*, it is a widely distributed species and, as it is a generalist herbivore, it occupies several types of environments (riparian forests, map groves), and as synanthropic animals, they can be found in agricultural areas, public parks, and residential areas, with low sensitivity to human presence (Ferraz et al. 2007). Its intrinsic ability to adapt to anthropic habitats, its great reproductive potential, in addition to the absence of predators, enable the occupation of social groups in these environments, with a considerable increase in population density, causing the capybara to be characterized as a problem species in several areas of Brazil (Ferraz et al. 2010, Rosenfield 2019). Some of the more obvious human-capybara conflicts are associated with economic damage due to crop destructions (Ferraz et al. 2003), or ending up as roadkill on highways and rural roads (Bager and Fontoura 2013), and most importantly, capybaras participate in the transmission cycle of Brazilian Spotted Fever (BSF), a zoonosis that is caused by the bacteria *R. rickettsii* and transmitted by vector ticks (Labruna 2009).

Thus, developing species-specific population management strategies, respecting public opinion, and considering animal welfare are all required for the effective mitigation of these tick-borne zoonotic diseases. One consensus on how to potentially control disease transmission is by directly controlling the host's population, while indirectly removing the dynamics that allow for the pathogens to be maintained. In a polygynous society with a strong hierarchal organization that is upheld by a dominant male, the integrity of the hormone-driven secondary sexual characteristics and courtship behavior is crucial. For an intervention by contraceptive strategies to be a viable management tool, it is imperative to preserve the alpha male's phenotypic and agonistic characteristics. Losing the dominant status would allow for the opportunistic entrance of a competitive male, consequently leading to a failure of the intended population management—the reason why reproductive technologies become an essential part of population control strategies (Rosenfield 2019).

Therefore, the understanding of the relationship between the morphological, physiological, and behavioral aspects of the capybara is essential to the understanding of the adaptive and evolutionary mechanisms that culminate in a successful survival and reproduction strategy of the species.

REPRODUCTIVE MORPHOPHYSIOLOGY OF THE MALE

Sexual dimorphism in capybaras is very limited, especially until sexual maturity. Alpha males have more prominent secondary sexual characteristics, such as an enlarged nasal gland (morillo), used for territorial marking; its pronounced development is believed to be androgen-dependent (Moreira et al. 2013) and aid in the identification of an alpha male within a group. Despite the difficulty of visualization, the testes in these males can also be seen when they are present in the upper medial region of the thigh (Rosenfield 2019) (Figure 15.1).

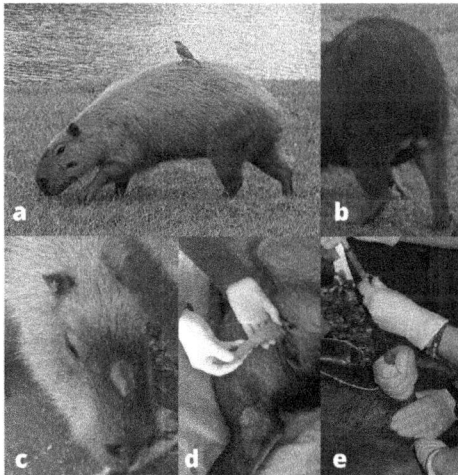

Figure 15.1 Images showing male capybaras. (a) Adult male capybara. (b) Visible alpha male testis. (c) Alpha male nasal gland. (d) Exposed penis. (e) Semen collection by urethral catheterization. *Source:* Rosenfield (2019).

Male capybaras lack a scrotum, and the testes are located subcutaneously in the inguinal region (Paula and Walker 2013) and can be retracted into the abdomen. Despite little information regarding the size, weight, and volume of the testes, Rosenfield et al. (2019) reported that alpha males have respective testicular weights and volumes of 45.76 ± 8.47 g and 62.29 ± 11.71 mL. In adult males, the average testis weight was 32 g, with small variations (Paula and Walker 2013) and the volume was 16.97 ± 9.67 cm^3 (Moreira et al. 1997). This variation in testicular weight and volume found in alpha males agrees with the findings of Herrera and Macdonald (1993), who found a positive correlation between the serum concentration of testosterone and a proportional increase in the volume of the nasal gland, in addition to the volume of Leydig cells.

Regarding spermatogenesis, the duration of the spermatogenic cycle was observed to be 11.9 ± 0.1 days over eight stages within the seminiferous epithelium, such as the morphology of spermatid nuclei and the presence of meiotic divisions (Paula et al. 1999). Testis histology confirmed the presence of occasional spermatogonia, primary spermatocytes, spermatids (initial and final), spermatozoa, Sertoli cells, and Leydig cells. Leydig cells were the most numerous, followed by germ cells and Sertoli cells. There was a correlation between the number of germ cells and Sertoli cells, and a strong negative correlation between the number of germ cells and Leydig cells (Paula et al. 1999).

The accessory sex glands of the capybara include the seminal vesicles and the prostate. The seminal vesicles are in the pelvic cavity, dorsal to the urinary bladder and parallel to the vas deferens (seminal colliculus). According to Paula and Walker (2013), they were paired and tubular in shape, with multiple ducts converging into a duct on each side, flowing into the urethra, forming the ejaculatory ostium with the vas deferens. The prostate is a paired structure, divided into several lobes (multi-lobed tubular gland). The lobes are surrounded by the serous tunic, with ducts opening through two folds, next to the ejaculatory ostium on the lateral side of the urethra. The vas deferens have been described as a tubular organ, found as a continuation of the tail of the epididymis, parallel to the pampiniform plexus, opening into the urethra and surrounded by the cremaster muscle (Fernandez 2003).

Furthermore, according to Paula and Walker (2013), the foreskin is connected to the anus to form an anogenital invagination, composed of the penis (flaccid), anus, and paranal glands (scent). The penis has penile bone, as described by (Rosenfield et al. 2020).

REPRODUCTIVE BIOTECHNOLOGY IN THE MALES

Assisted reproductive technology (ART) in neotropical wildlife rodents is still very limited. Besides intensive zootechnic breeding programs, similar to swine production that was adopted for capybara (Pinheiro 2005), only a few reports exist on tested biotechnologies, with the exceptions of several rodent species that are studied in laboratory settings (Mochida 2020, Lamas 2021, Yeonmi and Eunju 2012).

When we approach ART in males, the scarcity of information is very great. The advancement of these biotechnologies is limited only to the collection of semen, with practically no apparent studies yet on the advancement of the use of these methodologies. To date, Lall et al. (2022) have found only three semen collection methods in the literature for capybaras: electro-ejaculation (Rodríguez et al. 2012), urethral catheterization after administration of ketamine and dexmedetomidine (Figure 15.1d and 15.1e), and epididymal aspiration after hemi-orchiectomy or necropsy (Rosenfield 2019). For electro-ejaculation, it was possible to obtain an average volume of 135.5 ± 93.56 µL, with a pH of 8.14 ± 0.38, mass motility of $32.60 \pm 13.46\%$, individual motility of $34.0 \pm 19.81\%$, viability of $51.3 \pm 19.42\%$, and sperm concentration of $127 \pm 59.01 \times 10^6$ sperm/mL (Rosenfield et al. 2019); refer to Table 15.1 for detailed reproductive indices.

Table 15.1 Reproductive biology indexes of capybaras (*Hydrochoerus hydrochaeris*)

Life span in the wild/captivity	5–7 years/>10 years
Adult weight in the wild/synanthropic	+/– 50 kg/>105 kg
Puberty	6–12 months
Sexual maturity Male/female—age/body weight	12–15 months/30 kg/13–15 months/30–40 kg
Estrous	Seasonal (peak during rainy season), poly estrous, spontaneous ovulation.
Estrous cycle	7.5 +/– 1.2 days
Placentation	Chorioallantoic, hemochorial, discoidal, labyrinthine, lobular, subplacental.
In-heat duration	+/– 24 h
Post-partum estrous	15 days
Gestation period	147–156 days (average 150.6)
Number of litter (pups)	2–8 (average 4/litter)
Birth weight	1–2 kg
Number of births per female/yr.	Potentially two per year
Lactation period	1.3–4 months
Weaning period/weight	1.2–4 months
Testicular weight – adult male*	~ 45 g
Spermatogenesis period – adult male*	53.6 days (11.9 × 4.5 cycles)
Sperm concentration – adult male	127–146.25 ± 3.75 sperm × 10^6/mL

*Alpha male (higher testosterone concentration.
Sources: Rosenfield et al. (2019), Moreira et al. 2012, Miglino et al. (2002)

REPRODUCTIVE MORPHOPHYSIOLOGY OF THE FEMALE

Females have symmetrical ovaries, with an elliptical outline, in addition to an ovarian pouch (Bursa ovarica), with associated arteries and ligaments. The Fallopian tubes have a papilla, isthmus, ampulla, and infundibulum where the fimbriae are marked and radiate around the abdominal ostium of the tube. Then, two uterine horns are individualized; the uterine body is absent, characterizing a double uterus, with the horns flowing directly into the uterine cervix, with longitudinal striations. The vagina is a tubular organ with particular hyperpigmentation for each animal and stands out for presenting pleated mucosa in the transverse direction, more evident near the uterine cervix and gradually decreasing from the vaginal fornix. There is a cavity connected to the digestive and urogenital system, called the anal sac, where it is possible to visualize the vulva, clitoris, and anus—characteristics of the species (Carvalho 2011). All indices of reproductive biology can be seen in Table 15.1.

The mentioned reproductive biological indices are to be seen as a reference and not as a strictly defined value, as these indices suffer great variations based on the habitat characteristics, such as geography, climate, seasonality, food offerings, carrying capacity, disease, wildlife vs captivity, hunting pressure, and individual traits.

REPRODUCTIVE BIOTECHNOLOGY APPLIED TO THE FEMALES

Information regarding ARTs in females is practically non-existent in literature, leaving a big gap regarding the advancement of these techniques in reproduction. In the work by Miranda and Oba (2000), colpocytology described the types of vaginal cells present in capybaras. It was observed that an increase in the size of the cells would correspond to an increase in acidity and a decrease in the nucleus-cytoplasm ratio.

REPRODUCTIVE TECHNIQUES FOR POPULATION CONTROL

Contraception, methods of manipulating fertility in males and females to prevent pregnancy, are part of ARTs. Besides surgical procedures to completely remove the gonads, and with that eliminate a principal site of sex hormone synthesis, there are alternative techniques to preserve the gonads and their endocrinological function. In males, vasectomy prevents sperm from entering the urethra, and in females, tubal ligation obstructs the egg from traveling from the ovaries through the fallopian tubes, and sperm from traveling up the fallopian tubes toward the egg (Nunes et al. 2020, Passos-Nunes et al. 2022, Yanai et al. 2022)—these being the two most effective methods. Other non-surgical contraceptive methods include chemical castration (hormone application) and immunocontraception for males and females, potentially acting on various locations along the hypothalamic-pituitary-gonadal axis, rendering the animal temporarily infertile.

The problem with steroid hormone-based contraceptives is their potential for severe adverse effects, depending on the species to be treated; also, many products are developed for human application. Furthermore, many contraceptives are using synthetic hormones, that are very similar to endogenous hormones, thus more potent and with a broader affinity to different cell receptors, potentially provoking serious illnesses, like neoplastic tumors, epithelial hypertrophy, hyperplasia, endometrial cysts, and pyometra, to name a few. Heightened sensitivity to such collateral effects is reported especially with species of the Carnivora order (Asa and Porton 2005). Another concern is the reversibility of the contraceptive effect when conservational efforts demand.

Following is an illustration of the proposed mechanism of an anti-GnRH immunocontraceptive as a contraceptive method for capybara population control, provoking a similar response to an autoimmune disease, whereby antibodies are targeting endogenous hormones, with the intent to cease their biological activities (Figure 15.2).

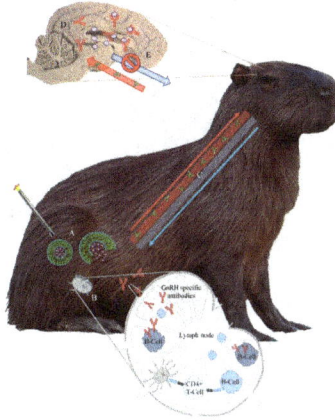

Figure 15.2 Mechanism of action of an anti-GnRH Immunocontraceptive. Illustrated mechanism of action of the immunocontraceptive: (a) Anti-GnRH Vaccine Dart injects the vaccine (adjuvant and GnRH-based antigens) into the large muscle group. (b) Provoking an immune response within a lymph node, synthesizing GnRH-specific antibodies. (c) Showing a normal passage of gonadotropin hormones (LH, FSH) through the hypothalamic-pituitary-gonadal axis, from the anterior pituitary gland in the direction of the gonads. After immunization, anti-GnRH antibodies are transported to the hypothalamic region. (d) Anti-GnRH antibodies capture endogenous GnRH peptide hormones, forming large immunocomplex molecules, thus inhibiting transfusion and ligation to GnRHR (receptors) in the pituitary region, ceasing LH and FSH synthesis and liberation. (e) Impeding LH and FSH biological activities in the gonadal region (steroidogenesis and gametogenesis).

As of 2022, the only successful study in the world on applied immuno-contraception in *Hydrochoerus hydrochaeris* under field conditions was presented as part of a Ph.D. thesis on capybara population control in the state of Sao Paulo, Brazil (Rosenfield 2019, Rosenfield et al. 2019). The results clearly showed the impact of the anti-GnRH vaccine on the fertility of male and female adult capybaras. In males (alpha male), the immunocontraceptive provoked a temporarily suppressed spermatogenesis to the level of infertility (Figure 15.3: a computer-assisted semen analysis CASA, pre- and post-treatment), sever atrophic testicular parenchyma, while preserving alpha male behavior. A very important characteristic is to maintain the integrity of the group's social structure. In females, the immunocontraceptive incited a temporary cessation of folliculogenesis, while maintaining alloparental behaviors. In general, the population control method of using an anti-GnRH vaccine proved successful in significantly driving down the group's birthrate during the observed reproductive seasons under treatment.

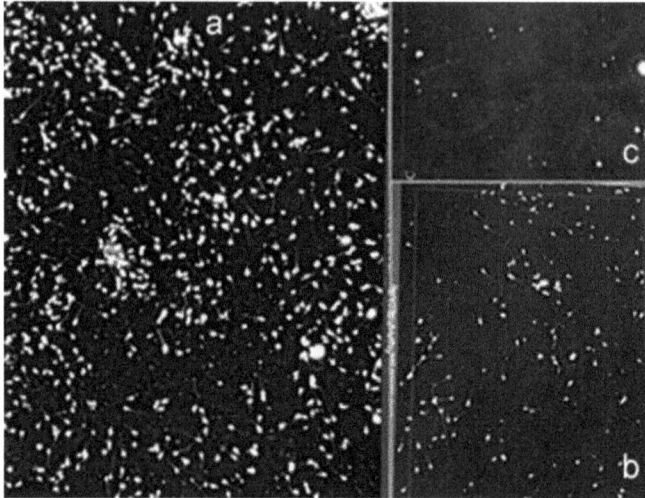

Figure 15.3 CASA images depicting (a) Semen of an intact male vs (b) Semen of an anti-GnRH vaccine treated male, 90 days post-vaccination, and (c) Semen of a treated male, 270 days post-vaccination. *Source:* Rosenfield (2019)

FINAL CONSIDERATIONS

Studies involving ARTs in neotropical rodent species are still at an embryonic stage, especially when it comes to capybaras. Due to the conservation status and their coexistence in very close proximities to humans in urban centers and agricultural activities, human-capybara conflicts are numerous and frequent, ranging from public health concerns to invasions of public and private spaces, and risks of car collisions, among others. Necessary studies on population control techniques in general are ongoing and advancing quickly. Despite this, capybara-specific research, including the morphophysiological aspects of reproduction and associated behaviors, are essential, both for conservation purposes as well as economic objectives to produce meat and other by-products.

REFERENCES

Asa, C.S. and I.J. Porton. 2005. Wildlife contraception: issues, methods, and applications. JHU Press, Baltimore, USA.

Bager, A. and V. Fontoura. 2013. Evaluation of the effectiveness of a wildlife roadkill mitigation system in wetland habitat. Ecol. Eng. 53: 31–38.

Carvalho, R.G. 2011. Morfologia e biometria do aparelhos reprodutor feminino da capivara (*Hydrochoerus hydrochaeris*). Ph.D. Thesis. Universidade Estadual Paulista Julio de Mesquita Filho, BR.

Fernandez, D.S. 2003. Morfologia do trato reprodutor masculino de capivara: estudo das glândulas anexas à uretra. Master. Dissertation, Universidade de São Paulo, BR.

Ferraz, K.M.P.M.B., S.F.B. Ferraz, J.R. Moreira, H.T.Z. Couto and L.M. Verdade. 2007. Capybara (*Hydrochoerus hydrochaeris*) distribution in agroecosystems: A cross-scale habitat analysis. J. Biogeogr. 34: 223–230.

Ferraz, K.M.P.M.B., B. Manly and L.M. Verdade. 2010. The influence of environmental variables on capybara (*Hydrochoerus hydrochaeris*: Rodentia, Hydrochoeridae) detectability in anthropogenic environments of Southeastern Brazil. Popul. Ecol. 52: 263–270.

Ferraz, K.M.P.M.B, M.A. Lechevalier, H.T.Z. Couto and L.M. Verdade. 2003. Damage caused by capybaras in a corn field. Sci. Agric. 60: 191–194.

Herrera, E.A and D.W. Macdonald. 1993. Aggression, dominance, and mating success among capybara males. Behav. Ecol. 4: 114–119.

IUCN. 2021. The IUCN Red List of Threatened Species. 2021. IUCN Red List of Threatened Species. Accessed September 30, 2021. https://www.iucnredlist.org/en.

Labruna, M.B. 2009. Ecology of Rickettsia in South America. Ann. N.Y. Acad. Sci. 1166: 156–166.

Lacher, T.E., R.D. Qualset Jr., L.M. Coburn and M.I. Goldstein. 1998. The role of agroecosystems in wildlife biodiversity. pp. 147–161. *In:* W.W. Collins and C.O. Qualset (eds). Biodiversity in Agroecosystems. CRC Press, Florida, USA.

Lall, K.R., K.R. Jones and G.W. Garcia. 2022. Reproductive technologies used in male neo-tropical hystricomorphic rodents. Animals 12: 34.

Lamas, S.C.S. 2021. Mouse Embryo Rederivation and Other Assisted Reproductive Techniques and Their Impact on Experimental Results. Ph.D. Thesis, Universidade do Porto, Portugal.

Miglino, M.A., A.M. Carter, R.H. Santos Ferraz and M.R.F. Machado. 2002. Placentation in the capybara (*Hydrochaerus hydrochaeris*), agouti (*Dasyprocta aguti*) and paca (*Agouti paca*). Placenta 23: 416–28.

Mochida, K. 2020. Development of assisted reproductive technologies in small animal species for their efficient preservation and production. J. Reprod. Dev. 66: 299–30.

Moreira, J.R. and D.W. Macdonald. 1996. Capybara use and conservation in South America. pp. 88–101. *In:* Taylor, V.J. and N. Dunstone (eds). The Exploitation of Mammal Populations. Dordrecht: Springer, Netherlands.

Moreira, J.R., J.R. Clarke and D.W. Macdonald. 1997. The testis of capybaras (*Hydrochoerus Hydrochaeris*). J. Mammal. 78: 1096–1100.

Moreira, J.R., K.M.P.M.B. Ferraz, E.A. Herrera and D.W. Macdonald. 2013a. Capybara: Biology, Use and Conservation of an Exceptional Neotropical Species. Springer Science & Business Media.

Moreira, J.R., M.R. Alvarez, T. Tarifa, V. Pacheco, A. Taber, D.G. Tirira, et al. 2013b. Taxonomy, natural history and distribution of the capybara. pp. 3–37. *In:* J.R. Moreira, K.M.P.M.B. Ferraz, E.A. Herrera and D.W. Macdonald (eds). Capybara: Biology, Use and Conservation of an Exceptional Neotropical Species. Springer, New York, USA.

Miranda, L.B. and E. Oba. 2000. Hormonal Evaluation in Capybara (*Hydrochoerus hydrochaeris*) by Squamous Epithelial Cell Types. 2000. IAAAM Archive, Davis, USA.

Nunes, F.B.P., A.Z. Nunes, M.P. Nunes, M.B. Labruna and C.S. Pizzutto. 2020. Reproductive control of capybaras through sterilization in areas at risk of transmission of Brazilian spotted fever. Cienc. Rural 50: e20200053.

Passos-Nunes, F.B., F.M.G. Jorge, M.P. Nunes, A.Z. Nunes, P.N. Jorge-Neto, A.C. Assis et al. 2022. Surgical sterilization of free-ranging capybaras (*Hydrochoerus hydrochaeris*): 'Passos Nunes' uterine horn ligature. Anim. Reprod. 19: e20220029.

Paula, T.A.R., H. Chiarini-Garcia and L.R. França. 1999. Seminiferous epithelium cycle and its duration in capybaras (*Hydrochoerus hydrochaeris*). Tissue Cell 31: 327–334.

Paula, T.A.R. and N.J. Walker. 2013. Reproductive morphology and physiology of the male capybara. pp. 107–29. *In*: J.R. Moreira, K.M.P.M.B. Ferraz, E.A. Herrera and D.W. Macdonald (eds). Capybara: Biology, Use and Conservation of an Exceptional Neotropical Species., Springer, New York, USA.

Pinheiro, M.S. 2005. Criação de Capivara em Sistema Intensivo. EMBRAPA.

Rodríguez, P.J., M. Peña J, A. Góngora O. and R. Murillo P. 2012. Obtención y evaluación del semen de capibara *Hydrochoerus hydrochaeris*. Rev. MVZ Córdoba. 17: 2991–2997.

Rosenfield, D.A. 2019. Study on the perspective of population control of capybaras (*Hydrochoerus hydrochaeris*) by reversible immunocontraceptive method. Ph.D. Thesis, Universidade de São Paulo, BR.

Rosenfield, D.A., M. Nichi, J.D.A. Losano, G. Kawai, R.F. Leite, A.J. Acosta., et al. 2019. Field-testing a single-dose immunocontraceptive in free-ranging male capybara (*Hydrochoerus hydrochaeris*): evaluation of effects on reproductive physiology, secondary sexual characteristics, and agonistic behavior. Anim. Reprod. Sci. 209: 106148.

Rosenfield, D.A., N.F. Paretsis, P.R. Yanai and C.S. Pizzutto. 2020. Gross osteology and digital radiography of the common capybara (*Hydrochoerus hydrochaeris*), Carl Linnaeus, 1766 for scientific and clinical application. Braz. J. Vet. Res. Anim. Sci. 57: e172323.

Yanai, P.R., M.A. Ferraro, A.F.K.T. Lima, S.R.G. Cortopassi and L.C. Silva. 2022. Surgical contraception of free-ranging female capybaras: description and comparison of open and minimally invasive techniques. Vet. Surg. 5: S1.

Yeonmi, L. and K. Eunju. 2021. Hormone induced recipients for embryo transfer in mice. J. Anim. Reprod. Biotechnol. 36: 247–252.

Chapter 16

The Cavies

Andréia Maria da Silva*, Samara Sandy Jerônimo Moreira,
and Alexandre Rodrigues Silva

Laboratory of Animal Germplasm Conservation, Department of Animal Sciences,
Federal University of Semiarid Region (UFERSA),
Mossoró, RN, 59625900, Brazil.
Email: andreia.m.silva@hotmail.com; samara.sandy@bol.com.br;
alexrs@ufersa.edu.br

INTRODUCTION

The cavies belong to the family Caviidae and sub-family Caviinae and are represented by four genera, namely as *Cavia*, *Galea*, *Microcavia*, and *Kerodon* (Adrian Sachser 2011). These animals are present in most of South America, with distribution in the most varied habitats, such as grasslands, savannas, swamps, forests, scrubland, deserts, and mountainous regions. The genera include semiaquatic (*Cavia magna*), semiescansorial (*Microcavia australis*), burrowing (*Microcavia spp.*), cursorial and terrestrial (*Cavia spp.* and *Galea spp.*) species (Adrian and Sachser 2011). Among the representatives, the Spix's yellow-toothed cavy (*Galea spixii* Wagler, 1831) stand out for being the best described, being ahead, in the studies, in relation to the other species.

The Spix's yellow-toothed cavy is a species native to the Caatinga biome, but they are also geographically distributed in other biomes of Brazil, Paraguay, and Bolivia East of the Andes (Cabrera 1961). These animals live in flocks, and feed leaves, stems, fruits, vines, roots, tubers, and bark of young trees (Mendes 1987). Moreover, the species presents daytime activities and a twilight habit, with activity being reduced at night (Reis et al. 2006). Its coat is dense, and the color of the hair on the dorsal region varies from gray to yellow (Reis et al. 2006).

*Corresponding author: andreia.m.silva@hotmail.com

They are small, with a length of 22.5 cm and a body mass of 375 g for males, and 23.5 cm and 405 g for females, respectively (Oliveira et al. 2008), with a stunted tail (Reis et al. 2006) (Figure 16.1). Despite environmental disturbance and intensive hunting, the Spix's yellow-toothed cavies' population is considered stable (Catzeflis et al. 2016). Then, as an alternative for its preservation and as a source of local income, its captive breeding has been developed in some regions of Brazil, with economic repercussions (Santos et al. 2011). Due to their small features, easy adaption to captivity, low maintenance costs and a short gestation period (Björkman et al. 1989, Oliveira et al. 2008), Spix's yellow-toothed cavies are adequate experimental models for the development of conservative strategies for endangered cavid rodents, as the Santa Catarina's guinea pig (*Cavia intermedia*) (Roach 2016a), the Patagonian cavy (*Dolichotis patagonum*) (Roach 2016b), and the Shipton's mountain cavy (*Microcavia shiptoni*) (Jayat and Ojeda 2008).

Figure 16.1   Adult male of Spix's yellow-toothed cavies.

The common yellow-toothed cavy (*Galea musteloides,* Meyen, 1833) is a species native to Bolivia, Chile and Peru (Dunnum 2015). These animals inhabit areas of grassland and scrub (Roach 2016c). They are terrestrial herbivores with mainly diurnal habits (Lacher 2016). When adults, they weigh between 400–700 g and unlike species of the genus *Cavia*, they are smaller animals and more resistant to adverse conditions (Weir 1970).

The Brazilian Guinea pig (*Cavia aperea* Erxleben, 1777) is native to Argentina, Bolivia, Brazil, Colombia, Guyana, Paraguay, Peru, Suriname, Uruguay, and Venezuela (Bernal 2016). They inhabit areas, such as savannas, grasslands, and rocky areas (Bernal 2016). In Brazil, they occur from Pernambuco to Rio Grande do Sul, whose size and color vary according to the region, being small with dark brown dorsal color in the Northeast region or even larger and lighter animals, with gray dorsal color in the South region country (Cherem and Ferigolo 2012).

The Santa Catarina's Guinea pig (*Cavia intermedia* Cherem, Olimpio and Ximenez, 1999) is an herbivore native to Southern Brazil. It is considered one of the rarest mammals among the cavies (Salvador and Fernandez 2008). It is the species with the smallest distribution, with the population restricted to a 10 ha island call "Moleques do Sul island", off the coast of Santa Catarina state, in Brazil (Salvador 2021).

The greater Guinea pig (*Cavia magna* Jiminez, 1980) is a resident species in Brazil and Uruguay (Gonzalez 2016). They are present in areas of forests and grasslands (Gonzalez 2016). They are larger animals in relation to *C. aperea*, having an orange to black dorsal coloration (Cherem and Ferigolo 2012).

The Southern Mountain Cavy (*Microcavia australis* I. Geoffroy and d'Orbigny, 1833) are native to Argentina and Chile, inhabiting areas of forests, savannas, and deserts (Rocha 2016). It is a diurnal herbivore, considered one of the smallest caviomorph rodents, weighing on average 200 g (Tognelli et al. 2001). These animals live in social groups of a maximum of eight males and five females, in addition to their young (Andino et al. 2011).

The ecological and economic importance of these rodents has aroused the interest of researchers. In this context, some studies focused on describing the main aspects of the reproductive biology of males and females. The aim of this chapter is to present such information regarding characterization of the germplasm of cavy species native to South America, as well as the main assisted reproduction techniques (ARTs) available at the time of writing this review.

MALE REPRODUCTION

Regarding male reproduction, the information available so far reports on the period of sexual maturation, in addition to some morphological characteristics of the male reproductive system. Among the cavy species, Spix's yellow-toothed cavy is the most reported species in the literature, with description of reproductive physiology and ARTs.

The Spix's yellow-toothed cavy' mating system is classified as polygamous (Lacher and Lacher 1981), with reproductive activity throughout the year. The male presents the reproductive system, composed of penis, scrotum, foreskin, testes, epididymis, and accessory genital glands, these represented by the prostate, vesicular and bulbourethral glands, as well as the ampullae of the vas deferens. Their testes are ovoid, and can be in the inguinal canal, in the abdominal cavity, or in an inguinal position, with a well-defined scrotum. The penis is cylindrical with a cranial flexure, curving caudally, being composed of root, body and glans, the latter presenting spicules (Rodrigues et al. 2013). The spermatogenesis in Spix's yellow-toothed cavies starts at 30 days of life, but only at 45 days there is the presence of sperm in the lumen of the seminiferous tubules. When the animals reach 120 to 150 days of life, all the seminiferous tubules have luminal areas with cells in different stages of division (Santos et al. 2012).

Regarding the other South American cavy species, the studies are very restricted, with little information about their reproductive biology. In this context, it is known that the *G musteloides* are a promiscuous species (Touma et al. 2001, Velez et al. 2010), and the males have a spherical testis, with a poorly vascularized epididymis that is arranged in a spiral (Cooper et al. 2000). The sexual maturity in these animals occurs around 90 days of age. Spermatogenesis occurs between 21–25 days of age, but it's only at 31–35 days that the first sperm appear in the seminiferous tubules, occurring more frequently after 60 days, when the seminiferous tubules are fully mature (Holt 1977). As for sperm cells, epididymal sperm are smaller, with narrow and small heads, when compared to *C. aperea* sperm (Cooper et al. 2000).

In *C. aperea*, sexual maturity occurs around 70–80 days of age, when males show adequate testicular mass (g) and relative testicular mass, both with 0.49% of body mass. Furthermore, the elevated testosterone levels in the serum are observed at 82 days (6.6 ± 1.8 ng/mL), in which males aged between 69 and 82 days tended to increase testosterone levels when in the presence of females (Trillmich et al. 2006). The first motile spermatozoa in the epididymis are observed at 55 days of age, and more frequently when starting from 74 days. *C. aperea* sperm are larger than those of *G. musteloides* and, unlike these, are arranged along the epididymis in the form of a rouleaux (Cooper et al. 2000, Trillmich et al. 2006). Guenther et al. (2014) observed that *C. aperea* males may have the time of maturation as well as their early growth occurring under photoperiod influence. These animals present a shorter maturation and growth time when exposed to simulated climatic conditions, referring to spring, with high concentrations of testosterone from 45 to 66 days of age.

Like *G. musteloides*, *M. Australis* are also promiscuous species (Touma et al. 2001, Velez et al. 2010). Velez et al. (2010) found that males from the Argentina region suffer the effects of environmental conditions, with changes in the weight of the testicles and epididymis of the animals, as well as in spermatogenic activity. The highest values of testicular weight (3.02 ± 0.26 g) and epididymal weight (0.67 ± 0.10 g) were acquired during the summer. In this case, precipitation was primarily responsible for 89% of the variation in testicular weight and 61% of the variation in epididymal weight (Andino et al. 2016).

For males, in general, there are studies focused on the physiology and anatomies of some species of wild cavies, but the application of ARTs is restricted to the Spix's yellow-toothed cavy. Therefore, more efforts are needed to deepen knowledge about the physiology and adequacy of assisted reproduction for this group of species.

SPERM RECOVERY AND CHARACTERIZATION

The first reported use of ARTs in Spix's yellow-toothed cavy was by Silva et al. (2016) who recovered epididymal sperm by different methods—flotation and retrograde flushing. Both methods provide similar values for all the sperm parameters (Table 16.1). In relation to the spermatozoa, it was described with a total length of 48.87 ± 0.1 μm, and the sperm head presented 9.4%, on average, a notable characteristic that was the prominent acrosome found in the sperm.

The next step was the evaluation of longevity of epididymal sperm from Spix's yellow-toothed cavies using the TES (2-[(2-Hydroxy-1,1bis(hydroxymethyl) ethyl) amino] ethanesulfonic acid) or TRIS (tris-hydroxymmethyl-aminomethane) extenders in a thermal resistance test (TRT), as well as fluorescence analysis as a complementary method to predict the plasma membrane integrity and mitochondrial activity of these gametes (Silva et al. 2017). In this study, the samples were subjected to a TRT which involved incubation in a water bath at 37°C for 3 h. During incubation, sample parameters were assessed at 0, 15-, 30-, 60-, 90-, 120-, 150- or 180-min intervals. In general, the TRIS resulted in the extender with the best performance, due to its capacity of preserving motility, vigor, membrane integrity, mitochondrial activity, and sperm morphology. Among the morphological defects observed, tail defects such as coiled and bent coiled tail were more predominant, especially when TES was used (Silva et al. 2017).

Table 16.1 Epididymal sperm of Spix's yellow-toothed cavy recovered by flotation and retrograde washing method

Sperm characteristics	Flotation	Retrograde flushing
Number of sperm recovered ($\times$ 10^6)	514.8 $\pm$ 202.3	345.1 $\pm$ 86.3
Motility (%)	72.2 $\pm$ 9.1	56.1 $\pm$ 7.1
Vigor (1–5)	3.2 $\pm$ 0.4	2.4 $\pm$ 0.4
Estructural membrane integrity (%)	64.2 $\pm$ 4.7	56.1 $\pm$ 6.1
Functional membrane integrity (%)	51.8 $\pm$ 6.3	61.5 $\pm$ 6.9
Normal (%)	55.3 $\pm$ 5.2	58.8 $\pm$ 4.4
Total defects (%)	44.7 $\pm$ 5.2	41.2 $\pm$ 4.4
Head defects (%)	1.7 $\pm$ 0.6	2.6 $\pm$ 0.4
Midpiece defects (%)	2.1 $\pm$ 1.0	0.8 $\pm$ 0.3
Tail defects (%)	40.6 $\pm$ 5.2	36.6 $\pm$ 4.2
Distal droplet (%)	0.1 $\pm$ 0.1	0.8 $\pm$ 0.4
Proximal droplet (%)	0.2 $\pm$ 0.1	0.4 $\pm$ 0.2

Source: Silva et al. (2016).

To evaluate plasma membrane integrity and mitochondrial activity of sperm of Spix's yellow-toothed cavies, the fluorescent probes propidium iodide, CMXRos and Hoechst 342 were used. The sample containing sperm was incubated at 37°C for 10 min and evaluated through an epifluorescence microscope (×400 magnification) by counting 200 cells. Sperm marked in blue (H-342) were classified as having an intact sperm membrane, and those fully or partially marked in red (PI) were classified as non-intact. Cells with midpiece marked in red were classified as showing mitochondrial activity. The authors showed the effective protocol by fluorescence analysis as a complementary method to evaluate sperm quality (Silva et al. 2017).

SPERM CRYOPRESERVATION

After Silva et al. (2017) proposed the TRIS-based extender as the most adequate medium for the recovery of epididymal sperm using the retrograde flushing method, the same authors conducted the first sperm cryopreservation attempt in Spix's yellow-toothed cavies epididymal spermatozoa (Silva et al. 2018). Briefly, samples were diluted in TRIS plus 20% egg yolk and stored in a water jacket (30 mL) at 27°C, following an equilibrium time of 40 min into an isothermal box at 15°C. Samples were then transferred to an incubator for 30 min to reach 4°C, when they were added to glycerol or dimethyl sulfoxide (DMSO) as cryoprotectants at a final concentration of 3%, 6% or 9%. The final dilution resulted in a sperm concentration of 100×10^6 sperm/mL. Samples were then packed into 0.25 mL plastic straws and placed horizontally in a thermal insulated box for 5 min, 3 cm above liquid nitrogen (LN_2); The straws were plunged into LN_2, and finally, samples were thawed by immersing the straws in a water bath at 37°C for 1 min and sperm was immediately evaluated, as reported for fresh sperm. As the main results, glycerol provided significantly better results than DMSO. Moreover, the use of 3%, 6%, and 9% glycerol was similar in osmotic response (40.66 ± 6.3%, 42.5 ± 7.1%, and 39.5 ± 5.0%, respectively), and membrane integrity (55.17 ± 5.5%, 68.4 ± 4.1%, and 59.1 ± 4.9% respectively). However, the use of 6% glycerol resulted in the greatest post-thaw values for total motility (60.9 ± 4.4%), rapid sperm with subpopulation (27.7 ± 3.1%) and sperm-binding ability (227.0 ± 20.2). Then, the authors recommended TRIS extender with 6% glycerol and 20% egg yolk for the cryopreservation of epididymal sperm from Spix's yellow-toothed cavies (Silva et al. 2018).

In sequence, TRIS- and powdered coconut water-based (ACP®-116c) extenders were supplemented with egg yolk or *Aloe vera* at a 10% or 20% concentration as the external cryoprotectants for the sperm cryopreservation in the same cavy species. The most effective preservation of sperm motility (68.1 ± 5.9%) and membrane integrity (48.2 ± 7.4%) was achieved by TRIS extender supplemented with 10% egg yolk (Moreira et al. 2021).

TESTICULAR TISSUE PRESERVATION

The first attempt at preserving testicular samples of the Spix's yellow-toothed cavy was conducted with a solid-surface vitrification method. The testes were recovered from adult individuals and further dissected in small fragments (3.0 mm³). They were then immersed in a minimum essential medium (MEM) based solution with 10% fetal bovine serum plus 0.25 M sucrose for 5 min before exposure to either 3 or 6 M of DMSO or ethylene glycol (EG) for 5 min. Then, the samples were placed on a solid aluminum surface floating on liquid nitrogen, in which they were subsequently immersed for vitrification to occur. After a week, samples were then warmed in a water jacket 37°C, and then washed in decreasing sucrose concentrations (0.50, 0.25 and 0 M) for cryoprotectant removal. Fresh and warming tissues were evaluated for proliferative activity by the simple silver staining technique (AgNOR), in which fresh control presented 3.73 ± 0.09 nucleolar organizing regions for spermatogonia, 3.72 ± 0.11 for spermatocytes, 1.68 ± 0.08 for spermatids, 2.55 ± 0.11 for Leydig and 3.89 ± 0.11 for Sertoli cells, these parameters being efficiently preserved after freezing/warming of the samples using all cryoprotectants. Besides, samples were processed for classic histology and evaluated for morphology according to the following scores: 3 – adequate; 2 – regular; 1 – poor, using the following criteria: separation of the basal membrane, structure integrity, cell swelling, cell loss, and rupture. Therefore, the scores for fresh samples were: 2.97 ± 0.02 for membrane separation, 2.49 ± 0.04 for structure integrity, 2.81 ± 0.04 for swelling, 2.97 ± 0.01 for cell loss, and 2.97 ± 0.01 for rupture. After warming, the results suggested that 3 M EG is the optimal cryoprotectant for testicular tissue vitrification in adult Spix's yellow-toothed cavy, because this group presented scores similar to fresh controls for cell swelling (2.71 ± 0.04), cell loss (2.98 ± 0.01) and membrane rupture (2.98 ± 0.01) (Silva et al. 2019).

FEMALE REPRODUCTION

The information available so far reports on the gestational period, the estrous cycle, and some information about the morphological characteristics of the female reproductive system. In general, studies involving Spix's yellow-toothed cavy show more information about the reproduction of these animals; also, among the species of cavy, it is the only one, so far, in which there are reports of ovarian tissue cryopreservation.

In the females of Spix's yellow-toothed cavy, ovaries are dorso-ventrally flattened ovoid structures, with 2.7 cm length on average. In the ovarian cortex, there are follicles at several developmental stages, with the mean follicular population estimated to be 416.0 ± 342.8 follicles. Interestingly, most of the follicular population is composed of 63.7% primary follicles, besides the presence of 32.3% primordial follicles, and 4.1% secondary follicles, from which 94.6% preantral follicles are morphologically normal (Praxedes et al. 2015). The primordial follicles form aggregates in the ovarian cortex, being characterized by an oocyte surrounded by one layer of squamous or squamous–cuboidal granulosa cells. The primary follicles consisted of an oocyte with a single layer of cuboidal granulosa cells; some already had a zona pellucida. Secondary follicles contained two or more concentric layers of cuboidal granulosa cells around the oocyte and had a distinct zona pellucida. Another interesting fact is that polyovular preantral follicles are found in ovaries cavies, being more frequent in primordial follicles (Praxedes et al. 2015). Other genital organs of Spix's yellow-toothed cavy comprise in uterine tubes whose epithelium and muscular layer thickness modify themselves throughout the isthmus, ampulla and infundibulum regions; double uterus and uterine horns with uterine glands that open to a single cervix which is linked to the vagina by the fornix; a variegated vaginal epithelium from different animals; a vulva with a clitoris trespassed by the urethra which features a lack of vaginal vestibule and the presence of a vaginal closure membrane (Santos et al. 2014).

The female of Spix's yellow-toothed cavy reaches sexual maturity at 55–90 days on average (Lacher and Lacher 1981), presenting a continuous polyestrous cycle ranging between 14 and 19 days with a mean of 15.8 ± 1.4 days (Santos et al. 2015). When evaluating the estrus cycle through vaginal cytology, there is a predominance of superficial cells, which goes from nucleated to enucleated. In metestrus, there is a predominance of parabasal cells with presence of large numbers of neutrophils. Superficial and intermediate cells are also present in this phase. In diestrus, there is predominance of small and large intermediate cells, with presence of neutrophils and parabasal cells. In proestrus, there is predominance of large intermediate cells and superficial cells, with little or no presence of neutrophils (Santos et al. 2015). The Spix's yellow-toothed cavy has continuous poliestral cycle, developing a type of inverted choriovitellinic placenta during pregnancy, and the gestational period has an average duration of 48 days and can generate 2 to 4 offspring (Oliveira et al. 2008, 2012).

Similar to Spix's yellow-toothed cavy, the females of *G. musteloides* have a gestational period of 55 days and can produce litters of 2 to 7 pups (40 g) (Lacher 2016). The first estrus in these animals occurs around 48.5 ± 1.2 days, and the estrus can be induced by the presence of a male (Weir 1970, 1971, Touma et al. 2001).

Females of *C. aperea* have a short estrous cycle, with an average duration of 20.6 ± 0.8 days. The first estrus occurs on average at 58.6 ± 3.0 days; however, they almost always do not conceive at the first estrus, with the first conception occurring only at 84.4 ± 9.9 days (Weir 1970), although there are reports of conception at 27 days (Kraus et al. 2005a). Gestation in these females takes a little longer, occurring around 59 to 74 days, with litter size reaching up to 6 pups (Lacher 2016). Weaning of pups can occur about a week after birth; however, it is recommended that pups have 2–3 weeks of contact with their mother (Kraus et al. 2005a). As in males, the onset of puberty in females of *C. aperea* can be influenced by photoperiod (long and short days). In pups born on long days, maturation can occur in about 47 days, almost half the time compared to those born on short days (79 days) (Trillmich et al. 2009). In addition to the photoperiod, females are also influenced by the presence of the male. Females in contact with adult males can show first estrus at 26.8 ± 3.7 days and sexual maturity at 42.0 ± 15.1 days (Trillmich et al. 2006).

The sexual maturity of *C. intermedia* females is very similar to the other species mentioned above, occurring around 59 days. The gestational period lasts 60 days, and it can generate up to 2 pups, whose weaning can occur in up to 30 days (Salvador and Fernandez 2008). As for the external morphology of the genitalia of *C. intermedia*, these females have an elongated clitoris, with a penniform shape and pink color, resembling the male penis (Furnari 2013).

The *C. magna* females are semiaquatic animals, and unlike the cavy's species mentioned above, they are seasonally breeding animals, reproducing mainly during the austral winter (Kraus et al. 2005b). They have a gestational period of 60 days, generating up to 3 pups (Kraus et al. 2005b), and can conceive 3–4 litters per year (Gonzalez 2016). The offspring of this species are born immature, weighing 18% of the maternal body mass, with mass gain and faster initial growth in captive offspring when compared to wild animals (Kraus et al. 2005b). Like *C. intermedia* females, *C. magna* also has a vaginal membrane that breaks during estrus, not being observed in *C aperea* females (Furnari 2013).

CRYOPRESERVATION OF OVARIAN TISSUES

In relation to ARTs, the first cryopreservation of ovarian tissues from Spix's yellow-toothed cavies was conducted through the use of a solution composed of 3 M DMSO in MEM, supplemented with 0.25 M sucrose and 10% fetal calf serum.

For the vitrification using solid-surface system, the entire ovary was exposed to 1.8 mL solution for 5 min, dried and placed on a metal cube surface partially immersed in liquid nitrogen (LN_2). Once vitrified, the sample was transferred to cryovials for storage in LN_2. After two weeks, the samples were rewarmed at room temperature for 1 min and then immersed in a water bath at 37°C for 5 s. The DMSO was removed from the ovarian cortex by three consecutive washes for 5 min each time in MEM supplemented with 10% fetal calf serum and decreasing sucrose concentrations (0.50, 0.25 and 0 M). A total of 91.2% morphologically normal preantral follicles was found for the non-vitrified group, after vitrification observed a significant decrease in this parameter was observed. However the protocol was efficient enough to provide the conservation of 69.5% morphologically normal preantral follicles after cryopreservation and rewarming. Moreover, transmission electron microscopy revealed preservation of oocytes and granulosa cell membranes and the morphological aspect of follicles; however, it was possible to verify the presence of vacuoles in the oocytes and granulosa cells cytoplasm and turgid mitochondria in vitrified samples (Praxedes et al. 2015).

In general, information about the characteristics of female cavy still has gaps that need to be filled, especially with regard to the follicular population. So far, this information is understood only in the yellow-toothed cavy and, therefore, this is the species that is at the forefront of works involving the conservation of gonadal tissue. However, this is just one of the many species that could benefit from the various proposed and potentially applicable ARTs for rodent conservation—the same species is just one to be reported by other ARTs.

OTHER ASSISTED REPRODUCTION TECHNIQUES

Among the species of cavid rodents in South America, the Spix's yellow-toothed cavy is the species that has shown the most significant advances in terms of the conservation of its genetic material. In addition to sperm cryopreservation techniques (Moreira et al. 2021) and gonadal tissue (Praxedes et al. 2015, Silva et al. 2019), the first steps were taken to establish a database of biological resources of ear somatic tissue from Spix's yellow-toothed cavy (Pereira et al. 2021).

Pereira et al. (2021) characterized the ultrastructure of the ear skin and cartilage of Spix's yellow-toothed cavy, noting the notorious presence of a corneum layer in the epidermis. The occurrence of bundles of fibers and hair follicles was also observed, delimiting the dermis and the presence of gaps in the cartilaginous tissue. In the skin of these animals, the presence of two layers was observed- the epidermis (corneal, spinous, and basal) and the dermis. In the dermis, the connective tissue is rich in collagen fibers, with no distinction between papillary dermis and reticular dermis. In the somatic samples, the presence of cartilaginous tissue was observed, with the occurrence of gaps in its matrix, in addition to the perichondrium. More details of the morphometric analysis of the skin and cartilaginous tissue are described in Table 16.2.

As for the cellular composition of the ear skin and cartilaginous tissue, an average of 29.9 ± 6.2 melanocytes, 29.8 ± 7.6 keratinocytes, 24.8 ± 2.4 epidermal cells, 112.2 ± 11 .3 fibroblasts and 27.5 ± 4.7 chondrocytes, with a higher incidence of skin collagen fibers, as well as a lower incidence of elastic fibers. Finally, a lower proliferative activity of fibroblasts was found in the skin, with a mean proliferative potential of 1.36 ± 0.15 AgNOR/cell and 1.01 ± 0.13 μm^2 of AgNOR/cell area. Based on this information, it will be possible to adjust an efficient and specific protocol for the conservation of tissues and somatic cells, according to their characteristics and composition, thus allowing the development of cryobanks of Spix's yellow-toothed cavy ear skin and cartilage that can serve as possible banks of genetic resources for the species.

Table 16.2 Morphometric analysis of the skin and cartilage tissue of Spix's yellow-toothed cavy

Morphometric analysis (µm)	
Corneum	3.9 ± 0.6
Spinosum	23.4 ± 8.1
Basal	4.8 ± 0.5
Dermis	258.2 ± 22.9
Total skin	290.3 ± 23.7
Perichondrium of the cartilage	10.5 ± 1.8

Source: Pereira et al. (2021).

FINAL CONSIDERATIONS

Despite the numerous species of cavid rodents found in South America, there is an immense lack of important information about the reproductive physiology of these animals. This makes progress in the development of ARTs for their conservation extremely difficult. In general, the Spix's yellow-toothed cavy is the most studied species, providing important results to be applied and extrapolated to other cavid rodents. It should be noted that efforts should be directed towards the study of all species, as knowledge is essential for future applications in sustainable production systems, as well as for the development and improvement of protocols that guarantee the maintenance of multiplication and preservation of biodiversity.

REFERENCES

Adrian, O. and N. Sachser. 2011. Diversity of social and mating systems in cavies: a review. J. Mammal. 92: 39–53.

Andino, N., L. Reus, F. Cappa, V. Campos and S. Giannoni. 2011. Social environment and agonistic interactions: strategies in a small social mammal. Ethology 117: 992–1002.

Andino, N., M. Nordenstahl, M. Fornés, J. Priotto and S. Giannoni. 2016. Precipitation drives reproductive activity in male Microcavia australis in the monte desert. Mastozool. Neotrop. 23: 17–24.

Bernal, N. 2016. *Cavia aperea*. A Lista Vermelha de Espécies Ameaçadas da IUCN. 2016: e.T86257782A22189256. https://dx.doi.org/10.2305/IUCN.UK.2016-2.RLTS.T862577 82A22189256.en. Accessed on 11 August 2022.

Björkman, N., V., Dantzer and R. Leiser. 1989. Comparative placentation in laboratory animals: a review. Scand. J. Lab. Anim. Sci. 16: 129–158.

Catzeflis, F., J. Patton, A. Percequillo and M. Weksler. 2016. The IUCN Red List of Threatened Species. 2016. e.T8825A22189453. (http://www.iucnredlist.org/details/ 8825/0). Accessed on 23 January 2018.

Cherem, J.J. and J. Ferigolo. 2012. Descrição do sincrânio de *Cavia aperea* (Rodentia, Caviidae) e comparação com as demais espécies do gênero no Brasil. Pap. Avulsos. Zool. 52: 21–50.

Cooper, T.G., S. Weydert, C.H. Yeung, C. Kunzl and N. Sachser. 2000. Maturation of Epididymal Spermatozoa in the nondomesticated guinea pigs *Cavia aperea* and *Galea musteloides*. J. Androl. 21: 154–163.

Cabrera, A. 1961. Catálogo de los mamíferos de América del Sur. Revista del Museo Argentino de Ciencias Naturales "Bernardino Rivadavia". Ciênc. Zool. 4: 309–732.

Dunnum, J. 2015. Família Caviidae G. Fischer, 1817. pp. 690–716. *In*: J.L. Patton, U.F.J. Pardiñas and G. D'Elía (eds). Mammals of South Americal, 2nd Ed. The University of Chicago Press.

Furnari, N. 2013. New findings on the origin of *Cavia intermedia*, one of the world's rarest mammals. Mammal Review 43: 323–326.

Gonzalez, E. 2016. *Cavia magna*. The IUCN Red List of Threatened Species. 2016: e.T4066A22188832. https://dx.doi.org/10.2305/IUCN.UK.20162.RLTS.T4066A221888 32.en. Accessed on 13 August 2022.

Guenther, A., R. Palme, M. Dersen, S. Kaiser and F. Trillmich. 2014. Photoperiodic effects on reproductive development in male cavies (*Cavia aperea*). Physiol. Behav. 123: 142–147.

Holt, W.V. 1977. Postnatal development of the testes in the euis, *Galea musteloides*. Lab. Anim. 11: 87–91.

Jayat, J. and R. Ojeda. 2008. *Microcavia shiptoni*. The IUCN Red List of Threatened Species. 2016. e.T8825A22189453. (http://www.iucnredlist.org/details/8825/0). Accessed on 23 January 2018.

Kraus, C., D.L. Thomson, J. Künkele and F. Trillmich. 2005a. Living slow and dying young? Life-history strategy and age-specific survival rates in a precocial small mammal. J. Anim. Ecol. 74:171–80.

Kraus, C., F. Trillmich and J. Kunkele. 2005b. Reproduction and growth in a precocial small mammal, Cavia magna. J. Mammal. 74: 763–772.

Lacher, T.E. and T.E. Lacher. 1981. The comparative social behavior of Kerodon rupestris and *Galea spixii* and the evolution of behavior in the Caviidae. Bulletin of Carnegie Museum of Natural History 17: 1–71.

Lacher, T.E. 2016. Family caviidae. *In:* D.E. Wilson, T.E. Lacher and R.A. Mittermeier, (eds). Handbook of Mammals of the World. Lagomorphs and Rodents: Part 1. Lynx Editions, Barcelona, Spanish.

Mendes, B.V. 1987. Plantas e animais para o Nordeste. Rio de Janeiro, Brazil.

Moreira, S.S.J., A.M. Silva, A.L.P. Souza, E.C.G. Praxedes, J.B.F. Souza-Junior, A.F. Pereira, et al. 2021. Cryopreservation of Spix's yellow-toothed cavy epididymal sperm using Tris- and coconut water-based extenders supplemented with egg yolk or *Aloe vera*. Cryobiology 99: 40–45.

Oliveira, M.F., A. Mess, C.E. Ambrósio, C.A. Dantas, P.O. Favaron and M.A. Miglino. 2008. Chorioallantoic placentation in *Galea spixii* (Rodentia, Caviomorpha, Caviidae). Reprod. Biol. and Endocrinol. 6: 1–8.

Oliveira, M.F., A.M. Vale, P.O. Favaron, B.G. Vasconcelos, G.B. Oliveira, M.A. Miglino, et al. 2012. Development of yolk sac inversion in *Galea spixii* and *Cavia porcellus* (Rodentia, Caviidae). Placenta. 33: 878–881.

Pereira, A.F., L.V.C. Aquino, M.B. Nascimento, F.V.F. Bezerra, A.A. Borges, E.A. Praxedes, et al. 2021. Ultrastructural and morphometric description of the ear skin and cartilage of two South American wild histricognate rodents (*Dasyprocta leporina* and *Galea spixii*). Pesq. Vet. Bras. 41: e06775.

Praxedes, E.C.G., G.L. Lima, A.M. Silva, C.A.C. Apolinário, J.A.B. Bezerra, A.L.P. Souza, et al. 2015. Characterisation and cryopreservation of the ovarian preantral follicle population from Spix's yellow-toothed cavies (*Galea spixii* Wagler, 1831). Reprod. Fertil. Dev. 29: 594–602.

Reis, N.R., A.L. Peracchi, W.A. Pedro and I.P. Lima. 2006. Mamíferos do Brasil. State University of Londrina, Londrina, Brazil.

Rocha, N. 2016. *Microcavia australis. A Lista Vermelha de Espécies Ameaçadas da IUCN* 2016: e.T13319A22189827. https://dx.doi.org/10.2305/IUCN.UK.2016-2.RLTS.T13319A22189827.en. Accessed on 4 Setempber 2017.

Roach, N. 2016a. *Cavia intermedia.* The IUCN Red List of Threatened Species. 2016: e.T136520A22189125. (http://www.iucnredlist.org/details/136520/0). Accessed on 4 September 2017.

Roach, N. 2016b. *Dolichotis patagonum.* The IUCN Red List of Threatened Species. 2016: e.T6785A22190337. (http://www.iucnredlist.org/details/6785/0). Accessed on 4 September 2017.

Roach, N. 2016c. *Galea musteloides.* A Lista Vermelha de Espécies Ameaçadas da IUCN. 2016: e.T86226097A22189593. https://dx.doi.org/10.2305/IUCN.UK.2016-3.RLTS.T86226097A22189593.en. Accessed on 4 September 2017.

Rodrigues, M.N., G.B. Oliveira, J.F.G. Albuquerque, J.D.A. Menezes, A.C. Assis Neto, M.A. Miglino, et al. 2013. Aspectos anatômicos do aparelho genital masculino de preás adultos (*Galea spixii* Wagler, 1831). Biotemas 26: 181–188.

Salvador, C.H. and F.A.S. Fernandez. 2008. Reproduction and growth of a rare, island-endemic cavy (*Cavia intermedia*) from southern Brazil. J. Mammal. 89: 909–915.

Salvador, C.H. 2021. *Cavia intermedia* (Green Status assessment). The IUCN Red List of Threatened Species. 2021: e.T136520A13652020221. Accessed on 4 September 2017.

Santos, P.R.S., T.V.B. Carrara, L.C.S. Silva, A.R. Silva, M.F. Oliveira and A.C. Assis Neto. 2011. Morphological characterization and frenquency of stages of the seminiferous epithelium cycle in captive bred Spix's Yellow-Toothed (*Galea spixii* Wagler, 1831). Pesq. Vet. Bras. 31: 18–24.

Santos, P.R.S., M.F. Oliveira, A.R. Silva and A.C.A. Neto. 2012. Development of spermatogenesis in captive-bred Spix's yellow-toothed cavy (*Galea spixii*). Reprod. Fertil. Dev. 24: 877–885.

Santos, A.C., B.M. Bertassoli, D.C. Viana, B.G. Vasconcelos, M.F. Oliveira, M.A. Miglino, et al. 2014. The morphology of female genitalia in *Galea spixii* (caviidae, caviinae). Biosci. J. 30: 1793–1802.

Santos, A.C., D.C. Viana, B.M. Bertassoli, G.B. Oliveira, D.M. Oliveira, F.V.F. Bezerra, et al. 2015. Characterization of the estrous cycle in *Galea spixii* (Wagler, 1831). Pesq. Vet. Bras. 35: 89–94.

Silva, A.M., J.A.B. Bezerra, L.B. Campos, É.C.G. Praxedes, G.L. Lima and A.R. Silva. 2016. Characterization of epididymal sperm from Spix's yellow-toothed cavies (*Galea spixii* Wagler, 1831) recovered by different methods. Acta. Zool. 98: 285–291.

Silva, A.M., P.C. Sousa, L.B. Campos, J.A.B. Bezerra, A.E.A. Lago, M.F. Oliveira, et al. 2017. Comparison of different extenders on the recovery and longevity of epididymal sperm from Spix's yellow-toothed cavies (*Galea spixii* Wagler, 1831). Zygote 25: 176–182.

Silva, A.M., E.C.G. Praxedes, L.B. Campos, L.G.P. Bezerra, S.S.J. Moreira, K.M. Maia, et al. 2018. Epididymal sperm from Spix's yellow-toothed cavies sperm successfully cryopreserved in Tris extender with 6% glycerol and 20% egg yolk. Anim. Reprod. Sci. 191: 64–69.

Silva, A.M., A.G. Pereira, E.C.G. Praxedes, S.S.J. Moreira, M.F. Oliveira, P. Comizzoli, et al. 2019. Vitrification of testicular tissue from adult Spix's yellow-toothed cavies (*Galea spixii*) using different cryoprotectants. *In:* Society for the Study of Reproduction 2019 Annual Meeting, San José. Proceedings. San José: Society for the Study of Reproduction.

Tognelli, M.F., C.M. Campos and R.A. Ojeda. 2001. *Microcavia australis*. Mamm. Species 648: 1–4.

Touma, C., R. Palme and N. Sachser. 2001. Different types of oestrous cycle in two closely related South American rodents (*Cavia aperea* and *Galea musteloides*) with different social and mating systems. Reproduction 121: 791–801.

Trillmich, F., C. Laurien-Kehnen, A. Adrian and S. Linke. 2006. Age at maturity in cavies and guinea-pigs (*Cavia aperea* and *Cavia aperea f. porcellus*): influence of social factors. J. Zool. 268: 285–294.

Trillmich, F., B. Mueller, S. Kaiser and J. 2009. Krause. Puberty in female cavies (*Cavia aperea*) is affected by photoperiod and social conditions. Physiol. Behav. 96: 476–480.

Velez, S., P.L. Sassi, C.E. Borghi, M.A. Monclus and M.W. Fornés. 2010. Effect of climatic variables on seasonal morphological changes in the testis and epididymis in the wild rodent *Microcavia australis* from the Andes Mountains, Argentina. J. Exp. Zool. 313: 474–483.

Weir, B.J. 1970. The management and breeding of some more hystricomorph rodents. Lab. Anim. 4: 83–97.

Weir, B.J. 1971. The evocation of oestrus in the cuis, *Galea musteloides*. J. Reprod. Fert. 4: 405–408.

Chapter **17**

The Pacas

Vânia Maria França Ribeiro*, Itacir Olivio Farikoski,
Patrícia Andrade dos Santos and Augusto Luiz Faino Alves

Programa de Pós-Graduação em Sanidade e Produção Animal Sustentável da
Amazônia Ocidental (PPGESPA), Universidade Federal do Acre (UFAC),
Campus Universitário Reitor Áulio Gélio Alves de Souza, BR 364, km 4,
Distrito Industrial, Rio Branco Acre, Brasil, 69915900.
Email: vania.ribeiro@ufac.br; itacir.farikoski@gmail.com;
patriciatlesantos@gmail.com; gutofaino1@hotmail.com

INTRODUCTION

Paca is a medium-sized mammal (Gast and Stevenson 2020) of the Cuniculidae family and *Cuniculus* genus. Currently, only two species of this genus are known: *Cuniculus taczanowskii* (Stolzmann 1885), popularly known as mountain paca restricted to mountainous areas in the Andean region, and lowland paca, *Cuniculus paca* (Linnaeus 1766), occupying a large habitat from South and Central America to Mexico and in Algeria and Cuba (Emmons 2016). Taxonomically, paca belongs to kingdom Animalia, phylum Chordata, class Mammalia, order Rodentia, suborder Hystricomorpha, and superfamily Cavioidea, and it is represented by the above-mentioned two species.

Owing to its wide geographic distribution, it is misunderstood and classified by the International Union for Conservation of Nature (IUCN 2022) as "Least Concern". In this classification, the number of individuals was not counted, and only the places where they are observed is monitored. However, population of paca is reducing leading to this animal becoming rare or non-existent in places where it was previously abundant (Chiarello et al. 2008, Emmons 2016).

*Corresponding author: vania.ribeiro@ufac.br

Pacas (Figure 17.1) are robust with short legs, have three fingers on the forelimbs, and have four fingers on the hind limbs, which are strong and adapted for digging. They have a vestigial tail, and the body is covered with short, bristly dark brown fur with a sequence of rounded longitudinal white spots on the sides of the back. The ventral part has a lighter shade, with the color approaching white (Bonvicino et al. 2008). Snouts have long tactile hairs seen developing in the fetal phase (El Bizri et al. 2017). They have territorial behavior and make burrows far away from each other (Harmsen et al. 2018).

Figure 17.1 Specimen of *Cuniculus paca*: a female with calf.

The animal's head is large, having an expansive zygomatic arch forming an internal cavity that amplifies sounds and favors greater muscle development, providing them a powerful bite. Upon onset of sexual maturity, males develop a larger zygomatic arch than females, and this is considered one of the unique features of visible sexual dimorphism (Hosken and Silveira 2001).

Similar to all rodents, paca has two pairs of incisor teeth that grow continuously and gets worn out because of consistent gnawing (Lange and Schmidt 2014). Its large eyes (Bonvicino et al. 2008) are physiologically adapted to survive in low-light environments. The diet of paca consists of vegetables, tubers, fruits, and seeds, but there is a marked preference for seeds of some leguminous plants, besides coconuts and chestnuts (Hosken and Silveira 2001, Ribeiro and Zamora 2008). Similar to other hystricomorph rodents, paca practices cecotrophy to retain microbial protein from its intestinal tract as a protein optimization strategy (Aldrigui et al. 2018).

However, their reproductive aspects have not been fully elucidated. According to Carretta-Junior (2012), pacas reach sexual maturity in 10 months. Females are in heat throughout the year, and sperm production by males is continuous across all seasons (Ribeiro and Zamora 2008, Urbina 2011).

Paca is the second most bred in legalized commercial farms in Brazil (Le Pendu et al. 2011). Growers choose them because they are docile and easily domesticated but face production constraints because these animals are less prolific and have a capacity to produce one calf in 150 days gestation. Twin births have rarely been reported (Oliveira et al. 2007, Guimarães et al. 2008, Ribeiro et al. 2012).

Most wild animals bred in captivity have reproductive difficulties due to stress and assisted reproduction techniques (ARTs) must be adopted to overcome this limitation (Micheletti et al. 2011). This review describes some of the ARTs that have been tried in pacas, as well as the anatomical structures of the male and female reproductive systems that help to successfully implement reproduction.

MALE REPRODUCTIVE ANATOMY

The male paca has a penis with foreskin, a pair of testicles surrounded by the scrotum, a pair of epididymis (composed of head, body, and tail), a pair of vas deferens and urethra, and four pairs of accessory genital glands (vesicular, prostate, coagulant, and bulbourethral) (Borges et al. 2013, 2014). The body of penis is cylindrical and is formed by two types of erectile tissues: the corpora cavernosa and corpus spongiosum, with the corpora cavernosa located dorsally and covered by the tunica albuginea from the root of the penis (Borges et al. 2013).

Furthermore, the glans presents a rounded dilatation, called as urethral torus or penile flower, at the time of erection (Figure 17.2a) (Carvalho et al. 2008, Mollineau et al. 2012). Pacas have rigid penis with serrated and sharp edges on the sides (Figure 17.2b) (Carvalho et al. 2008). Borges et al. (2013) observed the presence of a saccular structure on the dorsal aspect of a longitudinal section of the glans in which two rigid spurs with pointed free ends are housed.

Mollineau et al. (2006) mentioned that such structures damaged vagina during copulation, which prevented females from accepting new copulations. Borges et al. (2013) reported that no information is available in literature that indicates a relationship between these spicules and ovulation induction in females, as seen in other mammals such as rabbits, llamas, some rodents, and cats.

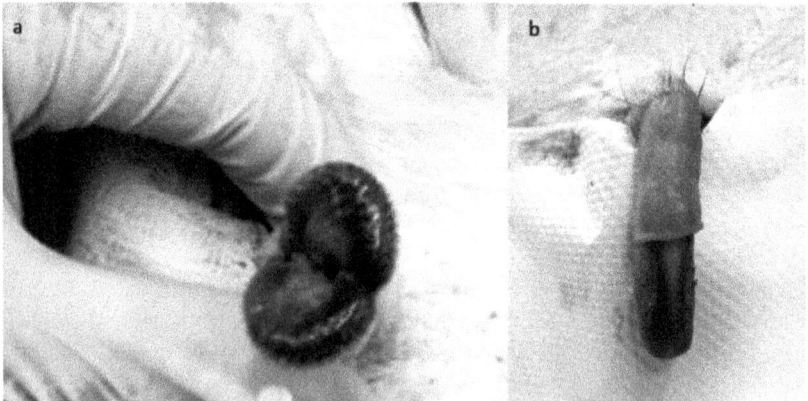

Figure 17.2 Male genital structure of *Cuniculus paca*. (a) Enlarged penile glans at the time of erection; (b) penis with serrations and spicules that are characteristic of the species.

Additionally, pacas do not have a well-defined scrotum. The testes are located subcutaneously in the inguinal region and have a well-developed cremasteric tunic. This location is associated with a wide inguinal canal and allows ample testicular movement into the abdominal cavity (Carretta-Junior 2012, Borges et al. 2013, Castelo et al. 2015). The testicle may be more evident during the breeding season when it relaxes and can be seen in the inguinal region next to the penis (Menezes et al. 2003, Costa 2009, Barros et al. 2016).

The epididymis, extending along the entire testes, has a firm consistency and is covered by tunic albuginea, except at the tail region. The epididymal tail is globular and large with a coiled epididymal duct (Borges et al. 2013). There is an interposed sinus of the epididymis, in which a serous membrane—derived from the peritoneum and distal mesorchium—can be observed between the body of the epididymis and epididymal margin of the testis. This fixes the epididymis to the epididymal margin of the testis (Borges et al. 2013).

The sperm cycle of pacas is 11.5 ± 0.16 days, and the duration of spermatogenesis is 51.6 ± 0.7 days (Costa 2009). Carretta-Junior (2012) stated that spermatogenesis in the adult has an 8.57 day cycle, and the daily sperm production per g of testis is 33.9×10^6. At the end of the spermiogenic phase, immature sperms are released in the seminiferous tubules and routed to the epididymis. As they pass through the epididymis, they undergo suitable modifications to become fit for fertilization. At the end of the process, mature sperms are stored in the tail of the epididymis (Shivaji 1988).

Finally, pacas have high spermatogenic efficiency, owing to the combination of high-volume density in the seminiferous tubules, high number of Sertoli cells, and short spermatogenesis. Because of these characteristics, this species is suitable for research aimed at increasing reproductive rates in captivity (Costa 2009).

FEMALE REPRODUCTIVE ANATOMY

Six organs, namely, the external and internal genitalia, vagina, cervix, uterus, oviducts, and ovaries, are part of the female reproductive system and play a vital role in reproduction (Hafez and Hafez 2004, Koning and Liebich 2004). The external genitalia or vulva have a large number of sensory nerve endings and are composed of the labia majora, labia minora, vestibule, and clitoris (Hafez and Hafez 2004).

These rodents have a unique urogenital tract, and females have anatomically distinct genital and urinary orifices. The genital orifice is closed most of the time by adhesive secretion, opening only during estrus or close to parturition (Ribeiro et al. 2012, Lange and Schmidt 2014). Additionally, pacas have a pair of inguinal and axillary breasts that are noticeable only close to childbirth when teats become swollen and easily recognizable (Bonvicino et al. 2008, Ribeiro and Zamora 2008).

ASSISTED REPRODUCTION

There exist only a few published reports regarding ARTs in pacas. In the subsequent sections, experiences of different research groups in Brazil investigating ARTs in paca have been reported.

Specifics of Sperm

Santos et al. (2020) observed the morphology, morphometry, and membrane integrity of sperm cells originating from the epididymal tail using the flotation method. This method involves cutting or slicing the tail of the epididymis longitudinally in Petri dishes containing collection media, where the spermatozoa migrate and are later recovered by filtration. In this study, the authors used ACP-123® medium (powdered coconut water) and sperm cells were stained with panoptic fast and eosin-nigrosin.

Rapid panotic staining 200 spermatozoa, to explore spermatozoa morphology, revealed an oval head with three vesicles in the acrosomal region, intermediate piece, and elongated tail, as well as 27% cellular defects. The morphometry of 100 sperm cells (observed under an optical microscope and visualized using EZ Leica LAS image acquisition software for Windows operating systems) showed the following measurements (mean ± standard deviation): total length (43.87 ± 4.91 μm), head (7.54 ± 0.82 μm), midpiece (5.35 ± 0.83 μm), tail (30.72 ± 2.55 μm), and head width (5.30 ± 0.68 μm). Of the 2,000 cells stained with eosin-nigrosin to assess membrane integrity, 83.8% showed intact membranes.

These results suggest that paca epididymal spermatozoa can be used in ARTs; however, further studies on the role of acrosomal vesicles in this species should be carried out, as due to their location, it is still unknown whether they play a role during the acrosomal reaction, a crucial process for fertilization. The aforementioned vesicles were also observed and reported in the sperm of this species, upon being obtained via electro-ejaculation and artificial vagina methods by Hoyos et al. (2001) and Ferreira et al. (2004), respectively, demonstrating that these are present even after passing through the vas deferens and ejaculatory canal.

Development of Methodology for Semen Collection

After a previous anatomical measurement of the male reproductive tract of the paca, in association with ejaculatory processes, Stradiotti et al. (2015) developed a specific electro-stimulator to test efficient techniques for semen collection and study. Four dead animals were dissected from the pelvic region for macroscopic visualization of the pelvic plexus and nerves associated with other structures. Measurements were taken to determine the distance from the pelvic plexus to the anal opening and the diameter of the rectum, to insert the probe of the electro-stimulation device. To test the device and the electro-stimulation protocol that is best suited, nine bulls (aged 18–36 months) were used. The procedures allowed the macroscopic evaluation of spermatozoa by removing the tails from the epididymis and subjecting them to the post-mortem sperm retrieval technique.

The typical dissection techniques, based on descriptions of the neuroanatomy of domestic mammals, are efficient in locating the pelvic plexus. The measurement taken up to the anal opening was also correct for inserting the probe of the electro-stimulation device, which was made of compact cylindrical material, 13 cm long and 1.6 cm in diameter, and tapered in the distal region, supporting two strips.

A 6 cm long longitudinal electrode was inserted laterally into the probe. To calibrate the device, a button was added to the electro-ejaculator (SA–200; Eletrogen AS 200, Champion co, Atlas Diagnóstico; Presidente Prudente, Brazil), which allowed changing the voltage from 1 to 5 V, in addition to the typical power button. Using a protocol that simulated natural copulation, and after mechanical and chemical restraint, it was observed that young animals presented ejaculatory responses with lower voltages than older animals. The evaluated seminal characteristics (volume, odor, color, appearance, and pH) were similar to those described by other authors for this species.

Collection, Evaluation, and Preservation of Semen

Alves (2018) evaluated the efficiency of two semen collection methods, namely, electro-ejaculation and collection of spermatozoa from the tail of the epididymis by the flotation method using two extenders (ACP-123® and Botusemen special®) and conserving sperm of this species by cooling (temperature of 5°C). Samples collected via electro-ejaculation registered an average volume, average concentration, motility, and mean vigor of 0.43 ± 0.33 mL, $45.5 \pm 42.44 \times 10^6$ spermatozoa/ mL, $33.33 \pm 32.14\%$, and 2.6 ± 1.15, respectively. The samples obtained directly from the epididymis tail had a mean volume and mean concentration of 1.5 mL ± 0.2 mL and $197.1 \pm 84.9 \times 10^6$ sperm/mL, respectively, with the mean motility and vigor of the ACP-123® diluent being $63.8 \pm 34.2\%$ and 4.2 ± 1.7, respectively. In contrast, samples diluted in Botusemen Special® showed a mean motility and mean vigor of $29.8 \pm 34.2\%$ and 2.4 ± 1.9, respectively. The membrane integrity of the spermatozoa when diluted with ACP-123® was preserved at $84.0 \pm 0.07\%$; the membrane viability was maintained at $53.9 \pm 3.78\%$, with Botusemen Special® at $73.0 \pm 0.21\%$ of the intact sperm and $39.0 \pm 17.9\%$ of viable membranes.

Epididymal collection recorded better results than the electro-ejaculation protocol in terms of sperm parameters. The evaluations of the different extenders indicated that the ACP-123® provided better sperm viability than the commercial Botusemen special®; however, neither of them preserved the sperm up to 24 h.

Pharmacological manipulation of estrous cycle with progestogen, prostaglandin, and equine chorionic gonadotropin (eCG) implants

Ribeiro et al. (2017) evaluated progestogen implants (Crestar® silicone auricular implant with 3 mg of Norgestomet plus an injectable 3 mg of Norgestomet with 5 mg of estradiol valerate) associated with two doses of equine chorionic gonadotropin (eCG) on the synchronization and induction of fertile estrus in pacas. For this, 18 non-pregnant females and nine males were divided into three groups (G1, G2, and G3), each with six females and three males. Females from G1 and G2 received implants with 1.5 mg of Norgestomet, and seven days later, 0.13 mg of prostaglandin intramuscularly (IM). On day 8 (D8), the implants were removed, and G1 and G2 received 25 IU and 50 IU of eCG IM, respectively, with G3 as control. Pairing of the three groups was performed on the same day.

G3 females showed estrus a few days after day 0 (D0). Females that received treatment (G1 and G2) showed estrus only after implant removal on day 8 (D8). The pregnancy rates for G1, G2, and G3 were 100%, 66%, and 50%, respectively. Regarding pups per birth, 100% of G1 and G3 produced one offspring, whereas 50% of G2 produced two offspring.

Therefore, it was concluded that: (i) the progestogen used in the form of subcutaneous implants was efficient in mimicking the luteal phase of the estrous cycle, suppressing the manifestation of estrus during the period during which they remained in the animal; (ii) the hormonal treatment protocol for estrus induction and synchronization favor the occurrence of fertile estrus and its physical and behavioral manifestations; (iii) Finally, a dose of 50 IU eCG presumably favored the occurrence of twin births.

This marked the first-time hormones were used for the induction and synchronization of fertile estrus in *Cuniculus paca*, and accurate observations were made in the two treated females (G2) in which estrus occurred, but adequate coverage was not provided for pregnancy. Therefore, ultrasound follow-ups were carried out at day 70 after inducing estrus and living with males; two pregnancies of approximately 40 days were observed. This demonstrated the non-interference of hormones in the natural estrous cycle.

Pharmacological Manipulation of Estrous Cycle using Melengestrol Acetate (MGA), Prostaglandin, and eCG

Ribeiro et al. (2017) and Farikoski (2021) chose melengestrol acetate, a known and extensively used oral progestogen that minimizes stress caused by handling. It is inexpensive and easy-to-orally-administer progesterone compared to other protocols, which require manipulation of the animals. This method also appears promising while working with wild species (Patterson et al. 1989, Quadros et al. 2020).

To observe the effect of this progestogen associated with prostaglandin eCG in the induction and synchronization of estrus in pacas, 18 adult animals (six males and 12 females) with proven reproductive efficiency were separated into three groups (G1: 0.22 mg of melengestrol acetate + 0.13 mg of prostaglandin and 50 IU of eCG; G2 0.22 mg of melengestrol acetate + 0.13 mg of prostaglandin and 75 IU of eCG; G3 control group) - each group containing four females and two males and confined to two stalls. The females in G1 received 0.22 mg of melengestrol acetate orally in palatable food for 8 days (D0 to D8). On day 8, they received 0.13 mg of prostaglandin and 50 IU of eCG IM. In G2, the females received a similar treatment with only a change in the dose of eCG, which became 75 IU, and G3 constituted the control group. The control animals received food, and saline solution was administered at the same time as rest of the groups receiving hormonal treatment. Blood samples were collected on D0, D4, and D8 to assess serum progesterone levels.

The animals studied had an average estrous cycle of 33–34 days with a minimum and maximum of 28 and 40 days, respectively, and a standard deviation of 3 ± 84; the average gestation time was 158.5 ± 12.7 days. Results of plasmatic measurement of progesterone via enzyme-linked immunosorbent assay indicated the following average concentrations during different periods: estrus (1.58 ± 1.0 ng/mL), metaestrus (7.18 ± 6.15 ng/mL), diestrus (5.14 ± 3.46 ng/mL), and proestrus (3.46 ± 2.07 ng/mL).

The application of eCG at a dosage of 50 IU demonstrated efficiency in inducing ovulation, and one of the animals from G1 had a triplet birth, a case observed in paca for the first time.

Laparoscopic Ovum Pick-up

Barros et al. (2016) studied the feasibility of laparoscopic follicular aspiration (LapOPU) and described the details of the surgical procedure, its complications, and oocyte recovery rate. Nine healthy, adult, non-pregnant females, kept in captivity, were used for a total of 39 procedures. When the surgical anesthetic plane was reached, the females were positioned in Trendelenburg with 20° of angulation. Three trocars were placed in the right and left inguinal and hypogastric regions, respectively. The abdomen was inflated with CO_2, and intra-abdominal pressure was maintained at 10 mmHg. Follicular punctures were carried out by manipulating the ovaries with atraumatic forceps. For follicular aspiration, an 18 G needle with a short bevel coupled to a vacuum system with pressure not exceeding 65 mmHg was used. Oocytes were recovered in 50-mL centrifuge tubes containing phosphate-buffered saline composite medium supplemented with 10 IU/mL heparin and maintained at 36°C.

Of the 39 video-laparoscopies, it was only possible to perform LapOPU in 30 (76.92%). The total surgical time for LapOPU was 37.34 ± 18.53 min. The total number of visualized follicles, aspirated follicles, and retrieved oocytes were: 502, 415, and 155, respectively, and the same parameters per animal were 14.34 ± 12.23, 11.86 ± 10.03, and 4.43 ± 4.69, respectively. The recovery rate was 32.56 ± 27.32%. This showed that the caudal positioning of portals, with slight triangulation, allows good visualization of the abdominal cavity and facilitates manipulation of the ovaries, which is easy to perform and feasible in pacas.

To improve the uptake rate and favor the *in vitro* maturation of paca oocytes, Barros et al. (2020) used three different hormonal protocols to evaluate follicular stimulating hormone (FSHp), eCG, and FSHp + eCG levels. Eight adult females that underwent a hormonal protocol were administered with a single dose of 45 mg of injectable progesterone and a single intramuscular injection of 0.075 mg of d-cloprostenol on Day 6.

Ovarian stimulation was performed on Day 6 after application of progesterone, in the following groups, as a single administered dose:

(a) 80 mg of FSHp and 200 IU of eCG IM in TFE group (FSHp and eCG)
(b) 80 mg of FSHp IM in group TF (FSHp), and
(c) 200 IU of eCG

The female used as control was administered with 1 mL saline solution (TC). LapOPU was performed at 22–26 h after gonadotropin treatment. All recovered oocytes were placed in maturation medium and incubated for 24 h.

There was no difference in the mean number of observed follicles, aspirated follicles, or oocytes recovered per treatment. Oocyte maturation rates did not differ between the groups, except that TF and eCG oocytes had higher maturation rates ($P = 0.043$ and $P = 0.048$, respectively) than TC oocytes. In this study, gonadotropin administration failed to superovulate the treated females and to increase the efficiency of oocyte retrieval. Despite the feasibility of this procedure, further studies are needed to develop and refine hormonal protocols for oocyte recovery and *in vitro* maturation in this species.

FINAL CONSIDERATIONS

Scarce information is available on the proper and efficient pharmacological manipulation protocols for improving fertility in paca. Therefore, it is essential to make adjustments to the known and established reproductive and behavioral biology-related methods used in domestic or wild animals. The following considerations are essential: (a) such information is still scarce, and paca have a unique physiology, which makes them likely to respond to drugs in an unexpected manner; (b) when stressed, pacas can influence responses to hormones; (c) past studies have been conducted with a small number of animals with reduced number of specific breeding sites; and (d) pacas have great genetic variability, which can lead to different individual responses when using appropriate drugs for ARTs.

Therefore, increased investigations in these areas are necessary to arrive at effective and appropriate protocols to favor pacas, which are known to have low reproductive rates, for meat consumption. Lastly, the gradual destruction of the habitat of pacas, on year-on-year basis, poses a risk of making this species disappear in due course of time.

REFERENCES

Aldrigui, L.G., S.L.G. Nogueira Filho, V.S. Altino, A. Mendes, M. Claus and A.S. Nogueira. 2018. Direct and indirect caecotrophy behaviour in paca (*Cuniculus paca*). J. Anim. Physiol. Anim. Nutr. 102: 1774–1782.

Alves, A.L.F. 2018. Methods of collection, evaluation, and preservation of paca (*Cuniculus paca* Linnaeus, 1766) (Cuniculidae) semen in two different diluents. Masters. Dissertation, Universidade Federal do Acre, BR.

Barros, F.F.P.C., P.P.M. Teixeira, M.E.B.A.M. Conceição, R.A.R. Uscategui, L.N. Coutinho, M.B.S. Brito, et al. 2016. Vasectomy in Spotted Paca (*Cuniculus paca*). Acta. Sci. Vet. 44: 175–181.

Barros, F.F.P.C., P.P.M. Teixeira, L.C. Padilha-Nakaghi, R.A.R. Uscategui, M.R. Lima, V.J.C. Santos, et al. 2020. Ovum pick-up and *in vitro* maturation in spotted paca (*Cuniculus paca*—Linnaeus, 1766). Reprod. Domest. Anim. 55: 442–447.

Bonvicino, C., J. Oliveira and O. D'Andrea. 2008. Guia dos Roedores do Brasil. Rio de Janeiro: Centro Pan-Americano de Febre Aftosa (OPAS/OMS).

Borges, E.M., É. Branco, A.R. Lima, L.M. Leal, L.L. Martins, A.C.G. Reis, et al. 2013. Morphology and topography of the external male genital organs of spotted paca (*Cuniculus paca* Linnaeus, 1766). Biotemas 26: 209–220.

Borges, E.M., É. Branco, A.R. Lima, L.M. Leal, L.L. Martins, M.A. Reis Miglino, et al. 2014. Morphology of accessory genital glands of spotted paca (*Agouti paca,* Linnaeus, 1766). Anim. Reprod. Sci. 145: 75–80.

Carretta-Junior, M. 2012. Comparative study of the spermatogenic process and duration of the cycle of the seminiferous epithelium by immunohistochemical technique with bromodeoxyuridine in three different species of rodents of the suborder Hystricomorfa: agouti (*Dasyprocta leporina*), paca (*paca Cuniculus*) and capybara (hydrochaeris). Ph.D. Thesis, Universidade Federal de Viçosa, BR.

Carvalho, M.A.M., A.A.N. Machado Junior, R.A.B. Silva, D.J.A. Menezes, A.M. Conde Jr. and D.A. Righ. 2008. Arterial supply for the penis in agoutis (*Dasyprocta prymnolopha,* Wagler, 1831). Anat. Histol. Embryol. 37: 60–62.

Castelo, T.S., A.L.P. Souza, G.L. Lima, G.C.X. Peixoto, L.B. Campos, M.F. Oliveira, et al. 2015. Interactions among different devices and electrical stimulus on the electroejaculation of captive agoutis (*Dasyprocta leporina*). Reprod. Dom. Anim. 5: 492–496.

Chiarello, A.G., L.M.S. Aguiar, R. Cerqueira, F.R. Melo and F.H.G. Rodrigues. 2008. Livro Vermelho da Fauna Brasileira Ameaçada de Extinção. 2: 680–880. Fundação Biodiversitas, Brasília.

Costa, K.L.C. 2009. Morphofunctional evaluation of the testis in brown brocket deer (*Mazama gouazoubira* Fischer, 1814). 2009. Masters. Dissertation, Universidade Federal de Viçosa, BR.

El Bizri, H.R., F.O.B. Monteiro, R.S. Andrade, J. Valsecchi, D.A. Guimarães and P. Mayor. 2017. Embryonic and fetal morphology in the lowland paca (*Cuniculus paca*): A precocial hystricomorph rodent. Theriogenology 104: 7–17.

Emmons, L. 2016. *Cuniculus paca.* The IUCN Red List of Threatened Species 2016: e.T699A22197347. http://dx.doi.org/10.2305/IUCN.UK.2016-2.RLTS.T699A22197347.

Farikoski, I.O. 2021. Reproductive behavior and pharmacological manipulation of the estrous cycle of paca (*Cuniculus paca*) using Melengestrol Acetate (MGA) in animals raised in captivity. Ph.D. Thesis, Universidade Federal do Acre, BR.

Ferreira, A.C.S., D.A. Guimarães, R.S. Luz-Ramos, L.V. Bastos and O.M. Ohashi. 2004. Morphologial and biometrics characteristics of *Agouti paca* semen, raised in captivity. Rev. Bras. Rep. Anim. 9: 224–224.

Gast, F. and P.R. Stevenson. 2020. Relative abundances of medium and large mammals in the Cueva de Los Guácharos National Park (Huila, Colombia). Biota. Neotrop. 20: e20160305.

Guimarães, D.A., L.V. Bastos, A.C.S. Ferreira and R.S. Luz-Ramos. 2008. Reprodutive characteristics of the female paca (*Agouti paca*) raised in captivity. Acta. Amaz. 38: 531–538.

Hafez, B. and E.S.E. Hafez. 2004. Anatomia e reprodução feminina. *In*: Reprodução Animal. 7ed. Manole, Barueri, Brazil. pp. 3–12.

Hoyos, D., J. López, A. Ramírez, F. Valencia, S. Molina, J. Sanchéz and M. Oliveira-Angel. 2001. Caracterización espermática de agouti paca y agouti taczanowskii. Rev. Colomb. Cienc. Pecu. 14: 86.

Hosken, F.M. and A.C. Silveira. 2001. Criação de Pacas. Viçosa: Aprenda fácil.

Harmsen, B.J., R.L. Wooldridge, D.M. Gutierrez, C.P. Doncaster and R.J. Foster. 2018. Spatial and temporal interactions of free-ranging pacas (*Cuniculus paca*). Mammal Res. 63: 161–172.

IUCN Red List Categories and Criteria, Version 3.1. 2022. IUCN, Gland, Switzerland and Cambridge, UK. Available on: www.iucnredlist.org. Accessed on: September 26th.

Koning, H.E. and H.G. Liebich. 2004. Órgãos genitais femininos. *In*: Anatomia dos animais domésticos - Texto e atlas colorido. 6ed. Artmed, Porto Alegre, Brazil. 856p.

Lange, R.R. and E.M.S. Schmidt. 2014. Rodentia – roedores selvagens (Capivara, Cutia, Paca e Ouriço). pp. 1137–1168. *In*: Z.S. Cubas, J.C.R. Silva and J.L. Catão- Dias (eds). Tratado de animais selvagens, 2nd Ed. Roca, São Paulo, Brazil.

Le Pendu, Y., D.A. Guimarães and Á. Linhares. 2011. State of the art on the commercial breeding of wildlife in Brazil. Rev. Bras. Zootec. 40: 52–59.

Menezes, D.J.A., M.A.M.D. Carvalho, A.C.D. Assis Neto, M.F. Oliveira, E.C. Farias, M.A. Miglino, et. al. 2003. Morphology of the external male genital organs of agouti (*Dasyprocta aguti* Linnaeus, 1766). Braz. J. Vet. Res. Anim. Sci. 40: 148–153.

Micheletti, T., Z.S. Cubas, W. Moraes, M.J. Oliveira, L.E. Kozicki, R.R. Weisset, et al. 2011. Assisted reproduction in wild felids – a review. Rev. Bras. Reprod. Anim. 35: 408–417.

Mollineau, W., A. Adogwa, N. Jasper, K. Young and G. Garcia. 2006. The gross anatomy of the male reproductive system of a neotropical rodent: The Agouti (*Dasyprota leporina*). Anat. Histol. Embryol. 35: 47–52.

Mollineau, W.M., T. Sampson, A.O. Adogwa and G.W. Garcia. 2012. Anatomical stages of penile erection in the agouti (*Dasyprocta leporina*) induced by electro-ejaculation. Anat. Histol. Embryol. 41: 392–394.

Oliveira, F.S., M.R.F. Machado, J.C. Canola and M.H.B. Camargo. 2007. Uniparity in pacas bred in captivity (*Agouti paca*, Linnaeus, 1766). Arq. Bras. Med. Vet. Zootec. 5: 387–389.

Patterson, D.J., G.H. Kiracofe, J.S. Stevenson and L.R. Corah. 1989. Control of the bovine estrous cycle with melengestrol acetate (MGA): a review. J. Anim. Sci. 67: 1895–1906.

Quadros, A.P., Y. Tanaka and J.M. Duarte. 2020. Assessment of melengestrol acetate on estrus synchronization in female of Brazilian dwarf brocket deer and Amazonian brown brocket deer: pilot study. Rev. Bras. Reprod. Anim. 44: 78–80.

Ribeiro, V.M.F. and L.M. Zamora. 2008. Pacas e capivaras, criação em cativeiro com ambientação natural. Bagaço, Rio Branco, Brazil.

Ribeiro, V.M.F., R. Rumpf, R. Satrapa, R.A. Satrapa, E.M. Razza, J.M. Carneiro-Junior, et al. 2012. Vaginal citology, serum progesterone concentration during pregnancy and fetal measurements in paca (*Cuniculus paca*, Linnaeus 1766). Acta. Amaz. 42: 445–454.

Ribeiro, V.M.F., R. Satrapa, J.V.A. Diniz, H.B. Fêo, L.M.M. Flórez, R.A. Satrapa, et al. 2017. Synchronization of estrus in paca (*Cuniculus paca* L.): possible impacts on reproductive and productive parameters. Braz. J. Vet. Res. Anim. Sci. 54: 27–35.

Santos, P.A., V.M.F. Ribeiro, A.L.F. Alves, V.L. Silva, B.K.F. Nascimento, R.A. Satrapa, et al. 2020. Morphology, morphometry, and membrane integrity of epididymal spermatozoa of spotted pacas (*Cuniculus paca,* Linnaeus 1766). Semin. Cienc. Agrar. 41: 181–190.

Shivaji, S. 1988. Seminal plasmin: a protein with many biological properties. Biosci. Reprod. 8: 609–618.

Stradiotti, C.G.P., J.F.S. Silva, I.C.N. Cunha, D.S. Júnior, A.C. Cóser, C.C. Rangel, et al. 2015. Development of methodology for collecting semen from pacas. Rev. Bras. Med. Vet. 37: 222–226.

Urbina, E.C.O. 2011. Plano de manejo tipo para tepezcuintle (*Cuniculus paca*) manejo intensivo. Relatório de referência. México D.F. Dirección general de vida silvestre.

Chapter 18

The Plains Vizcacha

Alfredo Daniel Vitullo[1,2]*, Pablo Ignacio Felipe Inserra[1,2],
Noelia Paola Leopardo[1,2], Julia Halperin[1,2],
and Verónica Berta Dorfman[1,2]

[1]Centro de Estudios Biomédicos Básicos, Aplicados y Desarrollo (CEBBAD),
Universidad Maimónides, C1405BCK, Buenos Aires, Argentina.
Email: vitullo.alfredo@maimonides.edu, inserra.pablo@maimonides.edu,
leopardo.noelia@maimonides.edu, halperin.julia@maimonides.edu,
dorfman.veronica@maimonides.edu

[2]Consejo Nacional de Investigaciones Científicas y Técnicas (CONICET),
C1425FQB, Buenos Aires, Argentina

INTRODUCTION

The South American plains vizcacha (*Lagostomus maximus* Desmarest, 1817) is a caviomorph fossorial rodent that inhabits the Southern area of the Neotropical region, mainly distributed in Argentina, from the North of Patagonia throughout the Pampean region and reaching the South of Bolivia and Paraguay (Jackson et al. 1996). The species (Figure 18.1) belongs to the family Chinchillidae, one of the more recently evolved from the Hystricomorpha clade, that radiated 21.3 million years ago (Voloch et al. 2013, Steppan and Schenk 2017).

Chinchillidae comprises three living species: the chinchilla (*Chinchilla lanigera*), the mountain vizcacha (*Lagidium viscascia*), and the plains vizcacha (*L. maximus*). Three subspecies of *L. maximus* are recognized according to their geographical distribution and morphological characteristics: *L. m. petilidens* Hollister, 1914, distributed in southern Buenos Aires, La Pampa, and Río Negro provinces in Argentina, *L. m. maximus* Desmarest, 1817, in Central Argentina and *L. m. immollis* Thomas, 1910, ranging from Northern Argentina to Paraguay and Bolivia (Llanos and Crespo 1952, Redford and Eisenberg 1992).

---

*Corresponding author: vitullo.alfredo@maimonides.edu

| Kingdom: Animalia |
| Phylum: Chordata |
| Class: Mammalia |
| Order: Rodentia |
| Family: Chichillidae |
| Genus: *Lagostomus* |
| Species: *Lagostomus maximus* |

Figure 18.1 Plains vizcachas (*Lagostomus maximus*) in the "Refugio Provincial de Bahía Samborombón", province of Buenos Aires, Argentina. Photo, courtesy of Bernardo Lartigau.

Diagnostic features that distinguish the plains vizcacha from other members of the family include a strong occipital crest, a noticeable facial pattern consisting of two parallel black bands—one passing through the eyes and the second across the nose separated by a white stripe—two pairs of bi-laminate molars, hind limbs with three digits, and a strong tail that could be used as a third leg (Pocock 1922, Jackson et al. 1996). Sexual dimorphism is very pronounced with males being much larger, having a bigger head, and a more pronounced facial mask than females. Neonates are relatively large (about 200 g) and precocious as they have teeth, hair, and nails, and can move independently from birth. In males, body mass increases rapidly until 18 months of age (around 2.5 kg), and then slows down. Adult males show a body weight of around 7.3 kg (30–32 months old); females gain body mass until 16–18 months of age reaching 2.5–4.3 kg at 65 months old. In both sexes, the hind foot length increases until animals reach one year of age, and the body length increases until 16–18 months of age (Jackson 1986, 1990, Jackson et al. 1996).

This chapter reviews the state of the knowledge on the biological characteristics of the germline of the male and female plains vizcacha whose reproductive strategy shows unusual characteristics that differentiate this species from other rodents and mammals. We will also examine the first steps in understanding and managing assisted reproductive techniques (ARTs) in this useful non-classical model for biomedical and biological basic research and commercial interest.

GENERAL REPRODUCTIVE TRAITS: GENERALITIES AND PECULIARITIES

The plains vizcacha is a large herbivore species that live in communal burrow systems and shows nocturnal foraging outings (Llanos and Crespo 1952). Night-time outputs expose them to predation, especially by puma (*Puma concolor*) and, eventually, by smaller felids like Geoffroy's cat (*Oncifelis geoffroyi*) (Branch 1995). The plains vizcacha is highly social and shows polygynous behavior (Llanos and Crespo 1952) and a biannual seasonal reproduction, with the main reproductive season occurring from March to April and births taking place by the end of August to the beginning of September (Weir 1970, Branch et al. 1993, Spotorno and Patton 2015). A second breeding season takes place by the end of October in females that have lost their progeny (Llanos and Crespo 1952).

Females have a vaginal closure membrane and form a copulatory plug after mating. The estrous cycle lasts 40–45 days releasing 200–800 oocytes, the highest polyovulation rate recorded so far for a mammal (Weir 1971a, b). After fertilization, gestation lasts around 155 days (Weir 1971b), representing the second longest gestation for a rodent, only surpassed by another caviomorph, the pacarana (*Dinomys branickii*), whose pregnancy lasts around 254 days (Merit 1984). In the plains vizcacha, embryo implantation takes place after an 18-day-long preimplantation period after which 5 to 6 blastocysts normally implant in each uterine horn (Roberts and Weir 1973). Despite an initial development of all implanted embryos during the first 30 days of gestation, by day 70 all embryos in each uterine horn, except those most caudally implanted, have been naturally resorbed. From the initial 10 to 12 implanted embryos, 1 or 2 pups (rarely 3) are usually delivered (Weir 1971b). Although the reproductive capacity is low, the precocial progeny exhibits a high survival rate. They remain in the burrow for the first two months of life, but from time to time the young make small trips outside to ingest solid food (Jackson et al. 1996). Females lactate for a minimum of 21 days; however, the lactation period may extend between 2–3 months. The male reaches sexual maturity at a year and a half and the female between eight months and a year of age (Jackson et al. 1996, Spotorno and Patton 2015).

THE MALE PLAINS VIZCACHA

The Spermatogenic Process:
Seasonal Cycling and Sperm Retrieval

Male plains vizcacha shows annual cycling of the seminiferous epithelium. During the long days of summer, testicles are completely active until winter days during which testis become entirely inactive with interrupted spermatogenesis (Fuentes et al. 1991, Aguilera-Merlo et al. 2005a, b). The period of testis inactivity is coincident with the female's gestation. Between testicular activity and inactivity stages, males undergo two transitional periods, a first one occurring by the end of summer/beginning of autumn in which regression of spermatogenesis takes place and a second transitional period in springtime characterized by a recrudescent stage in which full spermatogenesis is re-established before the mating season starts (Gonzalez et al. 2018).

In mammals, changes in reproductive function in seasonal breeding males are modulated by photic stimuli which regulate regression and recrudescence of the seminiferous epithelium. The underlying cellular processes that regulate the transition between testis activity and inactivity are mostly related to a balance between cell death, normally driven by apoptotic mechanisms, and proliferation as shown in many mammalian species (Young et al. 1999, 2000, Young and Nelson 2001, Morales et al. 2002, Strbenc et al. 2003). In some cases, apoptosis is the causative mechanism of testicular regression (Young et al. 1999, 2000) whereas, in other cases, proliferation rather than apoptosis is the responsible mechanism of testicular germ cell loss (Blottner et al. 2007, Dadhich et al. 2010). However, no significant changes in the number of apoptotic spermatogonia and meiotic germ cells are seen from the active to the inactive stage in plains vizcacha. Accompanying the invariable rates of apoptosis, proliferating spermatogonia are found at high levels in the recrudescent stage. Then, when full testicular activity is restored, its proliferation rate decays (González et al. 2018). However, contrary to what occurs in other mammals, neither apoptosis nor proliferation by themselves seems to be responsible for seasonal variation of the seminiferous epithelium in male plains vizcacha. Modulation of changes in testicular activity arises from the interplay between apoptosis and autophagy (González et al. 2018).

Mature spermatozoa of plains vizcacha show mammotypical morphology with a rounded head in which the flagellum implants centrally at the base. During the season of spermatogenic activity, mature spermatozoa displaying good vitality and motility together with acrosomal integrity can be easily obtained from cauda epididymis (Giacchino 2017). Males respond well to penis electro-stimulation by using a special device designed and produced by Giacchino (2017), enabling the recovery of mature spermatozoa of comparable vitality and motility as those from cauda epididymis. Electro-stimulation requires a two-step cycle promoting firstly the externalization of the penis and inducing ejaculation as the electrical stimulus increases. Collecting viable sperm samples requires the use of an artificial vagina in which the collecting tube is introduced to maintain the appropriate temperature and vitality of spermatozoa (Giacchino 2017). The possibility of obtaining sperm samples through electro-ejaculation together with protocols for inducing ovulation by the administration of exogenous hormones (see below) offers the first steps in managing ARTs in this species.

THE FEMALE PLAINS VIZCACHA

The Unusual Ovarian Anatomy and Function

The ovaries of adult plains vizcacha are irregular leaf-shaped structures approximately 1.5 cm long, 1.2 cm wide, and 0.6 cm thick (Jensen et al. 2006), located in the sublumbar area and fixed by extensions of the uterine ligament (Flamini et al. 2009). Unlike the solid ovaries exhibited by their close evolutionary species such as the chinchilla or the mountain vizcacha (Pearson 1949, Weir 1967, 1971a), those of the plains vizcacha have highly convoluted invaginations of the germinal epithelium and the tunica albuginea that greatly expand the surface of the ovary (Weir 1971a, Jensen et al. 2006). Regardless of the animal's reproductive status, the cortical surface bears follicles in all stages of development, the most abundant being always the primordial ones which are usually displayed in clusters (Flamini et al. 2020). In addition, it was reported that the interstitial tissue may have endocrine properties as it shows cytological aspects compatible with steroidogenic activity (Gil et al. 2007). In fact, after an *in vivo* induction protocol for ovulation using pregnant mare serum gonadotropin (PMSG) and human chorionic gonadotropin (hCG), the interstitial tissue becomes luteinized.

This atypical highly folded morphology of the ovary is interpreted as an adaptation to support the unusual number of follicles present in the ovary (Espinosa et al. 2011). Curiously, the size of the cohort of primordial follicles remains almost unchanged even though literally hundreds of oocytes are ovulated with each estrous cycle (Weir 1971b, Fraunhoffer et al. 2017). This aspect is related to another striking ovarian feature which is the suppression of the expression of the pro-apoptotic gene *Bax* and the overexpression of the anti-apoptotic gene *Bcl-2* in all follicular stages. This naturally suppressed apoptosis prevents the decline of the oocyte pool by suppressing follicular atresia (Jensen et al. 2006). Moreover, it has been shown that the elimination of altered follicles and remnant corpora lutea is most probably carried out by autophagy. This mechanism would provide the necessary space for the maturation of primordial follicles that continuously enter the growing follicular pool (Leopardo et al. 2020). All this would suggest that the suppressed follicular apoptosis, the active autophagy, and the permanent folding/remodeling of the ovarian architecture are key aspects to sustaining the availability of primordial follicles and the massive natural poly-ovulation that characterizes the species.

In the plains vizcacha, ovulation may take place spontaneously after the vagina has been open for several days (Weir 1971b). Once ovulation has occurred, primary corpora lutea form from the remnants of the ovulated follicles and measure, on average, 300 μm (Cortasa et al. 2022). If pregnancy takes place, in addition to the primary corpora lutea, and as other members of the suborder Hystricomorpha (Pearson 1949, Weir 1967), the ovary of the pregnant plains vizcacha shows luteinized unruptured follicles, i.e. accessory corpora lutea, which are considerably smaller than the primary ones (Weir 1971a). Accessory corpora lutea develop during mid-pregnancy upon reactivation of gonadotropin-releasing hormone (GnRH) and luteinizing hormone (LH) delivery (Dorfman et al. 2013). Precisely, LH stimulus is responsible for luteinizing those LH-receptor expressing-antral follicles present in the ovary at that stage.

In addition, such reactivation induces a restart of steroidogenesis in the primary corpora lutea (Cortasa et al. 2022). It is interesting to understand how this species overcomes, through this mechanism of reactivation of the hypothalamic-pituitary axis, the drop in progesterone levels that begins on day 70 of gestation, when it had not yet reached midterm. At this stage of gestation, the development of the embryos is not sufficient to allow them to live outside the mother's womb. The reboot of the steroidogenesis of the primary corpora lutea and the generation of a new set of steroidogenically active accessory corpora lutea during mid-pregnancy restore adequate levels of progesterone and allow the pregnancy to be maintained for another three months and complete embryonic development (Jensen et al. 2008, Dorfman et al. 2013, Cortasa et al. 2022).

NATURAL AND INDUCED MASSIVE POLY-OVULATION

Natural Ovulation

Natural ovulation in plains vizcacha shows great variability which ranges between 200 and 800 eggs per cycle, both in wild-caught and captivity-bred animals. Weir (1971b) was able to record up to 800 oocytes by counting the eggs found in serial ovary sections and flushing of oviducts and uterine horns during the oestrus phase in captivity-bred animals. In our laboratory, we were able to collect up to 400 eggs in wild-caught individuals (Willis 2012). Poly-ovulation is found in several other groups of mammals such as tenrecs and elephant shrews; however, eggs released in the tenrec (*Hemicentetes ecaudatus*) or the shrew (*Elephantulus myurus*) never exceed a hundred oocytes per cycle (Weir and Rowlands 1973). Therefore, the astonishing amount of oocytes released by the plains vizcacha places it as the mammal with the highest ovulatory rate described to date.

During the ovulatory cycle, the left and right ovaries show hemorrhagic stigmata suggesting that both ovaries ovulate at the same oestrus (Jensen et al. 2006, Willis 2012). This is not a common characteristic among hystricognatha species, as reported before (Pearson 1949, Weir 1967, 1971a). The number of hemorrhagic stigmata is very variable, and it does not show a correlation with the number of oocytes released or embryos implanted or born. Moreover, the presence of hemorrhagic stigmata does not always correlate with oocyte extrusion since in many cases, the stigmata correspond to follicles undergoing luteinization without effective oocyte release. In those structures, haematic material migrates into the antral cavity (Willis 2012, Dorfman et al. 2013, Fraunhoffer et al. 2017). Ovulation in the female plains vizcachas probably occurs spontaneously after the vagina has been open for several days, but only copulating females display luteinized follicles in the ovaries, suggesting that ovulation is a spontaneous process and luteinization is induced by copulation (Weir 1971a). Oocytes released during ovulation show heterogeneity with a predominance of immature, fragmented, activated, or altered morphology (Willis 2012). In addition, an ampullary region is not discernible and an unrestricted transit of ovulated eggs is observed between the oviduct and the uterine horn (Weir 1971a).

Induced Ovulation

Mature female plains vizcacha respond well to the administration of exogenous hormones to induce follicular recruitment and ovulation. Different variants and combinations of PMSG can be used to promote follicular recruitment, followed by the administration of hCG or porcine pituitary LH to trigger follicular rupture (Willis 2012). Our best results were achieved by the administration of 250 IU/day of PMSG for three consecutive days followed by 1,000 IU of hCG on the fourth day (Willis 2012, Charif et al. 2016, Inserra et al. 2020). By using this protocol, we were able to recover between 60 and 284 oocytes from a single animal, all of them with similar characteristics to those released during natural ovulation

(Willis 2012). Oocytes were found scattered throughout the reproductive tract, which seems to be a common characteristic in this species. Almost half of the oocytes were found in the oviduct and the rest in the uterine horns with variable percentages among different animals (Willis 2012). The oocyte quality obtained from stimulated females was as poor as in natural ovulation (Willis 2012).

OVULATION AND PSEUDO-OVULATION

Regulation of the Hypothalamic-pituitary-ovarian Axis

Female plains vizcachas exhibit a fine regulation of their hypothalamic-pituitary-ovarian (HPO) axis during pregnancy. From embryonic implantation until around day 70 of gestation, circulating progesterone gradually decreases as a result of the decay in the steroidogenic activity of the primary corpora lutea (Fraunhoffer et al. 2017, Cortasa et al. 2022). Approximately at day 90 of gestation, circulating progesterone reaches its minimum level (Fraunhoffer et al. 2017). At that time, the embryos from each uterine horn closest to the cervix are the only ones still viable; the remaining embryos proximally implanted at each uterine horn have already been completely resorbed (Barbeito et al. 2021, Acuña et al. 2022). However, the two surviving embryos are still too immature to be born and a hormonal boost turns essential to continue gestation.

The minimum level of progesterone reached at this point seems to enable the reactivation of the hypothalamic activity around 100–120 days of pregnancy which, through the delivery of GnRH, triggers pituitary activity (Dorfman et al. 2013, Inserra et al. 2017). So, the systemic delivery of the pituitary follicle-stimulating hormone (FSH) increases ovary estradiol secretion stimulating the surge of the pituitary LH (Dorfman et al. 2013, Fraunhoffer et al. 2017, Proietto et al. 2019). This results in the re-activation of the steroidogenic activity of the primary corpora lutea and the development of secondary corpora lutea from ovulatory follicles with oocyte retention (Dorfman et al. 2013, Fraunhoffer et al. 2017, Cortasa et al. 2022). This process was called "pseudo-ovulation" (Halperin et al. 2013, Dorfman et al. 2016) since although the formation of active corpora lutea occurs, follicle rupture with the release of the oocyte does not. During this process of HPO reactivation, resetting of primary corpora lutea and the newly added set of accessory corpora lutea provides the boost of progesterone necessary for fully developed fetuses to be born (Cortasa et al. 2022).

True Ovulation Versus Pseudo-ovulation During Gestation

In 2012, Willis described that before mating, the vast majority of oocytes are released devoid of cumulus cells; however, he showed that during the 3–7 days the ovulatory process lasts, a small number of oocytes that are released as cumulus-oocyte complexes (COCs) display good morphological quality. Moreover, those COCs can be recovered from females that were stimulated by the administration of exogenous hormones (Giacchino 2017).

It seems reasonable to suppose that COCs found in ovulating before mating represent a suitable germ cell population for fertilization (called "true ovulation"). The accompanying massive release of denuded poor-quality oocytes may well represent an adaptive mechanism for eliminating not viable germ cells since the plains vizcacha does not show intraovarian follicular removal through apoptosis-dependent atresia (Jensen et al. 2006, Leopardo et al. 2011, Inserra et al. 2014) as it occurs in most mammals so far studied (see above). Whether COCs mean a fertilizable oocyte cohort must await to be proven until *in vitro* fertilization techniques are fine-tuned for this species.

Contrary to what is found in ovulation before mating, the recovery of oocytes by flushing the oviducts in pseudo-ovulating females in mid-pregnancy is rare in some cases and null in most (as reviewed in Willis 2012). A few females who showed signs of ovulation released a very few morphologically abnormal oocytes to the oviduct. Hence, the level of circulating LH triggered when the PO axis re-activates at mid-pregnancy may be insufficient to promote follicular rupture and oocyte release but enough to induce follicular luteinization (Fraunhoffer et al. 2017). As mentioned above, pseudo-ovulation seems to be an adaptive mechanism to increase progesterone levels from mid-pregnancy rescuing the only viable fetuses that survived embryo resorption (Jensen et al. 2008, Dorfman et al. 2013).

Ovulation in mammals can occur spontaneously as in most rodents and primates or induced by copulation as in the rabbit and cats. Rowlands and Weir (1974) have proposed an intermediate stage of spontaneous/induced ovulation in the plains vizcacha. In agreement with this, the release of COCs at some moment during the 3–7 days lasting ovulation together with the description of spikes in the male penis (Giacchino 2017) argue in favor of this intermediate stage in which not viable oocytes are spontaneously released and viable ones need induction.

Implantation and Natural Abortion

Implantation occurs between 16- and 18-day post-coitus (dpc) and proceeds from the caudal to the cranial portion of each uterine horn (Barbeito et al. 2021). After implantation, a gradient in embryo growth retardation is observed. Embryos implanted in the caudal end of the uterine horns show a higher degree of development compared to those implanted in the cranial end. At 26dpc, signs of embryonic death are observed in the cranial and middle implantation sites (lymphocyte infiltration, hemorrhage, fibrin deposition, and tissue necrosis) (Acuña et al. 2020, 2022). This difference is even greater when embryos implanted at the caudal end are compared to those implanted at the cranial end of the horn (Leopardo et al. 2011, Giacchino et al. 2020). Moreover, these characteristics are more evident at mid-pregnancy (Giacchino et al. 2020). Finally, at day 70 of gestation, the development of the embryos implanted in the cranial and the middle portions of both uterine horns is completely arrested, and only the two embryos implanted in the caudal uterine horn nearest the cervix were gestated to term (Weir 1971a, b, Acuña et al. 2020, 2022, Giacchino et al. 2020). This process of partial embryonic resorption that results in the delivery of only two pups is a constitutive event that characterizes the species.

Acuña et al. (2022) proposed that uterine structural variations in the plains vizcacha could restrict growth, generating the loss of placental homeostasis and promoting embryo death. They searched for differences in the uterine glands, vasculature, and musculature along the uterine horns in non-pregnant females, finding that all these structures were more developed in the caudal region as compared to more cranial portions of the uterine horn wall (Acuña et al. 2020). These changes in the vasculature of the organ are thought to be established in prenatal life (Flamini et al. 2020, Barbeito et al. 2021). They also reported that the length and width of the implantation sites and their glandular and vascular areas significantly increased in a craniocaudal direction (Acuña et al. 2021), reinforcing the key potential role of these structures in the gestation of the species.

Similar studies focused on analyzing other possible causes of embryo loss in the plains vizcacha, suggesting that it could be related to placental and nutritional insufficiency produced by poor irrigation (Giacchino et al. 2020), and to progesterone decay (Jensen et al. 2008, Dorfman et al. 2013). The analysis of arterial architecture and blood supply of the female genital tract provided evidence that the uterine horns are irrigated through the uterine artery in an ascending way from the cervix and that segmental arteries irrigating the embryo vesicles become thinner as they approach the ovary. The ascending circulation pattern suggests that embryos implanted caudally benefit from receiving oxygen-, nutrient- and hormone-rich blood while, in contrast, embryos implanted cranially, that will be resorbed, are irrigated by segmental thinner branches born at more distal positions from the uterine artery (Giacchino et al. 2020). In addition, angiographies performed in mid-pregnant females revealed contrast solution accumulated in the placenta of embryos implanted caudally despite the dose of contrast solution administered, indicating that those embryos benefit from the greatest amount of bloodstream (Giacchino et al. 2020).

Using endoscopic inspection, incomplete partitions have been previously described around the embryonic vesicles in the uterine horns of early pregnant females constituting a physical barrier between contiguous embryos that could contribute to the formation of differential environments (Giacchino et al. 2018). The separation of each embryonic vesicle could prevent the spreading of the apoptotic and necrotic processes suffered in the middle and cranial embryonic vesicles to the caudal embryos which are the only ones that will be gestated to term. Moreover, once resorption has been completed, from day 70 of gestation onwards pseudo-ovulation occurs, promoted by the re-activation of the HPO axis (Dorfman et al. 2013, Proietto et al. 2019). As a result, a reboot of primary corpora lutea steroidogenesis occurs, and a considerable number of accessory corpora lutea are developed, all of which helps to recover progesterone levels and rescue the only two surviving caudally implanted embryos that escaped resorption (Jensen et al. 2008, Dorfman et al. 2013, Fraunhoffer et al. 2017, Cortasa et al. 2022). Further studies are needed to fully understand natural embryonic death in the plains vizcacha.

FINAL CONSIDERATIONS

Why and How to Assist the Plains Vizcacha in its Reproduction?

The South American plains vizcacha has proven to be a useful alternative model for biological and biomedical research due to its unusual reproductive strategy. The development of the female germ line occurs in an extremely opposite way to that found in murid rodents (Leopardo and Vitullo 2017, Leopardo et al. 2017), although the same gene network is used in both cases, thus opening the possibility to explore subtle differences in gene expression that can be targeted for therapeutic purpose in women ovarian insufficiency (Albamonte et al. 2008). Moreover, and in contrast to what occurs in classic rodent models, such as mice and rats, in plains vizcacha germ line develops in fetal life through the overlapping of primordial germ cell proliferation, meiosis entrance, primary oocyte, and primordial follicle establishment, and first meiotic arrest as in the human ovary (Inserra et al. 2014). In the same line, once primordial germ cells differentiate and specify early during gastrulation, colonization of the gonadal crest takes place by a differential pattern of gene expression comparable to humans and deviating from pathways found in mice (Leopardo and Vitullo 2017).

On the other hand, it is important to note that plains vizcacha's meat has a high nutritional value, because it is low in fat and cholesterol, low in carbohydrates, and is a significant source of protein and calcium. Although its commercialization is still incipient and mainly based on wild-trapped specimens, specialized companies have already started producing frozen and processed meat for general consumption (cf. https://pampacompaniadecarnes.com/vizcacha/). With all these ideas in mind, it could be stated that developing ARTs will enhance the chances of establishing breeding colonies of plains vizcacha which will promote the use of this species for both biomedical research and commercialization.

REFERENCES

Acuña, F., C.G. Barbeito, E.L. Portiansky, G. Ranea, F. Nishida, M.A. Miglino, et al. 2020. Early and natural embryonic death in *Lagostomus maximus*: association with the uterine glands, vasculature, and musculature. J. Morphol. 281: 710–724.

Acuña, F., C.G. Barbeito, E.L. Portiansky, M.A. Miglino and M.A. Flamini. 2021. Prenatal development in *Lagostomus maximus* (Rodentia, Chinchillidae): A unique case among eutherian mammals of physiological embryonic death. J. Morphol. 282: 720–732.

Acuña, F., C.G. Barbeito, E.L. Portiansky, G. Ranea, M.A. Miglino and M.A. Flamini. 2022. Spontaneous embryonic death in plains viscacha (*Lagostomus maximus*—Rodentia), a species with unique reproductive characteristics. Theriogenology 185: 88–96.

Aguilera-Merlo, C., E. Muñoz, S. Dominguez, L. Scardapane and R. Piezzi. 2005a. Epididymis of vizcacha (*Lagostomus maximus maximus*): morphological changes during the annual reproductive cycle. Anat. Rec. A Discov. Mol. Cell Evol. Biol. 282: 83–92.

Aguilera-Merlo, C., E. Muñoz, S. Dominguez, M. Fóscolo, L. Scardapane and J.C. De Rosas. 2005b. Seasonal variations in the heterologous binding of vizcacha spermatozoa. A scanning electron microscopy study. Biocell. 29: 243–251.

Albamonte, M.S., M.A. Willis, M.I. Albamonte, F. Jensen, M.B. Espinosa and A.D. Vitullo. 2008. The developing human ovary: immunohistochemical analysis of germ-cell-specific VASA protein, BCL-2/BAX expression balance and apoptosis. Human Reprod. 23: 1895–1901.

Barbeito, C.G., F. Acuña, M.A. Miglino, E.L. Portiansky and M.A. Flamini MA. 2021. Placentation and embryo death in the plains viscacha (*Lagostomus maximus*). Placenta 108: 97–102.

Blottner, S., J. Schon and H. Roelants. 2007. Apoptosis is not the cause of seasonal testicular involution in roe deer. Cell Tissue Res. 327: 615–624.

Branch, L.C., D. Villarreal and G.S. Fowler. 1993. Recruitment, dispersal, and group fusion in a declining population of the plains viscacha (*Lagostomus maximus*; Mammalia, Chinchillidae). J. Mammal. 74: 9–20.

Branch, L.C. 1995. Observations of predation by pumas and Geoffroy's cats on the plains vizcacha in semiarid scrub of central Argentina. Mammalia 59: 152–156.

Charif, S.E., P.I.F. Inserra, N.P. Di Giorgio, A.R. Schmidt, V. Lux-Lantos, A.D. Vitullo et al. 2016. Sequence analysis, tissue distribution and molecular physiology of the GnRH preprogonadotrophin in the South American plains vizcacha (*Lagostomus maximus*). Gen. Comp. Endocrinol. 232: 174–184.

Cortasa, S.A., P.I.F. Inserra, S. Proietto, M.C. Corso, A.R. Schmidt, A.D. Vitullo, et al. 2022. Achieving full-term pregnancy in the vizcacha relies on a reboot of luteal steroidogenesis in mid-gestation (*Lagostomus maximus*, Rodentia). PLoS One 17: e0271067.

Dadhich, R.K., F.M. Real, F. Zurita, F.J. Barrionuevo, M. Burgos and R. Jiménez. 2010. Role of apoptosis and cell proliferation in the testicular dynamics of seasonal breeding mammals: a study in the Iberian mole, *Talpa occidentalis*. Biol. Reprod. 83: 83–91.

Dorfman, V.B., L. Saucedo, N.P. Di Giorgio, P.F. Inserra, N. Fraunhoffer, N.P. Leopardo, et al. 2013. Variation in progesterone receptors and GnRH expression in the hypothalamus of the pregnant South American plains vizcacha, *Lagostomus maximus* (Mammalia, Rodentia). Biol. Reprod. 89(5): 115.

Dorfman, V.B., P.I.F. Inserra, N.P. Leopardo, J. Halperin and A.D. Vitullo. 2016. The South American plains vizcacha, *Lagostomus maximus*, as a valuable animal model for reproductive studies. JSM Anat. Physiol. 1: 1004–1006.

Espinosa, M.B., N.A. Fraunhoffer, N.P. Leopardo, A.D. Vitullo and M.A. Willis. 2011. The ovary of *Lagostomus maximus* (Mammalia, Rodentia): an analysis by confocal microscopy. Biocell 35: 37–42.

Flamini, M.A., C.G. Barbeito, E.J. Gimeno and E.L. Portiansky. 2009. Histology, histochemistry and morphometry of the ovary of the adult plains viscacha (*Lagostomus maximus*) in different reproductive stages. Acta. Zool. 90: 390–400.

Flamini, M.A., R.S.N. Barreto, G.S.S. Matias, A. Birbrair, T.H.C. Sasahara, C.G. Barbeito, et al. 2020. Key characteristics of the ovary and uterus for reproduction with particular reference to poly ovulation in the plains viscacha (*Lagostomus maximus*, chinchillidae). Theriogenology 142: 184–195.

Fraunhoffer, N.A., F. Jensen, N. Leopardo, P.I.F. Inserra, A.M. Abuelafia, M.B. Espinosa et al. 2017. Hormonal behavior correlates with follicular recruitment at mid-gestation in the South American plains vizcacha, *Lagostomus maximus* (Rodentia, Caviomorpha). Gen. Comp. Endocrinol. 250: 162–174.

Fuentes, L.B., N. Caravaca, L.E. Pelzer, L.A. Scardapane, R.S. Piezzi and J.A. Guzmán. 1991. Seasonal variations in the testis and epididymis of Vizcacha (*Lagostomus maximus maximus*). Biol. Reprod. 45: 493–497.

Giacchino, M. 2017. Caracterización del proceso de implantación en *Lagostomus maximus*: un mamífero con poliovulación y reabsorción embrionaria selective. PhD. Thesis. Universidad de Buenos Aires. Argentina.

Giacchino, M., P.I.F. Inserra, F.D. Lange, M.C. Gariboldi, S.R. Ferraris and A.D. Vitullo. 2018. Endoscopy, histology and electron microscopy analysis of foetal membranes in pregnant South American plains vizcacha reveal unusual excrescences on the yolk sac. J. Mol. Histol. 49: 245–255.

Giacchino, M, J.A. Claver, P.I.F. Inserra, F.D. Lange, M.C. Gariboldi, S.R. Ferraris, et al. 2020. Nutritional deficiency and placenta calcification underlie constitutive, selective embryo loss in pregnant South American plains vizcacha, *Lagostomus maximus* (Rodentia, Caviomorpha). Theriogenology 155: 77–87.

Gil, E., M. Forneris, S. Domínguez, A. Penissi, T. Fogal, R.S. Piezzi, et al. 2007. Morphological and endocrine study of the ovarian interstitial tissue of viscacha (*Lagostomus maximus maximus*). Anat. Rec. 290: 788–794.

Gonzalez, C.R., M.L. Muscarsel Isla and A.D. Vitullo. 2018. The balance between apoptosis and autophagy regulates testis regression and recrudescence in the seasonal-breeding South American plains vizcacha, *Lagostomus maximus*. PLoS One 13: e0191126.

Halperin, J., V.B. Dorfman, N.A. Fraunhoffer and A.D. Vitullo. 2013. Estradiol, progesterone and prolactin modulate mammary gland morphogenesis in adult female plains vizcacha (*Lagostomus maximus*). J. Mol. Histol. 44: 299–310.

Inserra, P.I.F., N.P. Leopardo, M.A. Willis, A.L. Freysselinard and A.D. Vitullo. 2014. Quantification of healthy and atretic germ cells and follicles in the developing and post-natal ovary of the South American plains vizcacha, *Lagostomus maximus*: evidence of continuous rise of the germinal reserve. Reproduction 147: 199–209.

Inserra, P.I.F., S.E. Charif, N.P. Di Giorgio, L. Saucedo, A.R. Schmidt, N. Fraunhoffer, et al. 2017. ERα and GnRH colocalize in the hypothalamic neurons of the South American plains vizcacha, *Lagostomus maximus* (Rodentia, Caviomorpha). J. Mol. Histol. 48: 259–273.

Inserra, P.I.F., S.E. Charif, V. Fidel, M. Giacchino, A.R. Schmidt, F.M. Villarreal, et al. 2020. The key action of estradiol and progesterone enables GnRH delivery during gestation in the South American plains vizcacha, *Lagostomus maximus*. J. Steroid. Biochem. Mol. Biol. 200: 105627.

Jackson, J.E. 1986. Determinacion de edad en la vizcacha (*Lagostomus maximus*) en base al peso del cristalino. Rev. Arg. Prod. Anim. 1: 41–44.

Jackson, J.E. 1990. Growth rates in vizcacha (*Lagostomus maximus*) in San Luis, Argentina. Vida Silvestre Neotropical 2: 52–55.

Jackson, J.E., L.C. Branch and D. Villarreal. 1996. *Lagostomus maximus*. Mamm. Species 543: 1–6.

Jensen, F., M.A. Willis, M.S. Albamonte, M.B. Espinosa and A.D. Vitullo. 2006. Naturally suppressed apoptosis prevents follicular atresia and oocyte reserve decline in the adult ovary of *Lagostomus maximus* (Rodentia, Caviomorpha). Reproduction 132: 301–308.

Jensen, F., M.A. Willis, N.P. Leopardo, M.B. Espinosa and A.D. Vitullo. 2008. The ovary of the gestating South American plains vizcacha (*Lagostomus maximus*): suppressed apoptosis and corpora lutea persistence. Biol. Reprod. 79: 240–246.

Leopardo, N.P., F. Jensen, M.A. Willis, M.B. Espinosa and A.D. Vitullo. 2011. The developing ovary of the South American plains vizcacha, *Lagostomus maximus* (Mammalia, Rodentia): massive proliferation with no sign of apoptosis-mediated germ cell attrition. Reproduction 141: 633–641.

Leopardo, N.P. and A.D. Vitullo. 2017. Early embryonic development and spatiotemporal localization of mammalian primordial germ cell-associated proteins in the basal rodent *Lagostomus maximus*. Sci. Rep. 7: 594.

Leopardo, N.P., P.I.F. Inserra and A.D. Vitullo. 2017. Challenging the paradigms on the origin, specification and development of the female germ line in placental mammals. *In:* Ahmed, R.G. (Ed.). Germ. Cell. IntechOpen, Egipt.

Leopardo, N.P., M.E. Velazquez, S. Cortasa, C.R. Gonzalez and A.D. Vitullo. 2020. A dual death/survival role of autophagy in the adult ovary of *Lagostomus maximus* (Mammalia-Rodentia). PLoS One 15: e0232819.

Llanos, A.C. and J.A Crespo. 1952. Ecología de la vizcacha (*Lagostomus maximus maximus* Blainv.) en el nordeste de la Provincia de Entre Ríos. Revista de Investigaciones Agrícolas 6: 289–378.

Meritt, D.A. 1984. The Pacarana, *Dinomys branickii*. pp. 154–161. *In:* O.A. Ryder and M.L. Byrd (eds). One Medicine. Springer-Verlag, New York, USA.

Morales, E., L.M. Pastor, C. Ferrer, A. Zuasti, J. Pallares, R. Horn, et al. 2002. Proliferation and apoptosis in the seminiferous epithelium photo inhibited Syrian hamsters (*Mesocricetus auratus*). Int. J. Androl. 25: 281–287.

Pearson, O.P. 1949. Reproduction of a South American rodent, the mountain viscacha. Am. J. Anat. 84: 143–173.

Pocock, R.I. 1922. On the external characters of hystricomorph rodents. Proc. Zool Soc. Lond. 25: 365–427.

Proietto, S., L. Yankelevich, F.M. Villarreal, P.I.F. Inserra, S.E. Charif SE, A.R. Schmidt, et al. 2019. Pituitary estrogen receptor alpha is involved in luteinizing hormone pulsatility at mid-gestation in the South American plains vizcacha, *Lagostomus maximus* (Rodentia, Caviomorpha). Gen. Comp. Endocrinol. 273: 40–51.

Redford, K.H. and J.F. Eisenberg. 1992. Mammals of the Neotropics. Volume 2: the southern cone (Chile, Argentina, Uruguay, Paraguay). The University of Chicago Press, Chicago, USA.

Roberts, C.M. and B.J. Weir. 1973. Implantation in the plains viscacha, *Lagosotomus maximus*. J. Reprod. Fertil. 33: 299–307.

Rowlands, I.W. and B.J. Weir. 1974. Reproductive characteristics of hystricomorph rodents. pp. 265–301. *In:* I.W. Rowlands and B.J. Weir (eds). The Biology of Hystricomorph Rodents. Symposia of the Zoological Society of London N° 34, London, UK.

Spotorno, A.E. and J.L. Patton. 2015. Superfamily Chinchilloidea Bennett 1833. pp. 762–764. *In*: J.L. Patton, U.F.J. Pardiñas and G.D. Elia (eds). Mammals of South America, Volume 2: Rodents. University of Chicago Press, Chicago, USA.

Steppan, S.J. and J.J. Schenk. 2017. Muroid rodent phylogenetics: 900-species tree reveals increasing diversification rates. PLoS One 12: e0183070.

Strbenc, M., G. Fazarinc, S.V. Bavdek and A. Pogacnik. 2003. Apoptosis and proliferation during seasonal testis regression in the brown hare (*Lepus europaeus* L.). Anat. Histol. Embryol. 32: 48–53.

Voloch, C.M., J.F. Vilela, L. Loss-Oliveira and C.G. Schrago. 2013. Phylogeny and chronology of the major lineages of New World hystricognath rodents: insights on the biogeography of the Eocene/Oligocene arrival of mammals in South America. BMC Res. Notes 6: 160.

Weir, B.J. 1967. Aspects of reproduction in some hystricomorph rodents. Doctoral thesis. University of Cambridge, UK.

Weir, B.J. 1970. The management and breeding of some more hystricomorph rodents. Lab. Anim. 4: 83–97.

Weir, B.J. 1971a. The reproductive organs of the female plains viscacha, *Lagostomus maximus*. J. Reprod. Fertil. 25: 365–373.

Weir, B.J. 1971b. The reproductive physiology of the plains viscacha, *Lagostomus maximus*. J. Reprod. Fertil. 25: 355–363.

Weir, B.J. and I.W. Rowlands. 1973. Reproductive strategies in mammals. Annu. Rev. Ecol. Syst. 4: 139–163.

Willis, M.A. 2012. Calidad y funcionalidad de las gametas femeninas en un modelo mamífero con poliovulación masiva y supresión de apoptosis folicular. Ph.D. Thesis. Universidad de Buenos Aires. Argentina.

Young, K.A., B.R. Zirkin and R.J. Nelson. 1999. Short photoperiods evoke testicular apoptosis in white-footed mice (*Peromyscus leucopus*). Endocrinology 140: 3133–3139.

Young, K.A., B.R. Zirkin and R.J. Nelson. 2000. Testicular regression in response to food restriction and short photoperiod in white-footed mice (*Peromyscus leucopus*) is mediated by apoptosis. Biol. Reprod. 62: 347–354.

Young, K.A. and R.J. Nelson. 2001. Mediation of seasonal testicular regression by apoptosis. Reproduction 122: 677–685.

Part V
Xenarthra

Part Editor: *Alexandre Rodrigues Silva*
Laboratory on Animal Germplasm Conservation
Federal University of Semiarid Region - UFERSA
Mossoró, Brazil
alexrs@ufersa.edu.br

Chapter 19

The Anteaters and Sloths

Rogério Loesch Zacariotti

Instituto Cuesta Selvagem, São Paulo, SP, Brazil 240, Avenida Marechal
Floriano Peixoto, Centro, Botucatu, SP, Brazil. 18603-970, P.O. Box 02.
Email: rogeriozacariotti@institutocuestaselvagem.com

INTRODUCTION

Sloths and anteaters are a group of placental mammals that belong to the order
Pilosa (superorder Xenarthras) (Table 19.1) and inhabits only the neotropical
region. The evolutive origins of the Pilosa are still unclear, but they can be traced
back in South America as far as the early Tertiary (Miranda et al. 2018, Abreu
et al. 2021).

Some of these species are threatened to extinction due to habitat loss and
fragmentation, roadkill, power line electrocutions, dog attacks, illegal wildfires,
poaching, illegal pet trade, among other causes (IUCN 2022).

Sloths species are restricted to neotropical rainforests of Central and South
America and divided into family Megalonychidae (two-toed sloth) genus *Choloepus*
spp. and family Bradypodidae (three-toed sloth) genus *Bradypus* spp. The species
are arboreal herbivores and feed mainly on leaves, although there are some reports
about *Choloepus* omnivore feeding habits (Chiarello 1998, Mesquita et al. 2021).

Anteaters, on the other hand, are insectivores that feed mainly on ants
and termites, with ground dwelling species as giant anteaters (*Myrmecophaga
tridactyla*), arboreal as silky anteaters (*Cyclopes* spp.) or semiarboreal as
tamanduas (*Tamandua* spp.) (Toledo et al. 2015).

The species within Pilosa order present some specific characters related to the
reproductive trait like absence of scrotum and intra-abdominal testicles, poorly

developed penis in males, a simple uterus without a cervix and a double vagina in females, among other features (Rossi et al. 2011, 2013, Fromme et al. 2021).

Table 19.1 Sloths and anteaters species from the Neotropical region

Species	Common name	Conservation status[1]
Bradypus pygmaeus	Pygmy Three-toed Sloth	Critically endangered
Bradypus torquatus	Northern Maned Sloth	Vulnerable
Bradypus crinitus	Southern Maned Sloth	Not evaluated
Bradypus tridactylus	Pale-Throated Three-Toed Sloth	Least concern
Bradypus variegatus	Brown-Throated Three-Toed Sloth	Least concern
Choloepus didactylus	Linnaeus's Two-Toed Sloth	Least concern
Choloepus hoffmanni	Hoffmann's Two-Toed Sloth	Least concern
Cyclopes catellus	Silky Anteater	Not evaluated
Cyclopes didactylus	Silky Anteater	Data deficient
Cyclopes dorsalis	Silky Anteater	Not evaluated
Cyclopes ida	Silky Anteater	Not evaluated
Cyclopes rufus	Silky Anteater	Not evaluated
Cyclopes thomasi	Silky Anteater	Not evaluated
Cyclopes xinguensis	Silky Anteater	Not evaluated
Myrmecophaga tridactyla	Giant Anteater	Vulnerable
Tamandua mexicana	Northern Tamandua	Least concern
Tamandua tetradactyla	Southern Tamandua	Least concern

(1) IUCN 2022. The IUCN Red List of Threatened Species. Version 2022-1. <https://www.iucnredlist.org>

These peculiar species also present between 40% to 60% of the metabolic rate expected for their body mass, one of the lowest metabolic rates among mammals. These characteristics could be part of explanation about why species within Pilosa order present a low reproductive output, even besides large body sizes in some species like giant anteater (up to 45 kg) (Aguilar and Superina 2015, Toledo et al. 2015).

In addition, the understanding of biology and biotechnology of reproduction in sloths and anteaters can be a powerful tool to be applied to *ex situ* and *in situ* conservation programs (Luna et al. 2014).

The present chapter brings a broad review about the biology of reproduction in sloths and anteater species, as well as information available for reproductive biotechnology techniques available for these species.

SLOTHS REPRODUCTIVE MORPHOLOGY AND PHYSIOLOGY

Although there are several studies on the anatomy and physiology of sloths, little information is available on reproductive morphology and biology for this group.

Sexual dimorphism can be difficult in two-toed sloths (*Choloepus* spp.) or immature individuals of all species. Females adult maned sloths (*B. torquatus*) are larger and heavier than males, as observed in other sloth species (*C. didactylus* and *B. tridactylus*). However, other characteristics of the external morphology are necessary to allow sex identification in sloths species. The penis size (> 1.5 cm in mature individuals) of male adult maned sloths is distinct, but external genitals of females can be problematic to identify because the clitoris resembles a small and undeveloped penis. A careful examination of species´ external genitalia is necessary to identify that only males present a terminal orifice (urogenital opening), while in females the cleft of the clitoris extends to the vaginal canal. Although both sexes present a mane, males' manes are larger, darker and denser in the mid dorsum than those of females, that usually resemble two lateral tufts of black hairs (Lara-Ruiz and Chiarello 2005).

Males of *B. tridactylus* and *B. variegatus* are easily identified by the presence of a speculum that is a patch of orange or yellowish hairs surrounded by black hairs located on the dorsum; females do not present this feature (Boscaini et al. 2019).

Sloth males present small intraabdominal testes, only around 0.13% of body weight, located medially in pelvic cavity. Testes are oval shaped, covered by albuginea tunica and connected to the adrenal glands by a peritoneum ligament. These gonads are not vascularized by a pampiniform plexus, but by one testicular artery and vein. In male *B. variegatus,* the gonads have an average length of 1.45 cm and volume of 1.42 cm³. Sexual accessories' glands described in sloths are vesicular glans (seminal vesicles), larger in *Bradypus* spp. and smaller in *Choloepus* spp., and prostate (Gilmore et al. 2000, Barretto et al. 2013, Rezende et al. 2013).

Female sloths present ovaries as two ovoid structures partially surrounded by a thin ovarian bursa and located ventral to the third lumbar vertebra in *B. variegatus.* In *B. variegatus,* the mean ovarian volume of non-pregnant females is around 70 mm³ and close to 110 mm³ in pregnant females, due to the presence of corpus luteum. As observed in other Pilosa, the ovaries microscopically consisted of an inner medulla and an external cortex surrounded by the tunica albuginea. Structures observed in the cortical region are primordial, primary, secondary, and antral follicles as well as corpus albicans or corpus luteum in pregnant females. Uterine tubes are narrow tubular and sinuous organs that extend supported by the mesosalpinx around the ovarian perimeter. The uterus is simplex, dorsoventrally flattened and suspended bilaterally by the broad ligament from the dorsal body. Uterus is divided into a cranial fundus, an elongated caudal uterine body, and a double cervix. As also observed in anteaters, the urogenital sinus is a distensible cavity with longitudinal mucosal folds, which connects the cervix directly to the vulva with a clitoris that can be bipartite in *B. variegatus* (Favoretto et al. 2016).

Sexual maturation is reported from two years in *B. torquatus* up to six years in other *Bradypus* species, and around four and a half years in males and three years for females *Choloepus* species. There is also a report about sexual maturity in *Choloepus hoffmanni* at 30 months of age. Breeding has been described as "slightly seasonal" in maned sloths and non-seasonal for other sloth species (Gilmore et al. 2000, Lara-Ruiz and Chiarello 2005).

Female three-toed sloths vocalize a high-pitched sound to attract males´ attention in the forest. On the other hand, two-toed sloths do not vocalize, but both sexes rub their anal glands against tree branches to mark their scent and "communicate" chemically their reproductive status (Gilmore et al. 2000).

For all sloth species, mating take place on the trees, and it is an event very difficult to be seen in wild. Some descriptions report that most of the mating initiatives seems to come from the males in three-toed sloths whereas female two-toed sloths in estrus seems to take initiative in mating. Scientific literature refers that mating can possibly occur as a face-to-face position or from behind (Taube et al. 2008).

Sloths are very conspicuous species and estrus signs are hard to detect. There is possibly vulva edema and bleeding during estrus phase, but this information needs to be confirmed. Estrus cycle duration is not described for sloth species, but one study about fecal sexual steroids suggests that the interval between ovulations, or duration of the estrous cycle, could be over 60 days in *B. variegatus* (Gilmore et al. 2000, Mühlbauer et al. 2006).

All sloth species present single offspring, although there is a report of a female taking care of two offspring simultaneously (Martins Bezerra et al. 2008). Estimated gestation in three-toed sloths is four to six months in *B. tridactylus*, five to six months in *Bradypus variegatus* and six months in *B. torquatus*, while two-toed sloths have longer gestations of around 10 months. *B. torquatus* and *B. tridactylus* exhibit an inter-birth interval of one year, *B. variegatus* from 10 to 12 months, while *C. didactylus* is between 14 and 26 months and *C. hoffmanni* is around 15 months. In general, young sloths becomes independent of its mother in about six months. However, field observations in *B. torquatus* reported that young sloths usually stayed with their mothers until 8 to 11 months old, started feeding on solid food as early as two weeks old, but continued suckling until four months of age (Lara-Ruiz and Chiarello 2005, Mühlbauer et al. 2006).

Compiled data about sloth reproduction parameters are provided next, to facilitate comparison among species (Table 19.2) and because the southern maned sloth (*B. crinitus*) was recently described (September/2022), no reproductive data is available.

Table 19.2 Reproduction parameters in sloth species

	Choloepus didactylus	*Choloepus hoffmanni*	*Bradypus torquatus*	*Bradypus tridactylus*	*Bradypus variegatus*
Gestation (months)	~10	~10	6	4–6	5–6
Inter-birth interval (months)	14–26	15		12	10–12
Litter size			1		
Lactation	–	–	2–8 weeks	–	1 month
Independence of young (months)	12	10	8–11	5	6–23
Age of sexual maturity (years)	3–4.5	2.5–4.5	~3	3	~3
Estrus cycle duration (days)	–	–	–	–	> 60
Placentation type			Discoidal, deciduate hemochorial		
Ovulation			Possibly spontaneous		

ANTEATERS REPRODUCTIVE MORPHOLOGY AND PHYSIOLOGY

The information provided here is mostly for giant anteater and tamandua species because very little is known about silky anteaters' reproduction.

Male giant anteaters and southern tamanduas do not present a scrotum (intraabdominal testes) and the penis is a short conical structure very similar in shape and size to the females' vulva. Because sexual dimorphism of anteaters species is not obvious, there are a few reports about incorrect sex identification of giant anteaters in captivity and some authors proposed the use of Polymerase Chain Reaction (PCR) of SRY gene related to chromosome Y in males to correctly determine the sex. However, an easier and cheaper way to determine the sex can be the close inspection of the external genital organs by a trained person (Takami et al. 1998, Bento et al. 2019, Fromme et al. 2021, Zacariotti et al. 2022).

Some authors discuss if intraabdominal testes in xenarthras could be linked to their low metabolism and body temperature (from 30 to 35.5°C), a more adequate temperature for normal spermatogenesis (Aguilar and Superina 2015).

The intraabdominal testes are ovoid structures, covered by tunica albuginea, medially located in the caudal abdominal cavity, and interconnected by dense connective tissue; consequently, the epididymis remain in contact with each other. The testicular histological arrangement is very similar to other mammals and presents a parenchyma with convoluted seminiferous tubules filled by germinal epithelium and a stroma with small groups of Leydig cells and blood or lymphatic vessels (Rossi et al. 2013, Oliveira et al. 2019, Fromme et al. 2021).

Anteaters present accessory sexual glands including vesicular glands, also called seminal vesicles, prostate, and bulbourethral glands. The bilobated prostate gland is located caudal to the urinary bladder and its glandular portion spreads around urethra and connected to its lumen. The non-glandular portion is formed by stroma with bundles of smooth muscle cells separated by dense connective tissue. The paired vesicular glands are elongated, and coiled structures located dorsally to the prostate and laterally to deferent duct. The bulbourethral glands are ovoid-shaped, situated dorsolateral to the bulb of the penis with the excretory ducts connected to urethra (Rossi et al. 2013, Fromme et al. 2021, Moura et al. 2021).

One peculiar feature recently described in giant anteaters are persisting Müllerian ducts (7 to 10 cm in length) in normal adult males. These developed structures have been characterized as male uterine tube, uterus, and vagina. Female giant anteaters also present persisting Wolf ducts and these structures extend from the ventral wall of vagina to the lateral pole of ovaries (Fromme et al. 2021). There are no descriptions of similar structures for any other anteater species.

Female anteaters present two oval shaped flattened ovaries, covered by tunica albuginea, and remain located caudal to the kidneys. Histologically, the ovaries present an external cortex and an internal medulla. The cortical portion presents follicles in different development stages, atretic follicles, corpora lutea and corpora albicans. The ovary length ranges from 15 mm in southern tamanduas up to 25 mm in giant anteaters. The uterine tubes are thin convoluted ducts connected from the fundic portion of uterus to the infundibulum close to the ovaries. Anteaters present a simple pear-shaped uterus, dorso-ventrally flattened, appearing grossly like a primate uterus. The uterus length is from around 19 mm in southern tamanduas to 50 mm in giant anteaters. The lumen of the uterus connects directly to the vagina without a cervix. In giant anteaters and southern tamanduas, the caudal portion of the vagina is separated by a longitudinal septum forming a double vagina closer to the vestibule region (Wislocki 1928, Rossi et al. 2011, Rezende et al. 2013, Fromme et al. 2021).

A hymen like structure in female tamanduas and giant anteaters has been described by different authors. Because penetration of penis seems to be short and shallow during copulation of these species, it is unlikely that this hymen-like structure would be perforated (Rossi et al. 2011).

Some reports describe a seasonal reproduction in giant anteaters from September to March in Argentina (Rezende et al. 2013). However, a 20 year study reports giant anteaters' births in captivity in virtually all months and another study presented estrous cycles year–round, which indicates a non-seasonal reproductive pattern (Leiva and Marques 2010, Knott et al. 2013).

For male southern tamanduas, seasonal spermatogenesis still needs to be confirmed, although variations of testicular size and weight have been recorded in different seasons for the species (Rossi et al. 2013).

Age of first reproduction in giant anteaters ranges from two to four years; however, there is one report of a single female giant anteater that gave birth at age of 19 months old and a male in captivity that sire offspring at 15 months old (Desbiez et al. 2020).

Mating behavior is rarely seen in wild species, especially in anteaters that are solitary animals, except when a female is carrying its offspring. Giant-anteaters' mating is described as males directly approach females, even when they are carrying its offspring on their back. Usually, the individuals produce loud nasal snarl, the male forces the female to her hind legs, strikes her with his forepaws to take her down, and after both fall down, male makes copulation movements in lateral position and mating is over in less than 30 seconds. Tamandua species seems to mate very alike giant anteaters (Matlaga 2006, Júnior and Bertassoni 2014).

Due to agonistic and potentially aggressive behavior during mating in giant-anteater, or even between two males, it is not recommended to house a male and a female together all the time in captivity or even more than one male in enclosure. In some cases of a female carrying a young over its back, there is a risk of offspring injuries and death during copulation (Leiva and Marques 2010).

Giant anteaters and tamanduas are referred as spontaneous ovulators with estrus cycles durations from 45 to 65 days for giant anteaters and around 43 days for tamandua species. In these species, vaginal bloody discharge can occur at the end of the luteal phase and estrus is usually expected approximately three weeks after first bleeding in southern tamanduas. Serosanguinous discharge can be used as an indicator of ovarian activity, but it is important to be differentiated from coagulation disorders derived from vitamin K deficiency, a frequent clinical disorder observed in anteaters in captivity. The length of gestation is around 180 days with return to estrus post-partum between 60 and 124 days for giant anteaters; meanwhile, southern tamandua presents an average gestation length of 160 days. There is evidence for delayed implantation in giant anteaters, although it needs to be confirmed (Hay et al. 1994, Knott et al. 2013, Hossotani and Silva e Luna 2016, Thompson et al. 2017, Amendolagine et al. 2018).

Giant anteaters, tamanduas and silky anteaters present a discoidal shape and villous placenta, classified as deciduate hemochorial type without remnants of the maternal vessel endothelium along the trabeculae, and with placental establishment at the fundic region of the uterus (Wislocki 1928, Mess et al. 2012).

The two mammary glands present in anteaters are located close to the axillar region next to the chest. Female southern anteaters carry their offspring for almost one year while female giant anteaters carry for up to nine months. One study in giant anteater reported that between third and eighth month, the juvenile increasingly begins to walk for itself, and by the ninth month the juvenile is carried on the mother's back only around 4.5% of the time. Giant anteaters can nurse up to nine months, but by the ninth month, the juvenile spent less than 3% of its time suckling (Valle and Halloy 2003, Smith 2007).

Compiled data about anteater reproduction parameters are provided next, to facilitate comparison among species (Table 19.3).

Table 19.3 Reproduction parameters in anteater species

	Myrmecophaga tridactyla	Tamandua tetradactyla	Cyclopes didactylus
Gestation (days)	174–190	130–190	160
Interbirth interval (months)	7–9	–	2 births/yr can occur
Litter size		1 offspring	
Age of sexual maturity (years)	2–4		–
Estrus cycle duration (days)	45–65	42–44	–
Estrus duration (days)	–	7–12	–
Return to estrus post–partum (days)	60 to 124	–	–
Placentation type	Discoidal, deciduate hemochorial		
Ovulation	Spontaneous ovulators		–

ASSISTED REPRODUCTIVE TECHNOLOGIES APPLIED FOR THE FEMALE

Assisted reproductive technologies (ART) applied to wild species can be an important tool to be used in conservation programs of xenarthras. These biotechnologies allow exchange of genetic material between *in situ* and *ex situ* populations through artificial insemination, *in vitro* fertilization, embryo transfer, cloning and even transgenics (Deco-Souza et al. 2021, Pizzutto et al. 2021).

Unfortunately, the knowledge about reproductive endocrinology in Pilosa order is scarce, although this information is crucial for the development of ARTs. Reproduction endocrinology was studied in two species of anteaters, *M. tridactyla* and *T. tetradactyla*, and one species of sloth, *B. variegatus*, by analyzing steroid metabolites in fecal and urine samples (Hay et al. 1994, Patzl et al. 1998, Amendolagine et al. 2018).

Monitoring of estrous cycle and gestation was carried out only in one species of sloth, *B. variegatus* was studied by measuring estrogen (E2) and progesterone (P4) fecal metabolites. The solid-phase ^{125}I radioimmunoassay kit (Coat-A-Count, DPC Medlab®, Los Angeles, CA, USA) was validated for evaluation of fecal steroid metabolites in the species (Mühlbauer et al. 2006). Fecal steroid metabolites were extracted in this species with approximately 0.3 g of wet feces boiled in 5 mL of 90% ethanol solution for 20 min. After centrifugation (500 g for 15 min), the supernatant was separated and the pellet resuspended in 5 mL 90% ethanol (Etanol, P.A., Merck S.A., Rio de Janeiro, RJ, Brazil), vortexed for 30 s, and then re-centrifuged. The ethanol supernatants were recombined, dehydrated under a stream of compressed air, and dissolved in 1.0 mL methanol (Metanol, P.A., Merck S.A., Rio de Janeiro, RJ, Brazil). This methanol solution was diluted in a buffered solution (13.8 g $NaPO_4$, 9.0 g NaCl, 1.0 g sodium azide, 1.0 g gelatin, and 1000 mL distilled water, pH 7.0). To optimize concentration detection during radioimmunoassay, these solutions can be diluted by 1/5 and 1/20 for metabolites of E2 and P4, respectively (Mühlbauer et al. 2006).

Southern tamandua estrous cycles were studied through urine samples analyzed for levels of pregnanediol-glucuronide (PdG), estrone conjugates (EC), using enzyme immunoassay (EIA) with rabbit anti pregnanediol-glucuronide and anti-estrone glucuronide antibody. Urine samples were diluted in EIA buffer (1:2) for the assays. Cortisol concentrations were measured in urine using a solid-phase solid-phase ^{125}I radioimmunoassay kit (Coat-A-Count, DPC Medlab®, Los Angeles, CA, USA). A volume of 0.5 mL of urine was combined with 1 ml methylene chloride, then it was shaken vigorously for 10 min before centrifuging (1,500 g for 5 min). The aqueous phase was removed, and the methylene chloride phase was dried down in antibody-coated tubes before samples were counted in a gamma counter (TM Analytic model 1290) (Hay et al. 1994).

Estrogen and progesterone metabolites in feces or urine were used to determine estrus cycle and gestation in giant anteaters (Patzl et al. 1998, Amendolagine et al. 2018). A report describes similar results in concentrations of giant anteaters estrogen and progestogen fecal metabolites obtained by three different extraction methods (80% ethanol, 40% methanol or Petroleum Ether: Methanol solutions) (Knott et al. 2013).

Other report described an efficient extraction of estrogens and progestogens metabolites from feces of giant anteaters utilizing Petroleum Ether: Methanol method. This extraction method vortexed 0.5 g of feces in methanol (4 mL) and distilled water (0.5 mL) for 30 min. Subsequently, petroleum-ether (3 mL) was added to the solution and re-vortexed for 15 s. After centrifugation, an aliquot of the methanol fraction was diluted with assay buffer (20 mmol

Trishydroxyaminomethan, 0.3 mol NaCl, 0.1% bovine serum albumin and 0.1% Tween 80, pH 7.5, with 1 mol HCl) for EIA analysis (Patzl et al. 1998).

Giant anteaters' estrogens and progestogens metabolites concentrations were determined in urine samples by EIA utilizing a mouse monoclonal antibody specific for progesterone (CL425, C. Munro, University of California, Davis, CA, USA), and a rabbit polyclonal antibody specific for estrone-3-glucuronide (R522-2, C. Munro, University of California, Davis, CA, USA) (Amendolagine et al. 2018).

Another method used for reproductive cycle monitoring in Pilosa is the vaginal cytology, using similar techniques to the ones described for domestic animals. However, vaginal cytology evaluation showed limited success in determination of estrous phase in *B. torquatus* and *T. tetradactyla* (Hay et al. 1994, Snoeck et al. 2011). The presence of red blood cells or vaginal "bleeding" seems to be related to the estrus phase in anteaters and sloths. However, it is very important to distinguish it from coagulation disorder related to vitamin K deficiency in xenarthras (Aguilar and Superina 2015).

Evaluation of vaginal cytology in *T. tetradactyla* revealed the presence of red blood cells about 2-3 days before the increase in estrogen metabolites concentrations in urine. Although the increase of superficial epithelial cells in vaginal cytology samples are usually related to estrus in the bitch and queen, in *T. tetradactyla* it is not related to increase in estrogen metabolites concentrations in urine (Hay et al. 1994). While cell types commonly observed in vaginal cytology in the bitches were recovered in *B. torquatus* samples, it is not clear if these cells are related to different reproductive phases in the species (Snoeck et al. 2011).

ASSISTED REPRODUCTIVE TECHNOLOGIES APPLIED FOR THE MALE

Semen was collected and evaluated in one species of sloth and two species of anteaters (Table 20.4). It is important to emphasize that semen samples of sloths and anteaters are evaluated similarly as in other mammal species, but scientific literature describes semen evaluation at temperatures of 34 or 37°C, although first one is more physiological for the species (Peres et al. 2008, Luba et al. 2015, Silva et al. 2019, Mendonça et al. 2021, Miranda et al. 2021).

Semen was collected from *B. tridactylus* by electroejaculation, using a probe size of 13 cm long and 1 cm wide, with three longitudinal electrodes. The lubricated probe was inserted approximately 4.5 cm into the rectum with the electrodes directed ventrally. Electrical stimulation was performed by three series of progressive intensities ranging from 20 to 60 mA (10 mA increment). Between each series of three intensities of stimulations (30 electrical stimulations), a three-minute interval is recommended (Peres et al. 2008).

Semen samples were obtained in giant anteaters using a bipolar rectal probe measuring 17 or 20 cm long and 1.3 cm wide, introducing approximately 16 cm into the rectum and by applying electrical stimulation that varied from 200 to 300 mA in five electrical stimuli or from 2 to 6 volts in 80 stimuli divided into three series (First series 10 electrical stimuli of 2 V, 10 of 3 V, and 10 of 4 V, 2 minute interval; Second series 10 stimuli of 3 V, 10 of 4 V, and 10 of 5 V, 2 minute interval; Third series 10 electrical stimuli of 5 V and 10 of 6 V). A large amount of lipid granules has been reported in giant anteaters' semen samples collected by electroejaculation (Luba et al. 2015, Mendonça et al. 2021).

Semen was collected in southern tamanduas by electroejaculation using a tripolar rectal probe of 1 cm in diameter using a protocol used for wild felids. However, no motility was observed, besides semen samples containing 37.5% of normal sperm. The major sperm cell abnormality found was distal droplets (16.3%) (Hay et al. 1994).

Pharmacological semen collection was accomplished in *M. tridactylus* using medetomidine (0.08 mg/kg) and ketamine (5 mg/kg) applied intramuscularly. After 15 to 20 minutes, a sterile urethral catheter was placed and connected to a 1 mL syringe, in order to collect semen (Silva et al. 2019). Additionally, pharmacological semen collection was also attempted in *C. didactylus* using medetomidine (0.02 mg/kg) and ketamine (5 mg/kg) applied intramuscularly, but no results about semen volume, sperm motility, vigor, and concentration were presented by authors, but only sperm cell measurements (Miranda et al. 2021).

Table 19.4   Semen parameters described for sloths and anteaters

	Semen volume (μL)	Motility (%)	Vigor (0–5)	Sperm concentration (× 10^6 spz/mL)	pH	Major defects (%)	Minor defects (%)
Myrmecophaga tridactyla[1]	200	70	4	630	–	34	14
Myrmecophaga tridactyla[2]	200– 2,600	06–85	1.5–4	1–302	7.2–10	19–91	4–35.5
Tamandua tetradactyla[2,3]	10–20	–	–	–	–	–	–
Bradypus Tridactylus[2]	20–80	~80	1	5–400	–	~17	~12

[1]Pharmacological semen collection, [2]Electroejaculation, [3]No motility observed and there was 62.5% of abnormal sperm cells

Cryopreservation of semen was described only for giant anteaters. One report describes that one part of semen sample was diluted in two parts of BotuCrio® extender without cryoprotectant (Botupharma, Botucatu, SP, Brazil) in 2-mL plastic tubes and then cooled at 0.05°C/min to 5°C. Cooled semen samples had sperm motility decreased continuously from 0 to 18 hours (Luba et al. 2015).

In the other report, semen samples were cooled after diluted to a concentration of 50 × 10^6 spz/mL in OptiXcell® (IMV Technologies, Campinas, SP, Brazil) media, packaged in mini straws (IMV Technologies, Campinas, SP, Brazil) and equilibrated for four hours at 5°C. Semen was frozen in liquid nitrogen vapor for 10 min and then plunged in liquid nitrogen for storage, no more details were provided. Sperm motility and vigor recovered after thawing was 30% and 3, respectively (Silva et al. 2019).

FINAL CONSIDERATIONS

As presented in this chapter, there are several gaps of knowledge about basic reproductive parameters for sloths and anteaters, and ARTs are still very limited for these species. No artificial insemination, *in vitro* fertilization or embryo transfers attempts are described in scientific literature.

Due to extinction processes' increasing speed, there is an urgent need to expand scientific studies about biology of reproduction of sloths and anteaters and the development of ARTs for these species.

ACKNOWLEDGMENTS

Author would like to acknowledge Dr. Marcelo Alcindo de Barros Vaz Guimarães (*In memorian*) and Dr. Barbara Durrant for their friendship and altruistic sharing of knowledge.

REFERENCES

Abreu, E.F., D. Casali, R. Costa-Araújo, G.S.T. Garbino, G.S. Libardi, D. Loretto, et al. 2021. Lista de Mamíferos do Brasil (2021–2) [Data set]. Zenodo.

Aguilar, R.F. and M. Superina. 2015. Xenarthra. pp. 355-369. *In*: R.E. Miller and M.E. Fowler (ed.). Fowler's Zoo and Wild Animal Medicine Volume 8. W.B. Saunders, St. Louis, USA.

Amendolagine, L., T. Schoffner, L. Koscielny, M. Schook, D. Copeland, J. Casteel, et al. 2018. In-house monitoring of steroid hormone metabolites in urine informs breeding management of a giant anteater (*Myrmechophaga tridactyla*). Zoo Biol. 37: 40–45.

Barretto, M.L.M., M.JA.A.L. Amorim and M.V.D. Falcão. 2013. Análise morfológica e morfométrica das gônadas de preguiça (*Bradypus variegatus,* Schinz, 1825). Pesq. Vet. Bras. 33: 1130–1136.

Bento, H.J., J.M.A. Rosa, T.O. Morgado, M.D.B. Granjeiro, M.A. Bianchini, G.A. Iglesias, et al. 2019. Sexagem em tamanduá-bandeira (*Myrmecophaga tridactyla*) por meio do teste da reação em cadeia da polimerase. Arq. Bras. Med. Vet. Zootec. 71: 538–544.

Boscaini, A., T.J. Gaudin, N. Toledo, B.M. Quispe, P.O. Antoine and F. Pujos. 2019. The earliest well-documented occurrence of sexual dimorphism in extinct sloths: evolutionary and palaeoecological insights. Zool. J. Linn. Soc. 187: 229–239.

Chiarello, A. 1998. Diet of the Atlantic Forest maned sloth *Bradypus torquatus* (Xenarthra: Bradypodidae). J. Zool. 246: 11–19.

Deco-Souza, T. de, C.S. Pizzutto, L.S.B. Souza, P.N. Jorge-Neto, A.C. Csermak-Jr and G.R. de Araújo. 2021. Desafios para a reprodução assistida em animais de vida livre. Rev. Bras. Repr. Anim. 45: 253–258.

Desbiez, A.L.J., A. Bertassoni and K. Traylor-Holzer. 2020. Population viability analysis as a tool for giant anteater conservation. Perspect. Ecol. Conserv. 18: 124–131.

Favoretto, S.M., E.G. da Silva, J. Menezes, R.R. Guerra and D.B. Campos. 2016. Reproductive System of Brown-throated Sloth (*Bradypus variegatus*, Schinz 1825, Pilosa, Xenarthra): Anatomy and Histology. Anat. Histol. Embryol. 45: 249–259.

Fromme, L., D.R. Yogui, M.H. Alves, A.L.J. Desbiez, M. Langeheine, A. Quagliatto, et al. 2021. Morphology of the genital organs of male and female giant anteaters (*Myrmecophaga tridactyla*). PeerJ 9: e11945.

Gilmore, D.P., C.P. Da-Costa and D.P.F. Duarte. 2000. An update on the physiology of two- and three-toed sloths. Braz. J. Med. Biol. Res. 33: 129–146.

Hay, M.A., A.C. Bellem, J.L. Brown and K.L. Goodrowe. 1994. Reproductive Patterns in Tamandua (*Tamandua tetradactyla*). J. Zoo Wildl. Med. 25: 248–258.

Hossotani, C.M.S. and H. Silva e Luna. 2016. Reproductive patterns of the lesser anteater (*Tamandua tetradactyla*, Linnaeus, 1758). Rev. Bras. Reprod. Anim. 40: 95–98.

IUCN. 2022. The IUCN Red List of Threatened Species. Version 2022-1. https://www.iucnredlist.org.

Júnior, J.F.M. and A. Bertassoni. 2014. Potential Agonistic Courtship and Mating Behavior between Two Adult Giant Anteaters (*Myrmecophaga tridactyla*). Edentata 15: 69–72.

Knott, K.K., B.M. Roberts, M.A. Maly, C.K. Vance, J. DeBeachaump, J. Majors, et al. 2013. Fecal estrogen, progestagen and glucocorticoid metabolites during the estrous cycle and pregnancy in the giant anteater (*Myrmecophaga tridactyla*): evidence for delayed implantation. Reprod. Biol. Endocrinol. 11: 83.

Lara-Ruiz, P. and A.G. Chiarello. 2005. Life-history traits and sexual dimorphism of the Atlantic Forest maned sloth *Bradypus torquatus* (Xenarthra: Bradypodidae). J. Zool. 267: 63–73.

Leiva, M. and M.C. Marques. 2010. Dados Reprodutivos da População Cativa de Tamanduá-Bandeira (*Myrmecophaga tridactyla* Linnaeus, 1758) da Fundação Parque Zoológico de São Paulo. Edentata 11: 49–52.

Lima, A.K.F., C.B. Felipe and G.M.L. Silva. 2018. Biologia reprodutiva de *Bradypus variegatus* Schinz (1825): desafios e perspectivas. Rev. Bras. Reprod. Anim. 42: 109–113.

Luba, C. do N., Y.L. Boakari, A.M. Costa Lopes, M. da Silva Gomes, F.R. Miranda, F.O. Papa, et al. 2015. Semen characteristics and refrigeration in free-ranging giant anteaters (*Myrmecophaga tridactyla*). Theriogenology 84: 1572–1580.

Luna, H.S., C.M.S. Hossotani and F.M.A. Moreira. 2014. Esforços para conservação da espécie *Myrmecophaga tridactyla* Linnaeus, 1758: tecnologias aplicadas à reprodução. Rev. Bras. Reprod. Anim. 38: 10–14.

Martins Bezerra, B., A. da Silva Souto, L.G. Halsey and N. Schiel. 2008. Observation of brown-throated three-toed sloths: mating behavior and the simultaneous nurturing of two young. J. Ethol. 26: 175–178.

Matlaga, D. 2006. Mating Behavior of the Northern Tamandua (*Tamandua mexicana*) in Costa Rica. Edentata 7: 46.

Mendonça, M.A.C., M. Nichi, R.H.F. Teixeira, F.R. Braga, R. Simões, J.D. de A. Losano, et al. 2021. Spermatic profile of captive giant anteaters (*Myrmecophaga tridactyla*): Knowing more to preserve better. Zoo Biol. 40: 227–237.

Mesquita, E.Y.E., P.C. Soares, L.R. Mello, E.C.B. Freire, A.R. Lima, E.G. Giese, et al. 2021. Sloths (*Bradypus variegatus*) as a polygastric mammal. Microsc. Res. Tech. 84: 79–88.

Mess, A.M., P.O. Favaron, C. Pfarrer, C. Osmann, A.P. Melo, R.F. Rodrigues, et al. 2012. Placentation in the anteaters *Myrmecophaga tridactyla* and *Tamandua tetradactyla* (Eutheria, Xenarthra). Reprod Biol. Endocrinol. 10: 102.

Miranda, F.R., D.M. Casali, F.A. Perini, F.A. Machado and F.R. Santos. 2018. Taxonomic review of the genus *Cyclopes* Gray, 1821 (Xenarthra: Pilosa), with the revalidation and description of new species. Zool. J. Linn. Soc. 183: 687–721.

Miranda, F., M. Kersul, A. Lopes, V. Gasparotto, K. Molina, P. Snoeck, et al. 2021. Pharmacological semen collection and sperm morphometric evaluation in Silky anteater (*Cyclopes didactylus*) Linneus, 1758. Proc. VIII Int. Symp. Anim. Biol. Reprod. Brazil.

Moura, F., L. Sampaio, P. Kobayashi, R. Laufer-Amorim, J.C. Ferreira, T.T.N. Watanabe, et al. 2021. Structural and Ultrastructural Morphological Evaluation of Giant Anteater (*Myrmecophaga tridactyla*) Prostate Gland. Biol. 10: 231.

Mühlbauer, M., D.P.F. Duarte, D.P. Gilmore and C.P. da Costa. 2006. Fecal estradiol and progesterone metabolite levels in the three-toed sloth (*Bradypus variegatus*). Braz. J. Med. Biol. Res. 39: 289–295.

Oliveira, F.R., F.R. Lima, M.J. Silvino, L.F. Pereira and F.G.G. Dias. 2019. Topography and syntopy of abdominopelvic viscera of the giant anteater (*Myrmecophage tridactyla*— Linnaeus, 1758). Arq. Bras. Med. Vet. Zootec. 71: 1961–1967.

Patzl, M., F. Schwarzenberger, C. Osmann, E. Bamberg and W. Bartmann. 1998. Monitoring ovarian cycle and pregnancy in the giant anteater *Myrmecophaga tridactyla* by faecal progestagen and oestrogen analysis. Anim. Reprod. Sci. 53: 209–219.

Peres, M.A., E.J. Benetti, M.P. Milazzotto, J.A. Visintin, M.A. Miglino and M.E.O.A. Assumpção. 2008. Collection and evaluation of semen from the three-toed sloth (*Bradypus tridactylus*). Tissue Cell 40: 325–331.

Pizzutto, C.S., H. Colbachini and P.N. Jorge-Neto. 2021. One Conservation: the integrated view of biodiversity conservation. Anim. Reprod. 18: e20210024.

Rezende, L.C., A.C. Galdos-Riveros, M.A. Miglino and J.R. Ferreira. 2013. Aspectos da biologia reprodutiva em preguiça e tamanduá: uma revisão. Rev. Bras. Reprod. Anim. 37: 354–359.

Rossi, L.F., J.P. Luaces, H.J.A. Marcos, P.D. Cetica, G. Gachen, G.P. Jimeno, et al. 2011. Female reproductive tract of the lesser anteater (*Tamandua tetradactyla*, Myrmecophagidae, Xenarthra). Anatomy and histology. J. Morphol. 272: 1307–1313.

Rossi, L.F., J.P. Luaces, H.J.A. Marcos, P.D. Cetica, G. Perez Jimeno and M.S. Merani. 2013. Anatomy and Histology of the Male Reproductive Tract and Spermatogenesis Fine Structure in the Lesser Anteater (*Tamandua tetradactyla*, Myrmecophagidae, Xenarthra): Morphological Evidences of Reproductive Functions. Anat. Histol. Embryol. 42: 247–256.

Silva, M., G. Araujo, M. Kersul, P.N. Jorge Neto, A. Aguiar, F. Miranda, et al. 2019. Pharmacological semen collection and cryopreservation of the Giant Anteater (*Myrmecophaga tridactyla*) in the wild. Rev. Bras. Reprod. Anim. 43: 705.

Smith, P. 2007. Smith P. 2007. FAUNA Paraguay Handbook of the Mammals of Paraguay Number 2. *Myrmecophaga tridactyla*. p. 1–21.

Snoeck, P.P.N., A.C.B. Cruz, L.S. Catenacci and C.R. Cassano. 2011. Citologia vaginal de preguiça-de-coleira (*Bradypus torquatus*). Pesq. Vet. Bras. 31: 271–275.

Takami, K., M. Yoshida, Y. Yoshida and Y. Kojima. 1998. Sex determination in giant anteater (*Myrmecophaga tridactyla*) using hair roots by polymerase chain reaction amplification. J. Reprod. Dev. 44: 73–78.

Taube, E., J. Keravec, J.-C. Vié and J.-M. Duplantier. 2008. Reproductive biology and postnatal development in sloths, *Bradypus* and *Choloepus*: review with original data from the field (French Guiana) and from captivity: Reproduction of sloths. Mamm. Rev. 31: 173–188.

Thompson, R., T.M. Wolf, H. Robertson, M.W. Colburn, A. Moreno, A. Moresco, et al. 2017. Serial ultrasound to estimate fetal growth curves in southern tamandua (*Tamandua tetradactyla*). J. Zoo Wildl. Med. 48: 294–297.

Toledo, N., M.S. Bargo, S.F. Vizcaíno, G. De Iuliis and F. Pujos. 2015. Evolution of body size in anteaters and sloths (Xenarthra, Pilosa): phylogeny, metabolism, diet and substrate preferences. Earth Environ. Sci. Trans. R. Soc. Edinb. 106: 289–301.

Valle, S. del and M. Halloy. 2003. El oso hormiguero, *Myrmecophaga tridactyla*: crecimiento e independización de una cría. Mastozool. Neotrop. 10: 323–330.

Wislocki, G.B. 1928. On the placentation of the two-toed anteater (*Cyclopes didactylus*). Anat. Rec. 39: 69–83.

Zacariotti, R.L., T. Zwarg, J.F. de P. Castro, M.D. Falbel and R. Aily. 2022. Avaliação sanitária e reprodutiva de tamanduá-bandeira (*Myrmecophaga tridactyla*) mantidos em cativeiro no estado de São Paulo. Proc. XIII Seminário de Pesquisa e XIV Encontro de Iniciação Científica do ICMBio, Brazil.

Chapter 20

The Armadillos

Patrícia da Cunha Sousa[1]*, Lívia Batista Campos[1],
Alexandre Rodrigues Silva[1] and Gabriela Liberalino Lima[2]

[1]Laboratory of Animal Germplasm Conservation (LCGA),
Department of Animal Sciences, Universidade Federal Rural do Semi-Árido (UFERSA),
Francisco Mota Avenue, 572, Mossoró, Rio Grande do Norte, 59625-900, Brazil.
Email: pattbio13@hotmail.com, livia_campos86@hotmail.com, alexrs@ufersa.edu.br

[2]Instituto Federal de Educação, Ciência e Tecnologia do Ceará (IFCE),
Campus Crato, CE – 292, S/N, Giselia Pinheiro, Crato,
Ceará, 63115-500, Brazil.
Email: gabriela.lima@ifce.edu.br

INTRODUCTION

Armadillos are mammals belonging to the Order Cingulata, which has a single family (Dasypodidae) with living taxa, totaling 21 species, of which 11 occur in Brazil, distributed in different biomes (Abba and Superina 2010, ICMBio 2022). Armadillo species are characterized by exhibiting some form of armored carapace, consisting of osteoderms covered by epidermal scales (Wetzel 1980, Engelmann 1985, Wetzel et al. 2008). They have low metabolic rates (McNab 1985) and other peculiarities that resemble their extinct ancestors (Fernicola et al. 2008). In addition, they are relatively solitary and antisocial animals, with rare social interactions, which usually occur during the breeding season (Loughry and McDonough 2013).

In Brazil, the armadillo population has been decreasing due to habitat loss (McDonough and Loughry 2001), traffic accidents and predatory hunting habits (Silva et al. 2015). Native communities in South America constantly exploit armadillos as a food source, even though biomedical research highlights their importance as natural reservoirs of the bacteria that cause leprosy (*M. leprae*)

---
*Corresponding author: pattbio13@hotmail.com

(Balamayooran et al. 2015). In addition to supporting biomedical research, non-threatened armadillos can be used as experimental models for the development of management techniques to conserve endangered species such as the three-banded armadillo (*Tolypeutes tricinctus*) and the giant armadillo (*Priodontes maximus*) (IUCN 2022).

Given these characteristics, it can be expected that armadillos have been the subject of intense and extensive scientific investigation. However, the semi-fossorial habits, the low population density, and the difficulty of capture (Silva and Henriques 2009) make *in loco* studies difficult. In addition, the small sample size results in nonspecific and general answers regarding knowledge about ecology, biology, and reproductive physiology in all species of Dasypodidae (Superina et al. 2013). The International Union for Conservation of Nature (IUCN) by its Red List of Threatened Species shows that there are considerable gaps in our knowledge about these neotropical mammals. Such knowledge is fundamental for the realistic assessment of the conservation status and for the planning of effective conservation and multiplication strategies, and the development of assisted reproduction techniques (Superina et al. 2013, Silva et al. 2015).

This chapter is dedicated to armadillos, highlighting important information about the reproductive anatomy and physiology of the main species studied in South America, associated with the state of the art regarding the experimental development of reproductive biotechnologies aimed at conservation strategies for these remarkable mammals.

GENERAL REPRODUCTIVE CHARACTERISTICS

Edentata (currently Xenarthra) are a group with a small to moderate degree of sexual dimorphism, in which males are generally larger than females (Ralls 1976). In some species, such as the large hairy armadillo (*Chaetophractus villosus*) the sexual dimorphism can be observed in the skull and mandible. These structures are larger in females compared to males (Squarcia et al. 2009). In this species, a slight sexual dimorphism was also noted based on the shape of the humerus and ulna, structures traditionally associated with digging capacity, with greater development in females (Acuña et al. 2017).

On the other hand, in the nine-banded armadillo (*Dasypus novemcinctus*), there is no obvious sexual dimorphism, but the sex can be easily determined by observing the genitalia (Loughry and McDonough 2001). Studies have found that some species of armadillos have a long penis. (Kelly 1997, Sousa et al. 2013). In other armadillo species like the six-banded armadillo (*Euphractus sexcinctus*), the penis can be two-thirds the length of the animal, being the male with one of the longest penises among mammals, extending approximately 2/3 of the body length, or the equivalent of 33% of the body length (Sousa et al. 2013) (Figure 20.1). On the ventral face, the average length of the penis can reach up to 10 cm, while on its dorsal face it can reach up to 14 cm in length. A rough region is observed on its dorsal surface (Silva et al. 2014). The testicles are intra-abdominal, not visible externally, which makes sex determination difficult.

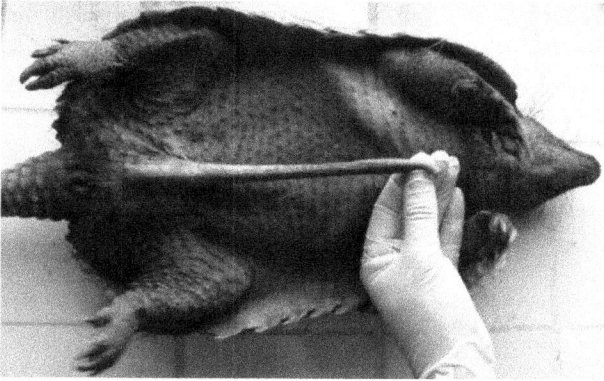

Figure 20.1 Six-banded armadillo, *Euphractus sexcinctus* (Linnaeus, 1758), male. Dorsal decubitus and penile extension. Medri (2008).

Sexual maturity varies between armadillo species, starting between 9 and 12 months of age in the wooded and hairy armadillos (Freitas et al. 2014) or later, around 6.5–8 years, as observed in the male giant armadillo (*P. maximus*) (Superina and Abba 2018, Soibelzon 2020).

Although some specific features are present between species, polyembryony is a feature of research interest. This characteristic is observed in the nine-banded armadillo, producing multiple embryos derived from a single fertilized egg, resulting in the production of four genetically identical offspring (Loughry et al. 1998).

Despite efforts to obtain information about the biology and physiology of armadillos, many studies focused on reproductive aspects are still needed to develop and adapt assisted reproduction techniques. Such studies are valuable for captive breeding programs aimed at increasing reproductive performance for commercial purposes and for conservation programs.

SEMEN COLLECTION

There are few publications focused on semen analysis and the development of reproductive biotechnologies in armadillos. Therefore, in this regard, this topic will address only the armadillo species with more recent studies in South America: *T. matacus, P. maximus* and *E. sexcinctus*.

In studies aimed at the andrological evaluation of armadillos, electroejaculation (EEE) has been the method of choice for semen collection, using different protocols between species. But, in general, the method consists of inducing the ejaculatory reflex by means of electrical stimuli on the floor of the rectal ampulla, whether the animal is anesthetized or not, by means of a probe connected to an electrical stimulator (Figure 20.2).

In a study conducted in the region of La Plata, Argentina, male three-banded armadillos (*T. matacus*) underwent the electroejaculation procedure, being previously anesthetized with isoflurane gas in an induction chamber and maintained (1–2%) for face mask. A lubricated rectal probe (diameter 10 mm), modified with a 30–401 mid-axis curve, was inserted into the rectum with the three longitudinal electrodes positioned ventrally (Herrick et al. 2002). Using an electrostimulator, three sets of 30 stimulations, separated by rest periods of 5 to 10 minutes, were delivered with incremental increases in voltage (range 2 to 6 V) (Beetem et al. 1989, Howard 1993).

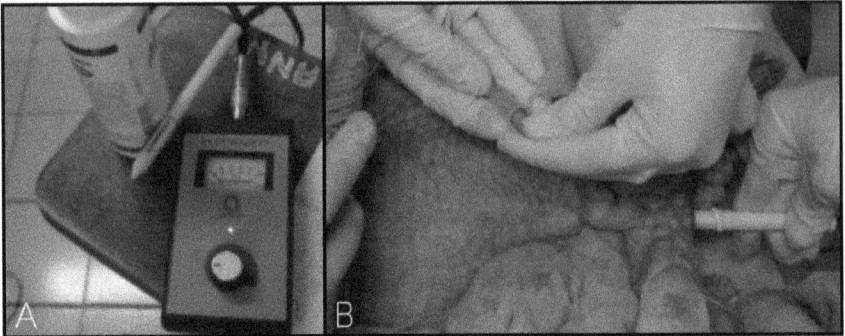

Figure 20.2 (A) Electroejaculator. (B) Electroejaculator probe inserted into the rectum for electrical stimuli on the floor of the rectal ampulla and semen collection in a plastic tube. Six-banded armadillo, *Euphractus sexcinctus* (Linnaeus, 1758).

In 2010, Serafim et al. were the first to describe the characteristics of the semen of the armadillo species (*E. sexcinctus*), which has, on average, a seminal volume of 0.3 ml, with a concentration of 450 million spermatozoa/ml. These samples were obtained by electroejaculation, using a stimulation protocol previously described for carnivores (Wildt et al. 1983). For semen collection in this species, an electroejaculator connected to a 12 V source is used to perform three successive cycles of electrostimulation with an interval of 5 minutes between cycles. The first cycle consists of 10 successive stimuli at 2, 3 and 4 mA; the second 10 consecutive stimuli at 3, 4 and 5 mA; and the third is 10 successive stimuli at 5 and 6 mA. The probe is 12.5 cm long, 1.0 cm in diameter and has two longitudinal electrodes. About 8 cm of the probe is inserted into the rectal canal of the animal, placed in dorsal decubitus, and the ejaculate is collected in graduated plastic tubes (Serafim et al. 2010).

Supporting the use of electroejaculation as a standard method for semen collection of Dasypodidae, the giant armadillo (*P. maximus*), the largest living armadillo, was another species subjected to the semen collection procedure, following a protocol previously proposed for giant anteaters (Luba et al. 2015). For this purpose, an electroejaculator with a probe (1.3 cm in diameter, 17 cm in length) containing two electrodes (7 cm in length each) (Eletrojet Premium, electrovet; Eletrovet LTDA, Valinhos, São Paulo, Brazil) was used. The probe was lubricated with carboxymethylcellulose and inserted into the rectum. Subsequently, 80 electrical stimuli were applied, divided into 3 series. The first series consisted of 30 electrical stimuli divided into 10 stimuli of 2 V, 10 of 3 V and 10 of 4 V, followed by a 2 min interval. The second series consisting of 30 stimuli, divided into 10 stimuli of 3 V, 10 of 4 V and 10 of 5 V, followed once more with a 2-min interval. The last series consisted of 20 electrical stimuli, divided into 10 stimuli of 5 V and 10 of 6 V. The stimulation procedure was interrupted as soon as the rich sperm fraction of the ejaculate was obtained.

Studies show that electroejaculation is considered a usual, reliable, and efficient method for semen collection in wild animals (Zimmerman et al. 2013). However, the method requires the use of anesthetic protocols that are species specific and may influence the efficiency of electroejaculation. This need for chemical restraint is evident, since the first studies on six-banded armadillos were conducted with mechanical restraint only, and the animals showed signs of stress and pain, such as intense vocalization, aversive behavior with attempts to escape (Serafim et al. 2010). In this sense, Sousa et al. (2016a) tested two anesthetic protocols to be used during electroejaculation in *E. sexcinctus*: the first consisted of administering intramuscular xylazine (1 mg/kg; Rompun, Bayer, São Paulo, Brazil) and ketamine (7 mg/kg, ketamine, Pfizer, São Paulo, Brazil), followed by intravenous (IV) bolus administration of 5 mg/kg of propofol (Propovan, Cristália, Fortaleza, Brazil). In the second protocol, subjects were premedicated with intramuscular butorphanol (0.4 mg/kg, Torbugesic-SA Zoetis, São Paulo, Brazil) and ketamine (7 mg/kg), followed by IV propofol in bolus (Figure 20.3). Although both protocols provide adequate semen parameters and excellent post-anesthetic recovery time, the combination of drugs that showed the best efficiency (91.6%–11/12 attempts) to obtain the ejaculate was xylazine/ketamine/propofol. This was significantly more efficient than the combined use of butorphanol/ketamine/propofol, which resulted in only 33.3% efficiency (4/12 trials). It is pointed out that such results are due to the use of xylazine, an alpha-adrenergic agonist, known to increase semen emission, since the muscles involved in this emission have alpha-adrenergic innervation (Sjöstrand 1965, Knight 1974).

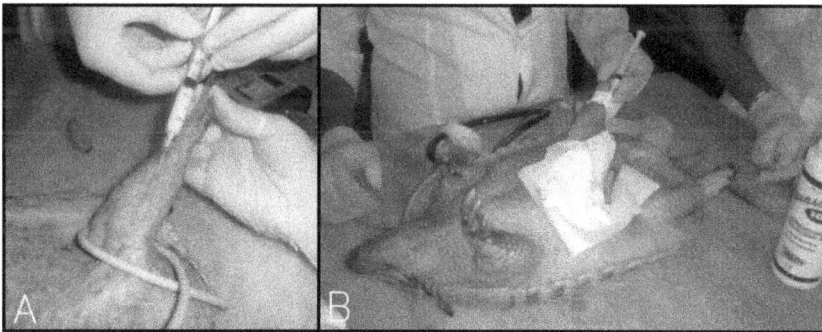

Figure 20.3   Anesthetic protocols: (A) intravenous (IV) administration of propofol. (B) Six-banded armadillo, *Euphractus sexcinctus* (Linnaeus, 1758).

SEMEN CHARACTERISTICS

Through semen analysis, it is possible to determine the basal reproductive characteristics of the species, including sperm concentration, motility, morphology, acrosomal state and gamete viability. In armadillos, *T. matacus* semen was recovered by electroejaculation, and showed high proportions (> 75%) of morphologically normal spermatozoa, observed in all analyzed samples (Herrick et al. 2002). In this study, sperm concentration and motility were generally low and urine contamination was a persistent problem—possibly due to anesthesia, reproductive anatomy and/or improper tube placement. Such aspects need to be considered before electroejaculation can be routinely used in this and other armadillo species.

In the giant armadillo (*P. maximus*), the semen collection through electroejaculation was used to identify the sexual maturity of males, and other semen parameters were not studied. Therefore, individuals presenting ejaculates containing sperm were classified as adults, while those azoospermic ones were categorized as subadults and sexually immature, specially when correlations with its body mass and morphometry were established (Luba et al. 2020). According to Superina et al (2013), the *P. maximus* is the most studied species of armadillos classified as vulnerable. However, information concerning more detailed basic analyzes of seminal aspects is practically non-existent. This is true for many other armadillo species, including those of little concern in the ecological conservation scenario, such as the *Cabassous unicinctus* (Anacleto et al. 2014). Table 20.1 presents the reference values for the main parameters of the semen of the armadillo evaluated after the electroejaculation procedure for semen collection.

Table 20.1 Semen characteristics (means ± SEM) from six-banded armadillos (*Euphractus sexcinctus*) and three-banded armadillo (*Tolypeutes matacus*) collected by electroejaculation

Semen parameters	Armadillos species	
	Tolypeutes matacus	*Euphractus sexcinctus*
Ejaculate volume (ml)	141.1 ± 45.3	353 ± 86
pH	7.8 ± 0.2	9.0 ± 0.0
Sperm concentration ($\times 10^6$/ml)	23.0 ± 14.8	45.0 ± 14.0
Motility (%)	24.5 ± 11.9	61.0 ± 7.0
Normal morphology (%)	86.9 ± 2.6	86.0 ± 2.0
Intact acrosome	42.8 ± 11.9	99.0 ± 0.3
References	Herrick et al. 2002	Serafim et al. 2010

The analysis of armadillo ejaculates revealed remarkable characteristics in the semen of this species: high viscosity of the seminal plasma (Serafim et al. 2010), presence of sperm aggregates called rouleaux (Figure 20.4) (Sousa et al. 2013), and large dimensions morphometric measurements of the sperm head (width 13 μm and length approximately 10.9 μm) (Sousa et al. 2013). So far, there is no explanation for the high viscosity, but it is likely that this characteristic may be related to the high concentration of glycosaminoglycans in the seminal plasma from accessory gland secretions, as described for other mammals (Cardoso et al. 1985). During semen analysis, high seminal viscosity interferes with motility and especially with sperm vigor, with a maximum value classified as 2, referring mainly to the beating of the flagellum without progressive displacement (Serafim et al. 2010, Sousa et al. 2013). This negative effect is aggravated by the presence of sperm rouleaux, grouping two or more sperm in a pile, also described for another species of armadillo, the *C. unicinctus* (Heath et al. 1987). In this a sense, a comparative study among different species of armadillos through transmission electron microscopy revealed an interesting fact that the sperm heads are very thin and have unusual spoon characteristics in mammalian spermatozoa (Cetica et al. 1997). The concave shape of the sperm head facilitates the fitting and permanence of the junction between these cells (Heath et al. 1987), which seems to protect the acrosome during epididymal transit as described in marsupials (Phillips 1974), in which separation occurs only in female genitalia (Rodger and Bedford 1982).

Santos et al. (2011) demonstrated marked individual variation in relation to the osmotic response of nine-banded armadillos and sperm longevity. The authors promoted different hyposmotic challenges to the spermatozoa and determined that a fructose solution of 50 mOsm/L is the most adequate to analyze the functionality of the armadillo sperm membrane. In addition, these authors diluted the semen of the three-banded armadillo in a Tris-based extender, after incubation at 34°C. They found that spermatozoa remain viable for up to 90 minutes on average, but in some individuals this time can extend up to 6 hours. Recently, Sousa et al. (2014) demonstrated that the Tris extender is even more efficient than the coconut water extender – ACP-119® – to maintain the viability of the three-banded armadillo semen at 34°C centrifuge semen samples before dilution, as this procedure has a negative effect on semen viability.

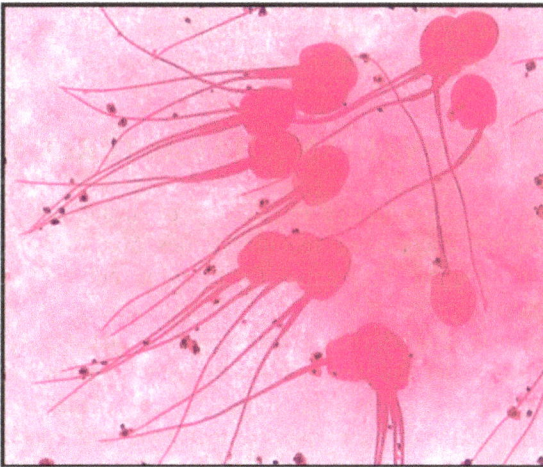

Figure 20.4 Sperm aggregated called rouleaux in semen smear stained with eosin-nigrosin. (Optical light microscopy – 40 × magnification). Six-banded armadillo, *Euphractus sexcinctus* (Linnaeus, 1758).

Using ultrastructural analysis, it was found that *E. sexcinctus* sperm commonly contains electrolucent spots in the nucleus, which may be associated with disturbances in chromatin condensation. In addition, the midpiece of the sperm usually has few mitochondrial whorls (~ 45), a characteristic that may be related to a low level of competition among the males of the species (Sousa et al. 2013).

SEMEN CONSERVATION

Among all armadillo species, the most significant advances in semen technology have been achieved in *E. sexcinctus*. In an initial attempt to preserve the semen of the three-banded armadillo, it was observed that its sperm is extremely sensitive to low temperatures, with a significant drop in sperm quality already after 3 h of cooling at 5°C, in Tris or ACP-119® (Amorim et al. 2012). Recently, the first attempt to freeze *E. sexcinctus* was published (Sousa et al. 2016b). In this study, a diluent based on Tris plus 20% egg yolk and 6% glycerol was used, associated with a freezing protocol previously described for other wild species (Silva et al. 2011), since there was no other protocol developed to Xenartras for that moment. After thawing, Sousa et al. (2016b) found a significant decline in all armadillo semen parameters with only 6.1% of motile spermatozoa found. However, the percentage of spermatozoa that remained with viable (13%) and functional (24.7%) membranes after thawing suggests that some cells may be alive but immobile. Analysis using fluorescent markers revealed that armadillo sperm mitochondria are highly sensitive to the freezing protocol and the findings of the ultrastructural analysis confirmed this statement. Additionally, the images obtained by transmission electronic microscopy reveal that the frozen-thawed spermatozoa presented damaged plasmatic membrane, nuclear modifications such as alterations in the chromatin and acrosomal alterations related to sperm capacitation. These results were extremely important, as it was possible to identify critical points in the freezing and thawing procedure, whose information may be useful to improve the protocols for cryopreservation of six-banded armadillo semen.

KNOWING THE OVARIAN CHARACTERISTICS

Many aspects of the reproductive physiology of armadillos still need to be elucidated, which has encouraged specific studies over the years (Table 20.2). Obtaining this information is essential for carrying out well-planned strategic conservation actions.

Regarding gamete characteristics, knowing their quantitative and qualitative aspects is fundamental for the development of growth and maturation protocols, cryopreservation and transplantation. It was observed that armadillo ovarian follicles are distributed in the ovarian cortical zone and consist of an oocyte containing scattered regions of chromatin and heterochromatin close to the nuclear membrane, surrounded by a flattened or cuboidal granulosa cell whose layers depend on the stage of the follicle, and large epithelioid cells internal to the theca (Rezende et al. 2013).

The estimation and characterization of the population of ovarian follicles was performed in some species of armadillos (Peppler et al. 1986, Rezende et al. 2013, Rossi and Solari 2021). In the hairy armadillo (*C. villosus*), for example, the ovarian follicles were categorized into primordial, intermediate, early primary, late primary, secondary, tertiary, and Graafian or pre-ovulatory, with diameters of 28.13 ± 5.01 μm, 38.64 ± 5.85 μm, 71.00 ± 12.97 μm, 134.52 ± 25.45 μm,

289.43 ± 55.40 μm and 489.58 ± 126.90 μm, respectively. A positive and linear correlation was observed between the diameter of the follicle and the oocyte, with the presence of fewer mitochondria in the tertiary follicle than in the other categories (Códon et al. 2001).

Table 20.2 Assisted reproduction techniques in different species of Armadillos.

Species	Reproductive Biotechniques		
	Estrous cycle monitoring	*Female genital tract histology description*	*Ovarian follicle characteristics*
Cabassous centralis	×	×	×
Cabassous chacoensis	×	√	×
Calyptophractus retusus	×	×	×
Chaetophractus vellerosus	√	√	√
Chaetophractus villosus	√	√	√
Chlamyphorus truncatus	×	√	×
Dasypus hybridus	×	√	√
Dasypus novemcinctus	√	√	√
Dasypus septemcinctus	×	×	×
Euphractus sexcinctus	√	√	√
Priodontes maximus	×	×	×
Tolypeutes matacus	√	√	×
Tolypeutes tricinctus	×	×	×
Zaedyus pichiy	√	√	√

In contrast, the diameter of *E. sexcinctus* follicles is smaller (primary: 14.59 ± 0.18 μm; primary: 23.43 ± 0.34 μm; secondary: 45.01 ± 1.45 μm), with a population of preantral follicles of 6175.8 ± 1923.5 follicles (Figure 20.5). In this species, it was observed that individuals infected with *M. leprae* had a smaller population of ovarian follicles, of 3,112.1 ± 615.1 follicles (Brasil 2020).

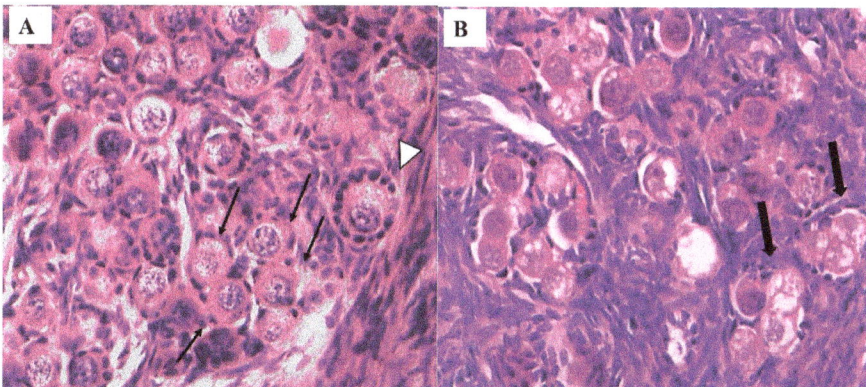

Figure 20.5 *Euphractus sexcinctus* preantral follicles. (A) Primordial (arrows) and primary follicles (arrow head). (B) Cell cysts containing a variable number of germ cells (full arrow)

A specific feature present in these species is the presence of ovarian germ cell cysts in adults (Rossi et al. 2020), containing a variable number of germ cells (2–30) interconnected by intercellular bridges. Also, in the hairy armadillo (*C. villosus*), using transmission electron microscopy, a distinct structure was described in the cytoplasm of oocytes in early growth, the multilamellar body. The authors described this structure as an organelle with a transitory function, during the initial stages of follicular development, from follicle growth to the beginning of zona pellucida formation (Rossi and Solari 2021). Given these data, oogenesis among Xenarthra remains little explored and its study can provide valuable tools for the development of assisted reproduction technologies.

ESTRUS CYCLE MONITORING

Regarding the adoption of estrus detection techniques and hormonal profile, different methods have been described (Table 20.3), allowing to know the duration of the estrous cycle and the hormonal levels in some species.

Table 20.3   Distinct techniques for estrous cycle monitoring in Armadillos

Species	Estrous cycle monitoring			
	Plasma hormone	*Fecal hormone metabolites*	*Ultrasound*	*Vaginal cytology*
Chaetophractus vellerosus	√	√	×	×
Chaetophractus villosus	√	√	×	×
Dasypus novemcinctus	√	×	×	√
Euphractus sexcinctus	√	×	√	√
Tolypeutes matacus	√	×	×	×
Zaedyus pichiy	×	√	×	×

Campos et al. (2016) demonstrated the characteristics of the six-banded armadillo estrous cycle by monitoring external signs of estrus, vaginal cytology, ultrasound, and measurement of blood hormones. On average, their estrous cycles last 23.5 ± 3.1 days, but some individuals can have cycles of 31 days, or as short as 16 days, highlighting individual variation within the species. The detailed identification of the different phases of the estrous cycle (proestrus, estrus, metestrus, diestrus) was not possible in this species, and the authors only distinguished the estrogenic phase from the progesterone phase. These results were equivalent to those described in three-banded armadillo (*T. matacus*), in which the authors found an average duration of the estrous cycle of 26.4 ± 1.3 days with no variation between individuals (Howell-Stephens et al. 2013).

In addition, the six-banded armadillo exhibits an estrogenic phase lasting 8.8 ± 1.4 days (range 4 to 16 days), in which females demonstrate external signs such as bloody vaginal discharge, vulvar edema, presence of mucus and ease of introduction of the vaginal swab. During this phase, estrogen peaks of 240.7 ± 13 pg/ml (202.1–322.5 pg/ml) have been identified as generally occurring 3–7 days before the onset of external signs of estrus. On this occasion, progesterone values were measured as 6.4 ± 1.2 ng/ml on average. Furthermore, vaginal cytology reveals that the proportion of vaginal epithelial cells does not differ during the estrogenic phase (Campos et al. 2016). These data were superior to those described by Luaces et al. (2011) who measured the peak of estrogen in *C. vellerosus* and *C. villosus* through fecal samples, over a year. In *C. villosus*, mean baseline estradiol concentrations are 3.04 ± 0.80 ng/g in dry feces, and in captive *C. vellerosus*, mean baseline estradiol concentrations are 3.46 ± 1.20 ng/g in dry stool.

In six-banded armadillo, the progesterone phase lasts an average of 15.6 ± 2.1 days (ranging from 7 to 22 days) with no bloody discharge and difficulty introducing the swab for vaginal cytology. In this phase, progesterone reaches values of 10.8 ± 1.9 ng/ml, but estradiol remains at baseline values of 111.7 ± 6.3 pg/ml on average. These data were superior to mean basal progesterone concentrations of C. villosus 28.72 ± 11.75 ng/g of dry feces and mean basal progesterone concentrations of a captive C. vellerosus 14.05 ± 3.03 ng/g of dry feces (Luaces et al. 2011). In nine-banded armadillos (Dasypus novemcinctus), a progesterone concentration indicative of ovulation ($\sim$10 ng/ml) occurred between 17 and 20 months of age (Peppler et al. 1986).

Regarding the use of vaginal cytology in six-banded armadillo, it was found that the proportion of parabasal cells gradually increases throughout the progesterone phase, reaching maximum values near the end of the cycle and decreasing as the new increase in estrogen increases (Campos et al. 2016). Apparently, females can synchronize their estrus when kept in groups, which may be influenced by an effect of the presence of males (Costa et al. 2014), which can easily detect their estrus (Fernandes et al. 2014).

An attempt to induce ovulation in D. novemcinctus was performed many years ago (Peppler and Stone 1980), using an antiestrogenic compound, clomiphene citrate. The majority of females (7/8) receiving 50 mg for 5 days had an increase in plasma progesterone to a level indicative of ovulation (>10 ng/ml) within 6 days of the last injection. In addition, the vaginal smear changed from a diestrus pattern (leukocytes and epithelial cells) before clomiphene administration to proestrus (rounded epithelial cells) and subsequently to estrus (cornified cells) 4–8 days after the last injection (6/7 animals). Currently, this study can be improved in other species of armadillos, and new drugs and associations can be evaluated, since new information about reproductive biology is available.

FINAL CONSIDERATIONS

Despite the ecological importance and perspectives of captive breeding for armadillos, the development of reproductive biotechniques is still scarce, especially for females. Knowing the characteristics of gametes is the first step towards the development of *in vitro* preservation and manipulation protocols, essential tools to increase reproductive performance.

REFERENCES

Abba, A.M. and M. Superina. 2010. The 2009/2010 armadillo Red List assessment. Edentata. 11: 135–184.

Acuña, F., N.S. Sidorkewicj, A.I. Popp and E.B. Casaneve. 2017. A geometric morphometric study of sex differences in the scapula, humerus and ulna of *Chaetophractus villosus* (Xenarthra, Dasypodidae). Sér. Zool. 107.

Amorim, R.N.L., K.D.M. Emerenciano, P.C. Sousa, T.S. Castelo, G.L. Lima and A.R. Silva. 2012. Short-term preservation at 5°C of Armadillo's (*Euphractus sexcinctus*) semen. International Symposium Animal Biology of Reproduction. Campinas. SP. 9: 968.

Anacleto, T.C.S, W.M. Tomas and M. Superina. 2014. *Cabassous unicinctus*. The IUCN Red List of Threatened Species 2014.

Balamayooran, G., M. Pena, R. Sharma and R.W. Truman. 2015. The armadillo as an animal model and reservoir host for *Mycobacterium leprae*. Clin. Dermatol. 33: 108–115.

Beetem, D.D., D.A. Schmidt, W.F. Swanson, R.W. Truman and R.A. Godke. 1989. Semen collection in the nine-banded armadillo. Procedings of the American Association of Zoological Parks and Aquariums. p. 615.

Brasil, A. 2020. Caracterização e estimativa da população folicular ovariana em tatus-peba (Euphractus sexcinctus) - Monografia (graduação em Medicina Veterinária) - Universidade Federal Rural do Semi-árido, Mossoró, Rio Grande do Norte, Brazil.

Campos, L.B., G.C.X. Peixoto, G.L. Lima, T.S. Castelo, C.I.A. Freitas and A.R. Silva. 2016. Monitoring the reproductive physiology of six-banded armadillos (*Euphractus sexcinctus*, Linnaeus, 1758) through different techniques. Reprod. Domest. Anim. 51: 736–742.

Cardoso, F.M., E.L. Figueiredo, H.P. Godinho and A.M. Cóser. 1985. Variação sazonal da atividade secretória das glândulas genitais acessórias masculinas de tatus *Dasypus Novemcinctus* Linnaeus, 1758. Rev. Bras. Biol. 45: 507–514.

Cetica, P., I.M. Rahn, M.S. Merani and A.Solari. 1997. Comparative spermatology in Dasypodidae II (*Chaetophractus vellerosus, Zaedyus pichiy, Euphractus sexcinctus, Tolypeutes matacus, Dasypus septemcinctus* and *Dasypus novemcinctus*). Biocell. 21(3): 195–204.

Codón, S.M., S.G. Estecondo, E.J. Galíndez and E.B. Casanave. 2001. Ultrastructure and morphometry of ovarian follicles in the armadillo *Chaetophractus villosus* (Mammalia, Dasypodidae). Braz. J. Biol. 61: 485–496.

Costa, T.O., W.O.B. Fernandes, C.L.C. Costa, R.M. Maia, A.R.S. Viana and C.I.A. Freitas. 2014. Aspectos reprodutivos em fêmeas de tatu-peba (*Euphractus sexcinctus*) mantidas em cativeiro. *In*: Congresso Norte Nordeste de Reprodução Animal, 7, 2014, Mossoró. Anais... Mossoró: Conera, 2014. Abstract.

Engelmann, G.F. 1985. The phylogeny of the *Xenarthra*. pp. 51–63. *In*: G.G. Montgomery (ed.). The Evolution and Ecology of Armadillos, Sloths and Vermilinguas. Smithsonian Institution Press, Washington, DC, USA.

Fernandes, W.O.B., T.O. Costa, S.S. Mendonça, S.L. Gomes, T.G. Coelho and C.I.A. Freitas. 2014. Condutas de diferentes categorias comportamentais em *Euphractus sexcinctus* (tatu-peba) em cativeiro associadas ao período reprodutivo. *In:* Congresso Norte Nordeste de Reprodução Animal, 7, 2014, Mossoró. Anais... Mossoró: Conera.

Fernicola, J.C., S.F. Vizcaíno and A.R. Fariña. 2008. The evolution of armored xenarthrans and a phylogeny of the glyptodonts. pp. 79–85. *In:* J. Loughry y SF Vizcaíno (Eds.). A Biologia do Xenarthra. University Press of Florida, USA.

Freitas, C.I.A, T.O. Costa, T.G. Coelho, M.O. Freitas, W.O. Barbosa, T.S. Lima, et al. 2014. Critérios de escolha e estratégias sexuais em tatu peba (*Euphractus sexcinctus* L., 1758). *In*: Encontro Anual de Etologia, 32, Simpósio Latino Americano de Etologia, 5, 2014, Mossoró. Anais... Mossoró: SLAE.

Heath, E., N. Schaeffer, D.A. Meritt Jr and R.S. Jeyendran. 1987. Rouleaux formation by spermatozoa in the naked-tail armadillo, *Cabassous unicinctus*. J. Reprod. Fertil. 79: 153–158.

Herrick, J.R., M.K. Campbell and W.F. Swanson. 2002. Electroejaculation and Semen Analysis in the La Plata Three-Banded Armadillo (*Tolypeutes matacus*). Zoo Biol. 21: 481–487.

Howard, J.G. 1993. Semen collection and analysis in carnivores. pp. 390–399. *In*: M.E. Fowler (ed.). Zoo and Wild Animal Medicine, Vol. III. Philadelphia: W.B. Saunders Co.

Howell-Stephens, J, D. Bernier, J.S. Brown, D. Mulkerin and R.M. Santymire. 2013. Using non-invasive methods to characterize gonadal hormonal patterns of southern three-banded armadillos (*Tolypeutes matacus*) housed in North American zoos. Anim. Reprod. Sci. 138: 314–323.

[ICMBio] Instituto Chico Mendes de Conservação da Biodiversidade. 2022. Centro Nacional de Pesquisa e Conservação de Aves Silvestres (Org.). Lista Xenarthra.

[IUCN] International Union for Conservation of Nature. 2022. Online version. Available in: <http://www.iucnredlist.org/search/details.php/8 306/all>. Accessed in: 12 August 2022.

Kelly, D.A. 1997. Axial orthogonal fiber reinforcement in the penis of the Nine-banded Armadillo (*Dasypus novemcinctus*). Journal of Morphology. 233: 249–255.

Knight, T.W. 1974. A qualitative study of factors affecting the contractions of the epididymis and ductus deferens of the ram. J. Reprod. Fertil. 40: 19–29.

Loughry, W.J., P.A. Prodöhl, C.M. McDonough and J.C. Avise. 1998. Polyembryony in armadillos. Am. Sci. 86: 274–280.

Loughry, W.J. and C.M. McDonough. 2001. Natal recruitment and adult retention in a population of nine-banded armadillos. Acta. Theriol. 46: 393–406.

Loughry, W.J. and C.M. McDonough. 2013. The nine-banded armadillo: a natural history. University of Oklahoma Press, Oklahoma.

Luaces, J.P., M. Ciuccio, L.F. Rossi, A.G. Faletti, P.D. Cetica, E.B. Casanave, et al. 2011. Seasonal changes in ovarian steroid hormone concentrations in the large hairy armadillo (*Chaetophractus villosus*) and the crying armadillo (*Chaetophractus vellerosus*). Theriogenology. 75: 796–802.

Luba, C.N., Y.L. Boakari, A.M.C. Lopes, M.S. Gomes, F.R. Miranda, F.O. Papa, et al. 2015. Semen characteristics and refrigeration in free-ranging giant anteaters (*Myrmecophaga tridactyla*). Theriogenology. 84: 1572–1580.

Luba, C.N., D. Kluyber, G.F. Massocato, N. Attias, L. Fromme, A.L.R. Rodrigues, et al. 2020. Size matters: penis size, sexual maturity and their consequences for giant armadillo conservation planning. Mammalian Biology. https://doi.org/10.1007/s42991-020-00065-3.

McDonough, C.M. and W.J. Loughry. 2001. Armadillos. pp. 796–799. *In:* McDonald, D.W. (Ed.). Encyclopedia of mammals. Oxford University Press, London, U.K.

McNab, B.K. 1985. Energetics, population biology, and distribution of xenarthrans, living and extinct. pp 219–232. *In*: G.G. Montgomery (ed.). The Evolution and Ecology of

Armadillos, Sloths and Vermilinguas. Smithsonian Institution Press, Washington, DC, EUA.

Peppler, R.D. and S.C. Stone. 1980. Clomiphene-induced ovulation in the 9-banded armadillo (*Dasypus novemcinctus*). Laboratory Animals 14: 329–330.

Peppler, R.D., F.E. Hossler and S.C. Stone. 1986. Determination of reproductive maturity in the female nine-banded armadillo (*Dasypus novemcinctus*). J. Reprod. Fertil. 76: 141–146.

Phillips, D.M. 1974. Spermiogenesis. Academic Press, New York, USA.

Ralls, K. 1976. Mammals in which females are larger than males. Q. Rev. Biol. 51: 245–276.

Rezende, L.C., S.A.S. Kückelhaus, A.C. Galdos-Riveros, J.R. Ferreira and M.A. Miglino. 2013. Vascularización, morfología e histología del ovario en el armadillo *Euphractus sexcinctus* (Linnaeus, 1758). Arch. Med. Vet. 45: 191–196.

Rodger, J.C. and J.M. Bedford. 1982. Separation of sperm pairs and sperm-egg interaction in the opossum Didelphis virginiana. J. Reprod. Fertil. 64: 171–179.

Rossi, L.F., S. Nottola, S. Miglietta, G. Macchiarelli, J.P. Luaces, V. Merico, et al. 2020. Germ cell cysts, a fetal feature in mammals, are constitutively present in the adult armadillo. Mol. Reprod. Dev. 87: 91– 101.

Rossi, L.F. and A.J. Solari. 2021. Large lamellar bodies and their role in the growing oocytes of the armadillo *Chaetophractus villosus*. J. Morphol. 282: 1330–1338.

Santos, E.A.A., P.C. Sousa, C.E.V. Dias, T.S. Castelo, G.C.X. Peixoto, G.L. Lima, et al. 2011. Assessment of sperm survival and functional membrane integrity of the six-banded armadillo (*Euphractus sexcinctus*). Theriogenology 76: 623–629.

Serafim, M.K.B., R.A. Lira, L.L.M. Costa, I.C.N. Gadelha, C.I.A. Freitas and A.R. Silva. 2010. Description of semen characteristics from six-banded armadillos (*Euphractus sexcinctus*) collected by electroejacula-tion. Anim. Reprod. Sci. v.118: 362–365.

Silva, K.F.M. and R.P.B. Henriques. 2009. Ecologia de população e área de vida do tatu-mirim (*Dasypus septemcinctus*) em um Cerrado no Brasil Central. Edentata. 8(10): 48–53.

Silva, M.A., G.C. Peixoto, T.S.Castelo, G.L. Lima, A.M. Silva, M.F. Oliveira, et al. 2011. Recovery and cryopreservation of epididymal sperm from agouti (*Dasiprocta aguti*) using powdered coconut water (ACP-109c) and Tris extenders. Theriogenology 76: 1084–1089.

Silva, R.M., P. Rocha, I.B.F.S. Silva, M.C. Cabral, A.T. Junior and A.Q. Santos. 2014. Anatomia dos órgãos genitais masculinos de tatus: *Dasypus novemcinctus* Linnaeus, 1758 (tatu-galinha) e *Euphractus sexcinctus* Linnaeus, 1758 (tatu-peba). Proceedings of the VII Encontro sobre Animais Selvagens—ENANSE and II Simpósio sobre Medicina e Conservação da Fauna do Cerrado, Uberlândia /MG- Brazil, 20–26.

Silva, A.R., P.C. Sousa and C.I.A. Freitas. 2015. Reprodução assistida em tatus-peba (Euphractus sexcinctus): peculiaridades e desafios. Rev. Bras. Reprod. Anim. Belo Horizonte. 39(1): 61–65.

Sjöstrand, N.O. 1965. The adrenergic innervation of the vas deferens and the accessory male genital glands. Acta. Physiol. Scand. 65: 1–81.

Soibelzon, E. 2020. Carrying offspring: An unknown behavior of armadillos Acarreo de crías: un comportamiento desconocido en los armadillos. Therya Notes 1: 101–105.

Sousa, P.C., E.A.A. Santos, J.A.B. Bezerra, G.L. Lima, T.S. Castelo, J.D. Fontenele-Neto, et al. 2013. Morphology, morphometry and ultrastructure of captive six-banded armadillo (*Euphractus sexcinctus*) sperm. Anim. Reprod. Sci. 140: 279–285.

Sousa, P.C., E.A.A. Santos, A.M. Silva, T.S. Castelo, G.C.X. Peixoto, C.I.A. Freitas, et al. 2014. Viabilidade do sêmen de tatus-peba (*Euphractus sexcinctus*) centrifugado e diluído em Tris ou agua de coco em pó. Cienc Rural. v.44: 1645–1650.

Sousa, P.C., R.N.L. Amorim, G.L. Lima, A.L.C. Paiva, V.V. Paula, C.I.A. Freitas and A.R. Silva. 2016a. Establishment of an anesthesic protocol for semen collection by electroejaculation in six-banded armadillos (*Euphractus sexcinctus* Linnaeus, 1758). Arq. Bras. Med. Vet. Zootec. 68(6): 1595–1601.

Sousa, P.C., E.A.A. Santos, A.M. Silva, J.A.B. Bezerra, A.L.P. Souza, G.L. Lima, et al. 2016b. Identification of ultrastructural and functional damages in sperm from six-banded armadillos (*Euphractus sexcinctus*) due to cryopreservation. Pesq. Vet. Bras. 36: 767–774.

Squarcia, S.M., N.S. Sindorkewicj, R. Camina and E.B. Casanave. 2009. Sexual Dimorphism in the Mandible of the Armadillo *Chaetophractus villosus* (Desmarest, 1804) (Dasypodidae) from Northern Patagonia, Argentina. Braz. J. Biol. 69: 347–352.

Superina, M., N. Pagnutti and A.M. Abba. 2013. What do we know about armadillos? An analysis of four centuries of knowledge about a group of South American mammals, with emphasis on their conservation. Mammal Review. 7: 33–44.

Superina, M. and A.M. Abba. 2018. Family Chlamyphoridae (*Chlamyphorid armadillos*). pp 48–73 in Handbook of the mammals of the world—Volume 8: insectivores, sloths and colugos (Wilson, D.E. and R.A. Mittermeier, Eds.). Lynx Editions. Barcelona, Spain.

Wetzel, R.M. 1980. Revision of the naked-tailed armadillos, genus *Cabassous* McMurtrie. Annals of Carnegie Museum, USA 49: 323–357.

Wetzel, R.M., A.L. Gardner,, K.H. Redford, J.F. and Eisenberg, 2008. Order cingulata. pp. 128–156. *In*: A.L. Gardner (ed.). Mammals of South America. Chicago, University of Chicago Press.

Wildt, D.E., M. Bush, J.G. Howard, S.J. O'Brien, D. Meltzer, A. Van Dyk, et al. 1983. Unique seminal quality in the South African cheetah and a comparative evaluation in the domestic cat. Biol. Reprod. 29: 1019–1025.

Zimmerman, D.M., M.A. Mitchell and B.H. Perry. 2013. Collection and characterization 323 of semen from green iguanas (Iguana iguana). A. J. Vet. Res. 74: 1536–1541.

Index

For Product Safety Concerns and Information please contact our EU
representative GPSR@taylorandfrancis.com
Taylor & Francis Verlag GmbH, Kaufingerstraße 24, 80331 München, Germany

www.ingramcontent.com/pod-product-compliance
Lightning Source LLC
Chambersburg PA
CBHW060330220326
41598CB00023B/2660